# GNSS-R Earth Remote Sensing from SmallSats

# GNSS-R Earth Remote Sensing from SmallSats

Editors

**Hugo Carreno-Luengo**
**Chun-Liang Lin**

MDPI

Basel • Beijing • Wuhan • Barcelona • Belgrade • Novi Sad • Cluj • Manchester

*Editors*
Hugo Carreno-Luengo
Climate and Space Sciences
and Engineering Department
University of Michigan
Ann Arbor, MI, USA

Chun-Liang Lin
Electrical Engineering
Department
National Chung Hsing
University
Taiwan

*Editorial Office*
MDPI
St. Alban-Anlage 66
4052 Basel, Switzerland

This is a reprint of articles from the Special Issue published online in the open access journal *Remote Sensing* (ISSN 2072-4292) (available at: https://www.mdpi.com/journal/remotesensing/special_issues/GNSS-R_earth_remote_sensing_SmallSats).

For citation purposes, cite each article independently as indicated on the article page online and as indicated below:

| Lastname, A.A.; Lastname, B.B. Article Title. *Journal Name* **Year**, *Volume Number*, Page Range. |
|---|

**ISBN 978-3-0365-9460-6 (Hbk)**
**ISBN 978-3-0365-9461-3 (PDF)**
**doi.org/10.3390/books978-3-0365-9461-3**

# Contents

# About the Editors

**Hugo Carreno-Luengo**

Hugo Carreno-Luengo received the degree of Ingeniero Aeronáutico (Plan Antiguo de Estudios, "bachelor+master"), with a specialization in spacecrafts, from the Escuela Técnica Superior de Ingenieros Aeronáuticos (ETSIA), the Universidad Politécnica de Madrid (UPM), Madrid, Spain, in 2010, and the PhD degree (Cum Laude) from the Department of Signal Theory and Communications (TSC), the Universitat Politècnica de Catalunya (UPC), Barcelona, Spain, 2016.

From 2009 to 2010, he carried out his final degree project at the Department of Aircrafts and Space Vehicles at UPM. In 2011, he undertook his master's degree in space science and technology at UPC, and was the recipient of a PhD fellowship from the Institut d'Estudis Espacials de Catalunya (IEEC), where he was involved in the design and development of the first-ever nanosatellite for Earth remote sensing using GNSS-R (2011–2016). From 2012 to 2015, he was the PI of the TORMES and TORMES 2.0 projects within ESA's REXUS/BEXUS and was a co-PI of the E-GEM FP7 project. From 2013 to 2014, he was a Visiting Researcher for the ESA/ESTEC, DLR, Esrange Space Center, and in the summer of 2016, he was invited by the China Great Wall Industry Corporation (CGWIC) to assist in the launch campaign of the 3Cat-2 CubeSat at Jiuquan Satellite Launch Center. From 2016 to 2017, he was a JPL Post-doctorate Researcher at NASA Jet Propulsion Laboratory, Pasadena, CA, USA, working on a GNSS-Reflectometry experiment, using the radar aboard NASA's SMAP mission. The experiment was successful, and it set the basis for the development of a new SMAP operational mode. Later, he was a JdC Post-doctorate Fellow on the NASA CYGNSS Science Team (2017–2019). Since 2019, he has been serving as a Research Scientist at the University of Michigan (UMich) College of Engineering, working directly with the Principal Investigator of NASA's CYGNSS mission, under a contract with the NASA Science Mission Directorate.

**Chun-Liang Lin**

Chun-Liang Lin was born in Tainan, Taiwan, in 1958. He received the PhD degree in aeronautical and astronautical engineering from the National Cheng Kung University, Tainan, Taiwan, in 1991. He was an Associate Professor and Professor with the Department of Automatic Control Engineering, Feng Chia University, Taichung, Taiwan, from 1995 to 2003. He is currently a Chair Professor with the Department of Electrical Engineering, National Chung Hsing University, Taichung, Taiwan. His research interests include guidance and control, intelligent control, network control, and robust control.

# Preface

SmallSats are changing the paradigm in Earth remote sensing, taking advantage of innovative payloads. As such, the operation of constellations of these instruments has the potential to observe the Earth's dynamic processes with a higher spatiotemporal sampling than traditional techniques. In particular, the so-called Global Navigation Satellite Systems Reflectometry (GNSS-R) is a type of L-band passive multi-static radar (as many transmitters as navigation satellites are in view) that provides a wide swath of up to 1500 km. GNSS-R spatiotemporal sampling properties could provide new process insights on mesoscale studies, wind speed determination, soil moisture content determination, vegetation water content monitoring, etc.

This reprint is intended to prompt the development of a potential virtual network of satellites, providing inter-comparable data to the scientific community, based on the new GRSS Standard for GNSS-Reflectometry. New and novel GNSS-R scientific applications, methodologies, and retrieval algorithms are the focus of this reprint, including contributions from academia, international space agencies, and private industry.

We thank all contributing authors for their interest in this project. It is our great pleasure to share this reprint with you and the wider GNSS-R community. We encourage all members of the community to contemplate and envisage the future of this remote sensing technique, and to continue working together in a coordinated manner.

**Hugo Carreno-Luengo and Chun-Liang Lin**
*Editors*

remote sensing

MDPI

*Review*

# The CYGNSS Mission: On-Going Science Team Investigations

**Hugo Carreno-Luengo [1,*], Juan A. Crespo [2], Ruzbeh Akbar [3], Alexandra Bringer [4], April Warnock [5], Mary Morris [2] and Chris Ruf [1]**

1 Climate and Space Sciences and Engineering Department, University of Michigan (UM), Ann Arbor, MI 48104, USA; cruf@umich.edu
2 Jet Propulsion Laboratory (JPL), California Institute of Technology, Pasadena, CA 91125, USA; juan.a.crespo@jpl.nasa.gov (J.A.C.); mary.g.morris@jpl.nasa.gov (M.M.)
3 Massachusetts Institute of Technology (MIT), Cambridge, MA 02139, USA; rakbar@mit.edu
4 Electrical and Computer Engineering, The Ohio State University, Columbus, OH 43210, USA; bringer.1@osu.edu
5 SRI International, Ann Arbor, MI 48105, USA; april.warnock@sri.com
* Correspondence: carreno@umich.edu; Tel.: +1-734-764-6561

**Abstract:** In 2012, the National Aeronautics and Space Administration (NASA) selected the CYclone Global Navigation Satellite System (CYGNSS) mission coordinated by the University of Michigan (UM) as a low-cost and high-science Earth Venture Mission. The CYGNSS mission was originally proposed for ocean surface wind speed estimation over Tropical Cyclones (TCs) using Earth-reflected Global Positioning System (GPS) signals, as signals of opportunity. The orbital configuration of each CYGNSS satellite is a circular Low Earth Orbit (LEO) with an altitude ~520 km and an inclination angle of ~35°. Each single Delay Doppler Mapping Instrument (DDMI) aboard the eight CYGNSS microsatellites collects forward scattered signals along four specular directions (incidence angle of the incident wave equals incidence angle of the reflected wave) corresponding to four different transmitting GPS spacecrafts, simultaneously. As such, CYGNSS allows one to sample the Earth's surface along 32 tracks simultaneously, within a wide range of the satellites' elevation angles over tropical latitudes. Following the Earth Science Division 2020 Senior Review, NASA announced recently it is extending the CYGNSS mission through 30 September 2023. The extended CYGNSS mission phase is focused on both ocean and land surface scientific investigations. In addition to ocean surface wind speed estimation, CYGNSS has also shown a significant ability to retrieve several geophysical parameters over land surfaces, such as Soil Moisture Content (SMC), Above Ground Biomass (AGB), and surface water extent. The on-going science team investigations are presented in this article.

**Keywords:** CYGNSS; GNSS-R; science investigations; ocean; land; atmosphere

**Citation:** Carreno-Luengo, H.; Crespo, J.A.; Akbar, R.; Bringer, A.; Warnock, A.; Morris, M.; Ruf, C. The CYGNSS Mission: On-Going Science Team Investigations. *Remote Sens.* **2021**, *13*, 1814. https://doi.org/10.3390/rs13091814

Academic Editor: Emanuele Santi

Received: 1 March 2021
Accepted: 1 May 2021
Published: 6 May 2021



## 1. Introduction

The CYGNSS Science Team [1–5] is composed of a wide variety of researchers, working in a collaborative and coordinated manner to further advance the use of CYGNSS for Global Navigation Satellite Systems Reflectometry (GNSS-R) Earth remote sensing [6–8].

GNSS-R is a type of L-band passive multi-bistatic radar, which uses the navigation spacecrafts which are in view as transmitters. GNSS-R provides global coverage, full temporal availability, and sampling of the Earth's surface over several tracks within a wide area up to ~1500 km. One potential disadvantage is the degraded spatial resolution under the incoherent scattering regime. The use of GNSS radio-navigation signals for Earth remote sensing has been investigated since it was originally proposed for mesoscale ocean altimetry by the European Space Agency (ESA) [9]. The first in-orbit proof-of-concept study was an experiment performed at the Jet Propulsion Laboratory (JPL) using the Space-borne Imaging Radar-C (SIR-C) on-board the Space Shuttle [10]. This experiment motivated the development of GNSS-R. At present, several GNSS-R space-borne studies have been performed with several satellites, including United Kingdom (UK) Disaster Monitoring System-1 (DMC-1) [11], UK Techdemosat-1 (TDS-1) [12], Soil Moisture Active Passive (SMAP) [13], and the CYGNSS 8 microsatellites constellation [3] which is the first-ever operational GNSS-R mission.

The CYGNSS Science Team (Figure 1, Table 1) holds two Science Team meetings every year to coordinate such activities, and to share our most recent results. The last 2020 conference was held in UM (virtual conference) in June 2020 (Figure 1). This conference was structured along six different sessions covering all of our research topics: Session 1 (CYGNSS Mission Overview and Data Products I), Session 2 (CYGNSS Data Products II), Session 3 (Land Processes I), Session 4 (Land Processes II), Session 5 (Altimetry and Tropical Cyclones and Tropical Convection I), and Session 6 (Tropical Cyclones and Tropical Convection II). This article provides a synoptic overview of the meeting.

**Table 1.** Some key-members of the CYGNSS Science Team. The full list is available in [14].

| **Member** | **Chris Ruf** |
|---|---|
| Home Institution<br>CYGNSS Role<br>Areas | University of Michigan<br>Principal Investigator<br>Earth environment remote sensing methods, instrumentation, atmosphere |
| Member | Mahta Moghaddam |
| Home Institution<br>CYGNSS Role<br>Areas | University of Southern California<br>Terrestrial Science Lead, Co-I<br>Inverse scattering, subsurface characterization, water resources |
| Member | Derek Posselt |
| Home Institution<br>CYGNSS Role<br>Areas | Jet Propulsion Laboratory, California Institute of Technology<br>Atmospheric Science Lead, Co-I<br>Clouds and precipitation, data assimilation, uncertainty quantification |
| Member | Ruzbeh Akbar |
| Home Institution<br>CYGNSS Role<br>Areas | Massachusetts Institute of Technology<br>Soil moisture sensor networks, calibration, and validation<br>Microwave remote sensing of Earth, hydrology, wireless sensor networks |
| Member | Alexandra Bringer |
| Home Institution<br>CYGNSS Role<br>Areas | The Ohio State University<br>CYGNSS Science Team member<br>Microwave remote sensing of the Earth, ocean and land applications |
| Member | Juan A. Crespo |
| Home Institution<br>CYGNSS Role<br>Areas | Jet Propulsion Laboratory, California Institute of Technology<br>Competed Science Team Member, CYGNSS ocean surface heat flux product<br>Extratropical cyclones & air-sea fluxes |
| Member | Mary Morris |
| Home Institution<br>CYGNSS Role<br>Areas | Jet Propulsion Laboratory, California Institute of Technology<br>CYGNSS Science Team member<br>Metereological and hydrological applications, Earth sciences |
| Member | April Warnock |
| Home Institution<br>CYGNSS Role<br>Areas | SRI International<br>CYGNSS Science Team member<br>Hydrology/storm surge modeling |
| Member | Hugo Carreno-Luengo |
| Home Institution<br>CYGNSS Role<br>Areas | University of Michigan<br>CYGNSS Science Team member<br>Surface scattering, Earth sciences |

**Figure 1.** Zoom-derived photo of some Science Team members attending the June 2020 CYGNSS meeting.

## 2. Methodology, Results, and Discussions

### *2.1. Data Products*

CYGNSS is within the NASA's Earth System Science Pathfinder (ESSP) program. The original goal of CYGNSS was to further advance extreme weather predictions with a focus on TCs inner core process studies. CYGNSS was designed to resolve the lack of accuracy with current TCs intensity forecasts, which lie in inadequate measurements and modeling of the inner core. The inaccurate measurements result from two main aspects:

- Traditional remote sensing techniques are blind to much of the inner core ocean surface when intense precipitation is in the eye wall and inner rain bands.
- Traditional high-inclination orbit and wide-swath surface wind imagers do not provide an enough temporal sampling of the dynamically evolving (genesis and rapid intensification) phases of the TCs life cycle.

CYGNSS was specifically designed to address these two limitations by combining the use of GNSS-R with the sampling properties of a constellation of eight microsatellites [15–19]. CYGNSS provides an unprecedented spatiotemporal resolution of the Earth's surface thanks to the multi-static nature of GNSS-R and the use of such a dense constellation of microsatellites. CYGNSS is the first operational GNSS-R mission launched into space, in 2016.

Since CYGNSS's launch in December 2016, various science data products (Figure 2) have been created, such as an expansion of its wind speed measurements, ocean surface heat flux measurements, and land observations [20]. These scientific data products expand CYGNSS's core mission beyond TCs observations and bolster its reach and influence within the Earth science community. The results presented here focus on some of these products that have been developed and published since launch. All CYGNSS data products available to the public are distributed by the Physical Oceanography Distributed Active Archive Center (PODAAC).

Session 1: CYGNSS Mission Overview and Data Products I

The current v2.1 CYGNSS mission science data products assume that the GPS transmit power is constant. The Climate Data Record (CDR) v1.0 was released in May 2020 to improve product performance. This is the most stable, and most accurate product currently available. Current v3.0 products incorporate real time monitoring of GPS power. At present, all eight CYGNSS spacecrafts are healthy and operating nominally. There was the open call "A.27 CYGNSS Competed Science Team" within NASA Research Opportunities in Space and Earth Science (ROSES) 2020 with the following research topics: Atmospheric River Generation and Development, Wetland Methane Emissions, Dynamic Inland Water Mask Development, Process, Coupling and Feedback Studies, and Weather and Storm Surge Data Assimilation Studies. This ROSES seeks to support the continued use of both the ocean and land data products through scientific investigations and end-user applications.

Several significant advances in ocean research activities were presented by the CYGNSS Science Team. By comparing the CYGNSS measured Mean-Square Slope (MSS) and modified Wave-Watch 3 (WW3) modeled MSS, it was shown that the mean MSS ratio has a dependence on both GPS Pseudo-Random Noise codes (PRNs) and CYGNSS Flight Models (FMs) (starboard and port antennas). A new end-to-end CYGNSS Level 1 calibration approach was proposed in [21] to: (1) improve the data quality of the CYGNSS Level 1 calibration and the Level 2 wind speed and MSS products; and (2) improve our understanding of the impacts of wind-wave and swell-wave on the GNSS-R ocean observations. The detailed performance assessments will be reported in the future. Sensitivity in CYGNSS data to wind direction was demonstrated by computing the kurtosis over areas in the glistening zone of different size and symmetry. CYGNSS raw Intermediate Frequency (IF) data processing at Institute of Space Sciences (ICE)—Institute of Space Studies of Catalonia (IEEC) demonstrated potential applications over land, inland water bodies, and in extreme events (e.g., hurricanes and floods) and regions of great geophysical interest (e.g., Himalayan glaciers). Finally, variational wind speed retrievals of uncalibrated CYGNSS data demonstrated a better response to high winds than the Advanced SCATterometer (ASCAT) radars and similar response as the Soil Moisture Ocean Salinity (SMOS) mission (Figure 3).

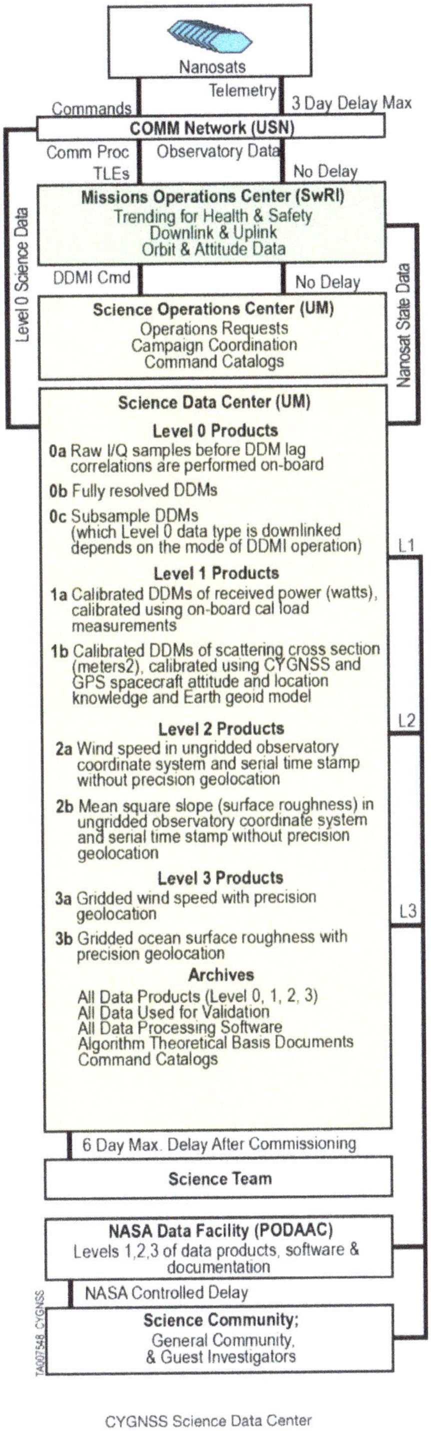


**Figure 2.** Data processing flow of the CYGNSS data products [19].

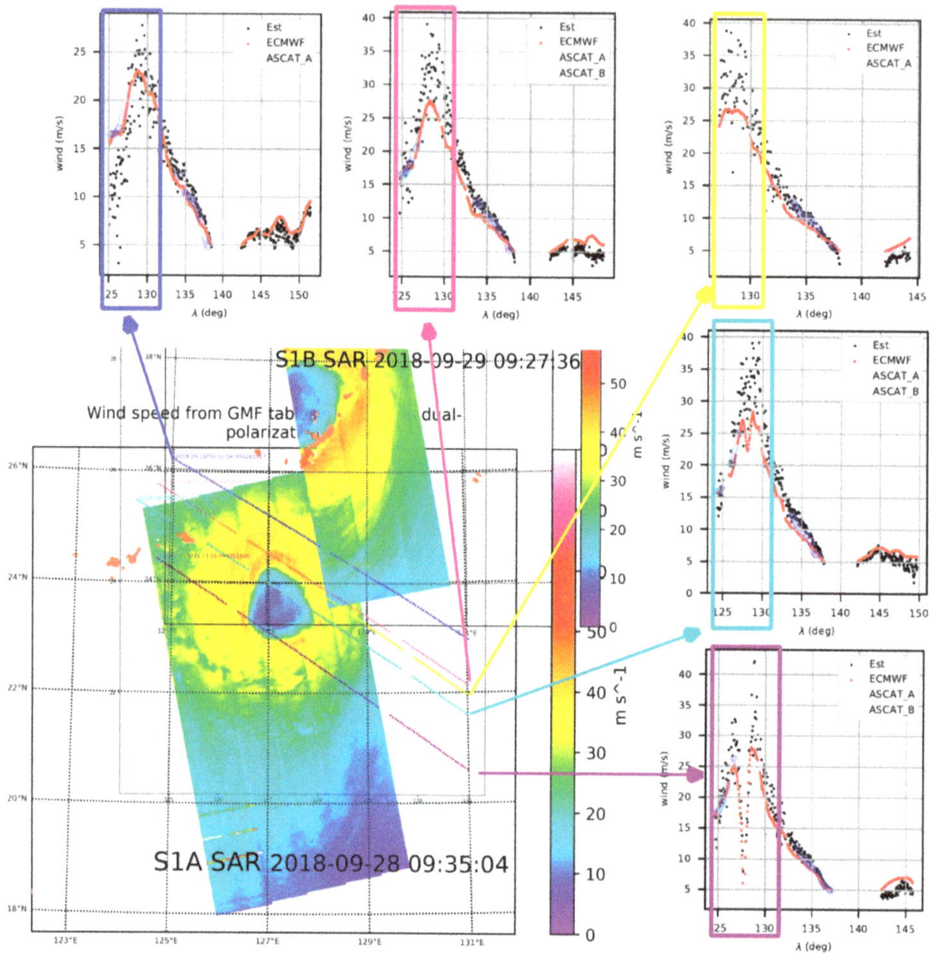


**Figure 3.** Sentinel-1A and -1B wind retrievals for Typhoon "Trami", acquired on Sep 28 9:35 (lower) and Sep 29 9:27 (upper) 2018, approximately 24 h apart (from https://cyclobs.ifremer.fr, 3 April 2021). The colored tracks are CYGNSS passes ~8 h before the first Synthetic Aperture Radar (SAR) wind field image (Sep 28, between 00:50 and 01:26) when the eye was further South. The CYGNSS wind retrievals for each of the tracks, estimated by A. Rius et al. [22], are shown in black dots ("Est" legend) in the panels surrounding the images. The red dots are for European Centre for Medium-Range Weather Forecasts (ECMWF) ERA-5 wind speeds interpolated to the CYGNSS tracks, and Advanced SCATterometer ASCAT-A/B are shown in blue circles when co-located.

Session 2: Data Products II

A CYGNSS L1 land product was developed by the University Corporation for Atmospheric Research (UCAR). This L1 land product [23] contains variables also found in the L1 ocean product, such as position, velocity, time, attitude, and antenna gain, while eliminates variables that were only applicable to ocean applications. The sandbox version of this product is available to all the Science Team members and includes almost all land bodies observed within CYGNSS coverage and a 50 km "skirt" of the ocean; however, very high-altitude regions, such as the Tibetan Plateau are not available due to on-board open loop tracker limitations.

In order to better assist with the CYGNSS wind speed product, several institutions including the National Oceanic and Atmospheric Administration (NOAA) have developed CYGNSS L2 wind speed products, which have been released through the PODAAC. As compared to the original CYGNSS L2 wind speed product, the NOAA version is gridded into 25 × 25 km grid cells, and it uses a custom Geophysical Model Function (GMF) with a track-wise debiasing algorithm. At the time of this meeting, the presenters highlighted some of the minor changes going into v1.1 of the product, such as, e.g., resolving an issue with the star tracker flag, which led to an 18% increase in daily data on some dates. Most importantly, v1.1 of the CYGNSS-NOAA winds have led to improvements in wind speed retrievals within the cores of TCs. This product (Figure 4) was released through the PODAAC during the last fall, with data availability from March 2017 through the present [24].

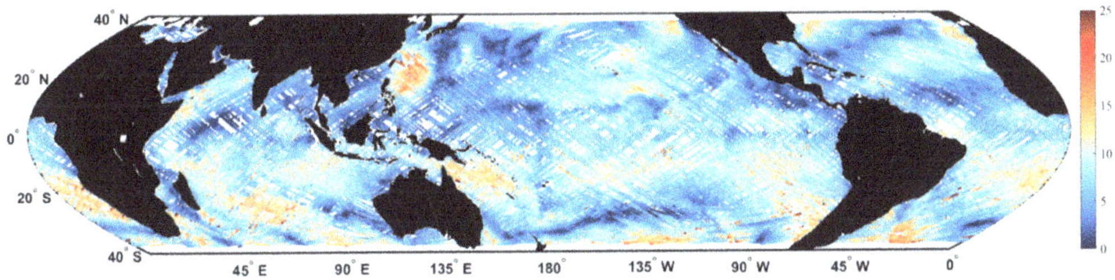


**Figure 4.** Full day of wind speed observations (m/s) on 30 August 2020 from the NOAA CYGNSS L2 Science Wind Speed 25-km v1.1 product.

An ocean surface heat flux product was released in August 2019, which combines CYGNSS L2 winds with reanalysis data from Modern-Era Retrospective analysis for Research and Applications (MERRA-2) in order to estimate the latent and sensible heat fluxes at each CYGNSS specular point [25]. An update of L2 CDR v1.0 of the CYGNSS ocean surface heat flux product was released in October 2020. As compared to the original version (Science Data Record (SDR) v1.0), the latest version of this product uses the CDR v1.0 L2 winds from CYGNSS, leading to a slightly improved performance as compared to buoy data and improved data availability for the Fully Developed Seas (FDS) versions of these products. Additionally, MERRA-2 variables are now matched in time and space to CYGNSS specular points using a tri-linear interpolation method rather than nearest neighbor, which has removed some biases and fixed issues near land. Similar to the CYGNSS-NOAA product, this product has been released through the PODAAC (Figure 5), with data availability from March 2017 through the present (with a 1-month data release latency) [26].

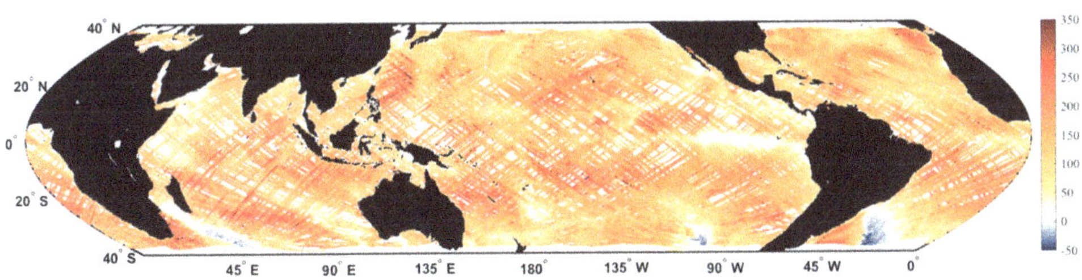


**Figure 5.** Full day of latent heat flux (W/m$^2$) observations on 30 August 2020 from the CYGNSS L2 Ocean Surface Heat Flux CDR v1.0 product.

NASA JPL serves as the primary data lead for CYGNSS within PODAAC, and continually provides updates to the CYGNSS Science Team regarding data publication. At the time of the meeting, the L1 ocean product remained the most popular CYGNSS product being downloaded, and the CYGNSS usage from October 2019 to May 2020 rebounded to pre-FTP retirement levels. It was also noted that by early 2021, cloud services for select datasets would become operational. Around the time of the meeting, PODAAC's web portal went through a revitalization to improve the user experience. This includes listing all of the CYGNSS's data products on a single organized page, allowing users to have a better access to the CYGNSS's scientific data products.

### 2.2. Land Surfaces

The use of GNSS-R for land-surface applications requires additional considerations because the dielectric properties of this scattering medium have significantly more variability than that of the ocean surface. Here, we provide some of the latest findings of the CYGNSS mission, including key-results, discussions, and on-going research activities. At present, a significant number of researchers within the CYGNSS Science Team are working on land surface applications, which remain less explored than ocean products.

Session 3: Land Processes I

The first Land Processes Session consists of a series of presentations and discussions focusing on the use of CYGNSS data and observations related to (a) surface SMC estimation, (b) creation of dynamic inland water body masks, and (c) advancements in forward GNSS-R-related electromagnetic scattering and forward models. GNSS-R-like observations have been shown to be sensitive to changes in surface SMC [27–31] as well as the presence (or absence) of inland water bodies [32,33]. More recently, a 3 km CYGNSS Signal-to-Noise SNR-based surface SMC estimation approach was developed in [30], with results comparable to those by the NASA SMAP mission [34] and unbiased Root-Mean-Squared (ubRMSE) on the order of 0.045 $m^3/m^3$. Given this background, the following briefly summarizes the key finding from each of the session's presentations.

Forward scattering properties of Earth-reflected GNSS signals were evaluated over land surfaces. The CYGNSS End-to-End Simulator (E2ES) was updated. A GNSS-R model capable of evaluating both the incoherent and the coherent scattering terms was developed based on the Huygens–Kirchhoff principle [35]. Results demonstrated the impact of higher order Fresnel zones on the spatial resolution of GNSS-R over heterogeneous areas, showing "ringing" fluctuations in the reflected power near high-contrast boundaries.

An inland water body detection method was also presented during this session. The method [36] demonstrates the utility of ~2 years CYGNSS Level-1 Delay-Doppler Maps (DDM) to create [1,3] km water masks (Figure 6). Inland water surfaces at L-band are presumed smooth enough such that specular scattering is the dominant features in the observed DDM. Therefore, by deploying a DMM coherence persistence and detection method [30], mapping inlands becomes possible. The derived inland water body masks were qualitatively and quantitively compared to, and validated against, historical (30 years) water body occurrence maps by Pekel et al. [37]. False-positive areas were identified and removed via comparison to the Pekel occurrence maps, and examination of high-quality Digital Elevation Maps (DEM). Areas with persistent coherence, unknown water bodies, and exceptionally flat surfaces were also removed. The current analysis was based on ~2 years of CYGNSS observations. Additional investigation to identify false-positive and improve the overall detection method is required.

Russo et al. [38] outlined an entropy-based CYGNSS coherence detector—via eigenvalue decomposition of the DDMs—which, overall showed good agreement with wetland maps derived from Advanced Land Observing Satellite-2 (ALOS-2). However, the current method does not discriminate between open water and flooded vegetation and merits further investigation. Additionally, Russo et al. outlined a newly initiated research project with the ultimate goal of fusing CYGNSS-derived wetland with future wetland products from the NASA-ISRO Synthetic Aperture Radar (NISAR) mission. The primary motivation of this work was to develop the GNSS-R/SAR framework for future CYGNSS/NASA-ISRO (NISAR) activities.

Santi et al. [39,40] presented a series of Artificial Neural Network (ANN) implementations to use CYGNSS SNR observations to estimate various geophysical properties. Specifically, a multi-layer ANN model was trained to estimate AGB and tree height using CYGNSS SNR observations which uses training and reference data from the Geocarbon Pan-tropical forest maps by Avitabile et al. and tree height maps derived by the Geoscience

Laser Altimeter System (GLAS) instrument aboard the Ice, Cloud, and land Elevation Satellite (ICESat-1)

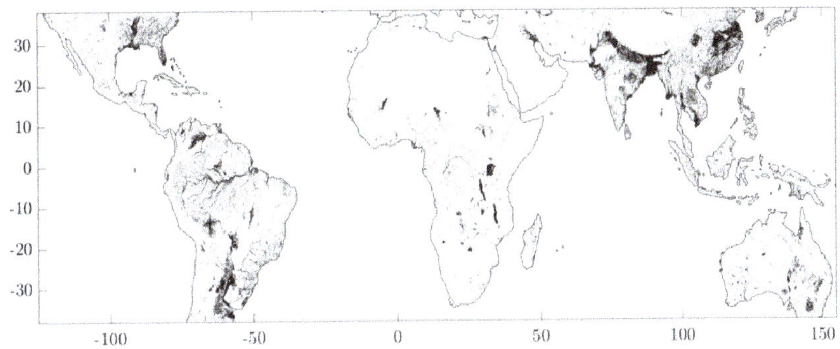


**Figure 6.** The 1−km land water body mask derived from CYGNSS L1 coherence-detection by M. Al-Khaldi et al. The map was generated using 1 year of CYGNSS data.

Recent theoretical developments in GNSS-R and scattering models over rough surfaces were also presented in this session [41]. Results and model validation efforts over the San Luis Valley, CO, showed promising matchups with the CYGNSS L1 Bistatic Radar Cross Section (BRCS). However, these studies also stressed the importance of accurately capturing and modeling multi-scale surface roughness and topographic effects.

This session also included several topics about retrieval and characterization of SMC and vegetation properties. A machine learning approach using random forest regression was presented to estimate SMC at a global scale [42]. The model was trained and generated using the reflectivity and relevant geophysical data layers, such as Normalized Difference Vegetation Index (NVDI) and DEM maps, along with SMC from the International Soil Moisture Network (ISMN). Results were validated over the Contiguous United States (CONUS) with consistent spatial patterns and magnitudes as those observed by the SMAP mission.

Similarly, by deriving a semi-analytical expression of vegetation transmissivity, and soil reflectivity, it was demonstrated how SMAP and CYGNSS observations can be concurrently leveraged to estimate either the Vegetation Optical Depth (VOD) and the SMC [43] (Figure 7). Additionally, a newly initiated study by Pu et al. sought to examine the effects of assimilation SMAP and CYGNSS SMC in near-surface weather forecasting models. This ongoing study showed that CYGNSS-derived SMC assimilation is on par with SMAP-based forecasts. However, further investigation is required to better quantify the added value by CYGNSS for SMC assimilation. Prior work by Pu et al. [44,45], however, demonstrates that strongly coupled land-surface assimilation frameworks which assimilate in situ, or SMAP, soil moisture can provide additional short-range and near-surface weather forecasting. The effects of CYGNSS SMC assimilation will be reported in future studies.

The use of CYGNSS high resolution ~3 km SMC maps was demonstrated by an experimental study supported by SERVIR [46]. This CYGNSS product, along with the ~3 km Land Information Systems (LIS) surface model, Integrated Multi-satellitE Retrievals for GPM (IMERG) rainfall, and Visible Infrared Imaging Radiometer Suite (VIRRS) vegetation information, was leveraged for locust monitoring applications in East Africa. SERVIR is a joint venture between NASA and the U.S. Agency for International Development. The SERVIR program is collaborating with regional entities in West and East Africa to evaluate the utility of Earth observations (EO), and their contribution to operational desert locust monitoring and tracking systems supported by the United Nations (UN) Food and Agriculture Organization (FAO).

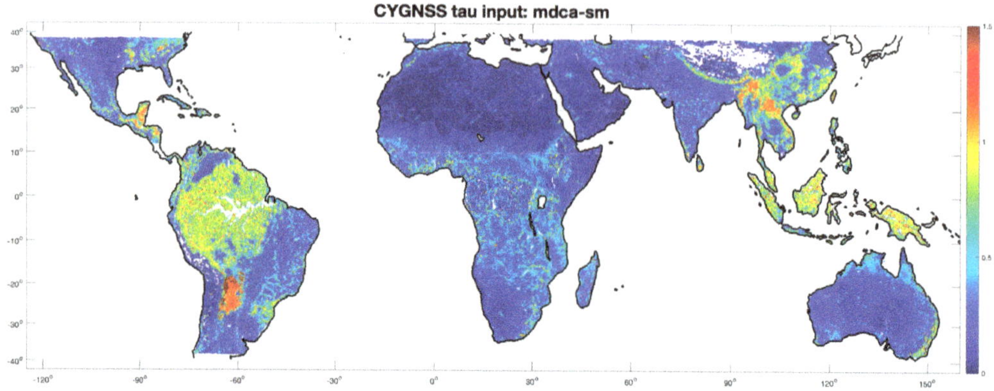


**Figure 7.** Example of CYGNSS-derived VOD by Xu et al.

Session 4: Land Processes II

Several studies have demonstrated the importance of coherent scattering of the GPS reflected signal over land [23,36,47–49]. In general, for coherent returns to exist, the GPS signal has to be reflected over very large flat areas with the surface RMSE height lower than ~10 cm at L-Band as presented in [50]. Recent analysis suggests that coherent returns are therefore mostly associated with the presence of inland water bodies [51,52]. This session will provide updates on current research using the SNR to detect inland water bodies. For land applications, surface roughness is a critical parameter for scattering models as detailed in the session. This surface roughness can be estimated using lidar airborne measurements [53]. Reflectometry has been used in various studies for land applications especially for retrieving SMC [53–56]. In this session, an update on the existing and potential calibration and validation sites for soil moisture applications is provided.

The first three presentations demonstrated the potential of the use of CYGNSS over land surfaces to detect and monitor water extent over short time scales [57]. The first method analyzed the distribution of time series of SNR for a given pixel to define dry and wet areas. With this method, a dynamic water extend mask can therefore be derived. Several case studies showed good correlation with rainfall events and the Pekel mask considered as the truth. Sensitivity of CYGNSS to water extent changes over short time scales was demonstrated by analyzing the spatial SNR variations with the seasonality of the Pascagoula River. Better detection results were obtained with a higher sampling rate. Finally, an investigation on the impact of land characteristics, such as SMC, surface water, topography and Vegetation Water Content (VWC) on the coherence of CYGNSS reflected signals was presented. The coherence was quantified using the tracked carrier phase from CYGNSS raw IF data. It was concluded that the coherent scattering term is most often present over inland water bodies.

The next topic of this session was assessing the impact of surface roughness, which is a critical parameter for correctly assessing the scattering of GNSS signals over land surfaces. A theoretical model was developed to decompose the surface into topography elevation and slopes, small-scale surface roughness, and surface correlation length. This new parametrization of the surface was tested using several classical forward-scattering models, including Geometrical Optics, Physical Optics, and Numerical Maxwell Model 3D (NMM3D). The modeled DDMs were then compared to CYGNSS-derived DDMs, showing a good agreement. In May 2020, experimental activities were performed along two cal/val sites over the San Luis Valley to characterize the small-scale surface roughness using lidar. Lidar-derived DEMs were generated at two different spatial resolutions, ~10 cm and ~30 cm.

The last three presentations dealt with actual and potential SMC cal/val sites for CYGNSS. First, an update was provided about the status of the SOILSCAPE in situ SMC sensor networks, that were installed in two cal/val sites in the San Luis Valley. All sensors were shown to be working properly. Calibration of the SMC measurements is in process. New potential cal/val sites have been identified in the USA (Walnut Glutch, AZ, White Sands, NM) and in New Zealand. New Zealand cal/val sites are particularly of interest because the next generation GNSS-R receiver [58] is going to be installed on a regional Air New Zealand commercial aircraft to complement CYGNSS data, and to test this new receiver in preparation of a potential CYGNSS follow-on mission. An extensive analysis of the coverage provided by one Air New Zealand commercial aircraft was performed to identify the best cal/val site locations for SMC and wetlands studies. These locations were cross-compared with the CYGNSS coverage, showing several overlapping sites. This project will advance terrestrial and coastal retrievals, by generating long-term datasets with high spatial resolution and high spatiotemporal sampling. In addition, in support of this project, the deployment of the next-generation GNSS-R receiver on one Air New Zealand aircraft was simulated to get a better understanding of the differences of the reflected signal acquired on an aircraft as compared to a satellite. Flight paths, flight frequency, and GNSS-R coverage were analyzed. It was found to have promising coverage, except for some mountain regions in the South Island (Figure 8). Installing this new receiver on one aircraft will provide a large amount of information over both ocean and several land cover types. It was concluded that the extension to a fleet of regional aircrafts will generate an unprecedented GNSS-R scientifically valuable dataset.

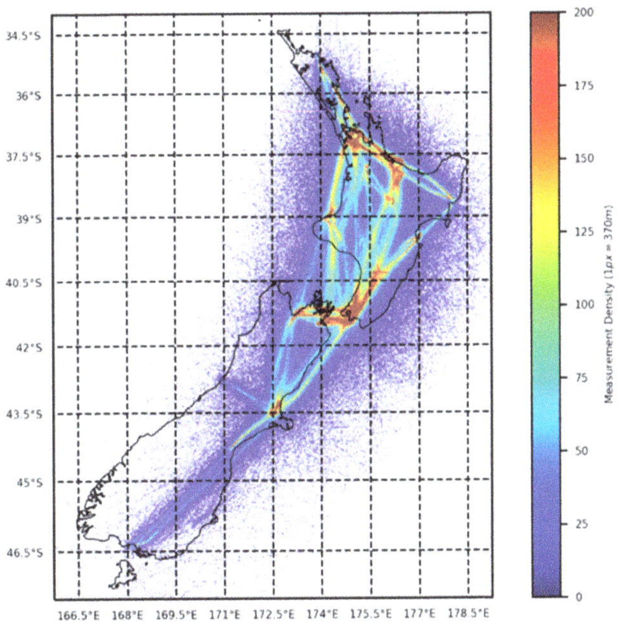


**Figure 8.** Simulated number of GNSS-R measurements over 1-year, considering just 1-single Air New Zealand aircraft by Linnabary et al.

### 2.3. Ocean Surfaces

The primary goal of the CYGNSS mission is to provide insight into the rapid intensification of TCs and to better measure their windspeeds. To this end, a number of studies since the mission's inception have focused on the application of CYGNSS data products to the study of TCs and tropical convection. Up until the release of the v3 winds data product, most of these studies have used simulated CYGNSS data for algorithm development. For example, Morris and Ruf developed parametric methods for filling in the gaps in the CYGNSS measurements [59], and for characterizing the size, structure, and strength of a TC [60]. Several studies have used simulated CYGNSS winds to develop assimilation methods for improving tropical cyclone track and intensity forecasts [61–64]. Simulated CYGNSS winds have also been used to demonstrate convective activity monitoring [65]. With the successful development of the v3 winds data products, these studies have transitioned to the application of actual CYGNSS observations for applications such as improving storm center estimates [66] and developing a surface heat flux product [25]. Sessions 5–6 of the Science Team meeting focus on the progress made for ocean-based applications: TCs and other science applications, improving weather forecasts, and ocean altimetry.

Session 5: Altimetry and Tropical Cyclones and Tropical Convection I

A new approach has been developed to incorporate CYGNSS winds into storm surge models, using machine learning to determine a predictive model for CYGNSS winds in the future, based on Global Forecast System (GFS) forecast data and parametric model representation of the CYGNSS inner core TCs winds. The use of CYGNSS in conjunction with ancillary satellite data to measure diurnal wind variations was presented, including also initial results from a "rapid change detector" that is being developed to detect rapid changes to the wind field, which could be used to signal convective activity. Preliminary results of identifying early cyclogenesis from easterly waves using CYGNSS data were highlighted. The modality of the identification comes from the findings of increased L2 latent heat flux and surface wind speeds leading up to tropical cyclone genesis. An update on comparisons of CYGNSS MSSs to those derived from buoy measurements and a coupled atmospheric-wave-ocean model, confirmed the importance of including short wave contributions to the MSS in the models to match the observed MSS values. Validation efforts have been performed for the CYGNSS CDR v.1 products using data from microwave radiometers, including SMAP, WindSat and Advanced Microwave Scanning Radiometer (AMSR-2). A comparison of the wind speeds indicates that the CDR v.1 Young Sea/Limited Fetch (YSLF) winds are more accurate than the v2.1 wind speeds but still are poorly correlated to the other microwave radiometer wind speeds. The CYGNSS-NOAA wind speeds however, correlated well with the SMAP, AMSR2 and WindSat winds. Results of data assimilation of CYGNSS L1 DDMs into ECMWF background winds, where the resulting wind speeds, with and without the CYGNSS L1 DDM, were compared to scatterometer winds from the ASCAT-A, and B and the OceanSat Scatterometer (OSCAT).

The assimilation of the CYGNSS data (Figure 9) improved the ECMWF background at specular points predominantly for wind speeds < 15 m/s. Matched filter retrievals were applied for maximum CYGNSS TCs winds, using the Willoughby–Darling–Rahn model [67]. After eliminating storms that violated the model's assumptions, good results were obtained for the matched filter output. An update of CYGNSS L1 data for ocean altimetry was presented, showing a number of improvements and corrections, including waveform pre-processing, delay compensation, and re-tracking. New results were presented for a case study in the Caribbean. Finally, an overview of several CYGNSS-based altimetry methods was presented, showing an accuracy on the order of several meters. Future improvements are likely to be achieved in accounting better for tides, and ionosphere and troposphere delays.

Session 6: Tropical Cyclones and Tropical Convection II

The final session of the June 2020 CYGNSS Science Team meeting focused on the use of the L2 wind speed products for improving forecasts and wind field analyses of TCs and tropical convection. CYGNSS provides wind speed data that can be used for storm surge predictions, which is a major source of destruction for communities lying in the paths of a storm. Simulations showed the storm surge from the Hurricane Harvey (2017), which were driven in part by wind speed observations from CYGNSS. Efforts were performed to diagnose the structure of TCs' wind fields using the CYGNSS wind speed observations and found that, given good quality data from CYGNSS, the size of the wind field can be determined. A methodology for creating storm wind fields that move with the storm over time was also presented (Figure 10). These datasets are available in PODAAC, alongside other CYGNSS products [68].

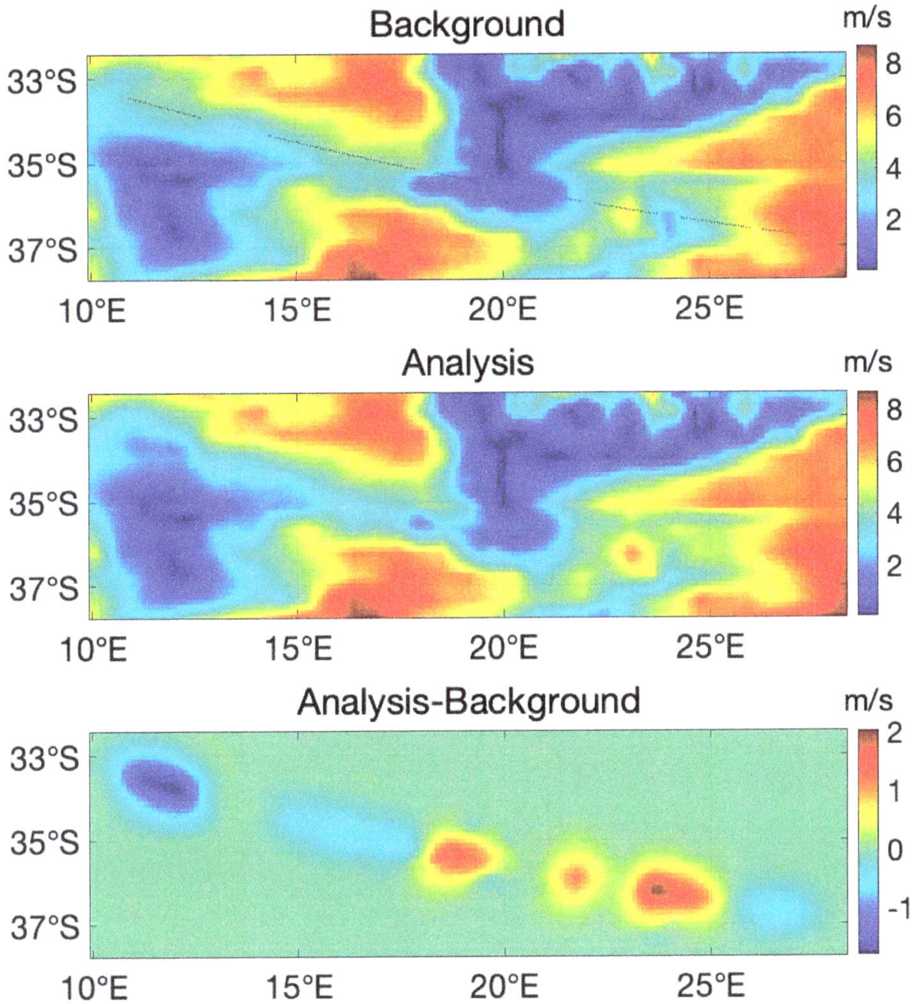


**Figure 9.** Assimilating CYGNSS DDMs into ECMWF background winds to improve surface wind forecasts: [**top**] demonstrating the ECMWF background at 150 km resolution, [middle] analysis field with the DDM assimilation, and [**bottom**] the difference between the two. Image by F. Huang et al.

Apart from analysis of storm wind fields, the observations from CYGNSS are also being used to improve weather forecasts. Efforts to improve the initialization of weather model simulations with CYGNSS observations were presented, including also information about where CYGNSS improves the forecasts of TCs. Throughout the session, team members compared techniques for quality control and optimal ingestion of wind speed data from CYGNSS.

CYGNSS is also playing a role in studies of tropical convection, where other observations suffer in heavy precipitation. Simulations of tropical convection with a coupled atmosphere, ocean, wave model were described, along with a comparison of simulated winds with those observed by CYGNSS. A number of on-going investigations examined CYGNSS observations near tropical oceanic thunderstorms, including an examination of the characteristics of the data in storms with and without lightning [69]. The importance of wind-driven fluxes on the Madden Julian Oscillation (MJO), using CYGNSS data was highlighted [70]. Finally, it is worth highlighting that several team members are using CYGNSS data to study the processes of air-sea interaction in weather and climate science [71].

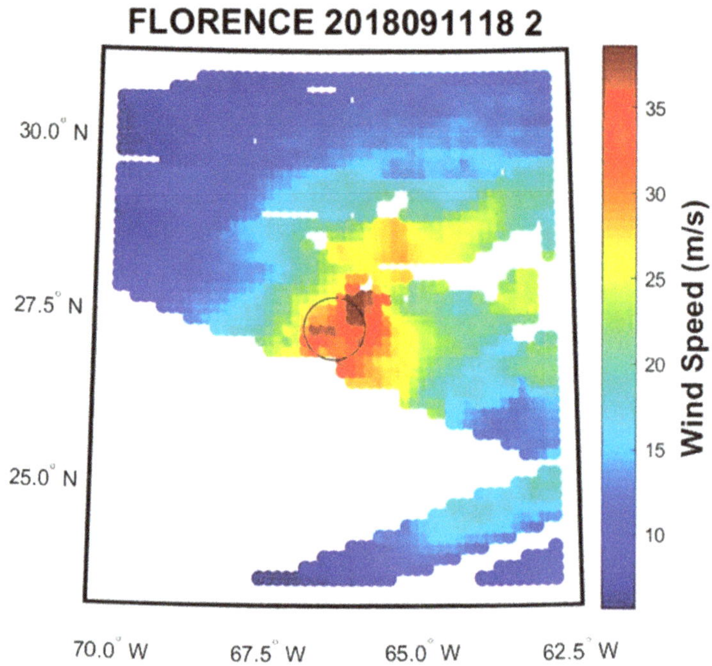


**Figure 10.** CYGNSS-derived wind speed retrievals over the Hurricane Florence by D. Mayers et al.

## 3. Conclusions and Final Remarks

NASA's CYGNSS 8 microsatellite-constellation is the first-ever operational GNSS-R mission, revolutionizing several scientific fields within our community, and attracting new members from different communities such as microwave radiometry and SAR. The number of GNSS-R scientific publications within our community continues to increase significantly thanks to the open-access CYGNSS products available at the PODAAC. This was a nice start, but the CYGNSS Science Team is a dynamic community always looking to improve products and apply them to new scientific applications. In this scenario, around 50% of the Science Team members are now involved with applications over land surfaces, including SMC, AGB, and water surface. The key enabling events were improving calibration procedures, product spatial resolution (e.g., reducing date integration to 0.5 s from 1 s), geolocation, metadata quality, and a more agile web for downlinking data.

The recent mission extension by NASA will enable the continuation of investigations by the CYGNSS Science Team. We will continue to contribute to the wider remote sensing community with our new and novel results, with the objective of advancing our understanding of key dynamic processes over the Earth's surface. With this article, we welcome new members to participate in the CYGNSS mission.

**Author Contributions:** Conceptualization, C.R. and H.C.-L.; methodology, C.R. and H.C.-L.; writing—original draft preparation, C.R., H.C.-L., R.A., A.B., J.A.C., M.M. and A.W.; writing—review and editing, C.R., H.C.-L., R.A., A.B., J.A.C., M.M. and A.W.; visualization, C.R. and H.C.-L.; supervision, C.R. and H.C.-L.; project administration, C.R.; funding acquisition, C.R. All authors have read and agreed to the published version of the manuscript.

**Funding:** This research was supported in part by the NASA Science Mission Directorate contract NNL13AQ00C with the University of Michigan.

**Institutional Review Board Statement:** Not applicable.

**Informed Consent Statement:** Not applicable.

**Conflicts of Interest:** The authors declare no conflict of interest.

## Abbreviations

Acronyms

Above Ground Biomass (AGB)
Advanced Land Observing Satellite-2 (ALOS-2)
Advanced Microwave Scanning Radiometer (AMSR-2)
Advanced SCATterometer (ASCAT)
Artificial Neural Network (ANN)
Bistatic Radar Cross Section (BRCS)
Climate Data Record (CDR)
Contiguous United States (CONUS)
Cyclone Global Navigation Satellite System (CYGNSS)
Delay Doppler Map (DDM)
Delay Doppler Mapping Instrument (DDMI)
Digital Elevation Model (DEM)
Disaster Monitoring System-1 (DMC-1)
Earth Observation (EO)
Earth System Science Pathfinder (ESSP)
End-to-End Simulator (E2ES)
European Centre for Medium-Range Weather Forecasts (ECMWF)
European Space Agency (ESA)
Flight Model (FM)
Food and Agriculture Organization (FAO)
Fully Developed Seas (FDS)
Physical Oceanography Distributed Active Archive Center (PODAAC)
Geophysical Model Function (GMF)
Global Forecast System (GFS)
Global Navigation Satellite Systems (GNSS)
Global Positioning System (GPS)
Institute of Space Sciences (ICE)
Institute of Space Studies of Catalonia (IEEC)
Intermediate Frequency (IF)
International Soil Moisture Network (ISMN)
Jet Propulsion Laboratory (JPL)
Land Information Systems (LIS)
Low Earth Orbit (LEO)
Madden Julian Oscillation (MJO)
Maxwell Model 3D (NMM3D)
Mean-Square Slope (MSS)
Modern-Era Retrospective analysis for Research and Applications (MERRA-2)
Multi-satellitE Retrievals for GPM (IMERG)
National Aeronautics and Space Administration (NASA)
NASA-ISRO (NISAR)
National Oceanic and Atmospheric Administration (NOAA)
Normalized Difference Vegetation Index (NVDI)
OceanSat Scatterometer (OSCAT)
Pseudo-Random Noise (PRN)Research Opportunities in Space and Earth Science (ROSES)
Science Data Record (SDR)
Signal-to-Noise Ratio (SNR)
Soil Moisture Active Passive (SMAP)
Soil Moisture Content (SMC)
Soil Moisture Ocean Salinity (SMOS)Space-borne Imaging Radar-C (SIR-C)

Synthetic Aperture Radar (SAR)
TechDemoSat-1 (TDS-1)
Tropical Cyclone (TC)
United Kingdom (UK)United Nations (UN)
University Corporation for Atmospheric Research (UCAR)
University of Michigan (UM)
Vegetation Optical Depth (VOD)
Vegetation Water Content (VWC)
Visible Infrared Imaging Radiometer Suite (VIRRS)
Wave-Watch 3 (WW3)
Young Sea/Limited Fetch (YSLF)

## References


1. Ruf, C.; Unwin, M.; Dickinson, J.; Rose, R.; Rose, D.; Vincent, M.; Lyons, A. CYGNSS: Enabling the Future of Hurricane Prediction [Remote Sensing Satellites]. *IEEE Geosci. Remote Sens. Mag.* **2013**, *1*, 52–67. [CrossRef]
2. Ruf, C.; Chang, P.S.; Clarizia, M.P.; Gleason, S.; Jelenak, Z.; Majumdar, S.; Morris, M.; Murray, J.; Musko, S.; Posselt, D.; et al. *CYGNSS Handbook. Cyclone Global Navigation Satellite System*; Michigan Publishing: Ann Arbor, MI, USA, 2016. Available online: https://clasp-research.engin.umich.edu/missions/cygnss/reference/cygnss-mission/CYGNSS_Handbook_April2016.pdf (accessed on 2 February 2021).
3. Ruf, C.; Atlas, R.; Chang, P.S.; Clarizia, M.P.; Garrison, J.L.; Gleason, S.; Katzberg, S.J.; Jelenak, Z.; Johnson, J.T.; Majumdar, S.J.; et al. New Ocean Winds Satellite Mission to Probe Hurricanes and Tropical Convection. *Bull. Am. Meteorol. Soc.* **2015**, *97*, 385–395. [CrossRef]
4. Ruf, C.; Cardellach, E.; Clarizia, M.P.; Galdi, C.; Gleason, S.T.; Paloscia, S. Foreword to the Special Issue on Cyclone Global Navigation Satellite System (CYGNSS) Early on Orbit Performance. *IEEE J. Sel. Top. Appl. Earth Obs. Remote Sens.* **2019**, *12*, 3–6. [CrossRef]
5. Ruf, C.; Asharaf, S.; Balasubramaniam, R.; Gleason, S.; Lang, T.; McKague, D.; Twigg, D.; Waliser, D. In-Orbit Performance of the Constellation of CYGNSS Hurricane Satellites. *Bull. Am. Meteorol. Soc.* **2019**, *100*, 2009–2023. [CrossRef]
6. Carreno-Luengo, H.; Luzi, G.; Crosetto, M. Sensitivity of CYGNSS Bistatic Reflectivity and SMAP Microwave Radiometry Brightness Temperature to Geophysical Parameters over Land Surfaces. *IEEE J. Sel. Top. Appl. Earth Obs. Remote Sens.* **2019**, *12*, 107–122. [CrossRef]
7. Carreno-Luengo, H.; Luzi, G.; Crosetto, M. Above-Ground Biomass Retrieval over Tropical Forests: A Novel GNSS-R Approach with CyGNSS. *Remote Sens.* **2020**, *12*, 1368. [CrossRef]
8. Warnock, A.; Ruf, C. Response to Variations in River Flowrate by a Spaceborne GNSS-R River Width Estimator. *Remote Sens.* **2019**, *11*, 2450. [CrossRef]
9. Martin-Neira, M. A Passive Reflectometry and Interferometry System (PARIS): Application to Ocean Altimetry. *ESA J.* **1993**, *17*, 331–355.
10. Lowe, S.T.; Labrecque, J.L.; Zuffada, C.; Romans, L.J.; Young, L.E.; Hajj, G.A. First spaceborne observation of an Earth-reflected GPS signal. *Radio Sci.* **2002**, *37*, 7-1–7-28. [CrossRef]
11. Gleason, S.; Hodgart, S.; Sun, Y.; Gommenginger, C.; Mackin, S.; Adjrad, M.; Unwin, M. Detection and Processing of bistatically reflected GPS signals from low Earth orbit for the purpose of ocean remote sensing. *IEEE Trans. Geosci. Remote Sens.* **2005**, *43*, 1229–1241. [CrossRef]
12. Unwin, M.; Jales, P.; Blunt, P.; Duncan, S. Preparation for the First Flight of SSTL's Next Generation Space GNSS Receivers. In Proceedings of the 6th ESA/European Workshop Satellite NAVITEC GNSS Signals Signal Processor, Noordwijk, The Netherlands, 5–7 December 2012; pp. 1–6.
13. Carreno-Luengo, H.; Lowe, S.; Zuffada, C.; Esterhuizen, S.; Oveisgharan, S. Spaceborne GNSS-R from the SMAP Mission: First Assessment of Polarimetric Scatterometry over Land and Cryosphere. *Remote Sens.* **2017**, *9*, 362. [CrossRef]
14. CYGNSS Mission Teams. Available online: https://clasp-research.engin.umich.edu/missions/cygnss/mission-teams.php (accessed on 3 March 2021).
15. Ruf, C.; Gleason, S.; Jelenak, Z.; Katzberg, S.; Ridley, A.; Rose, R.; Scherrer, J.; Zavorotny, V. The CYGNSS Nanosatellite Constellation Hurricane Mission. In Proceedings of the IEEE International Geoscience and Remote Sensing Symposium, Munich, Germany, 22–27 July 2012; pp. 214–216.
16. Dickinson, J.; Ruf, C.; Rose, R.; Ridley, A.; Walls, B. CYGNSS: The Cyclone Global Navigation Satellite System's CubeSat Foundations. In Proceedings of the 12th Annual Joint Agency Commercial Imagery Evaluation (JACIE) Workshop, St. Louis, MO, USA, 16–18 April 2013.
17. Ruf, C.; Lyons, A.; Ward, A. NASA Intensifies Hurricane Studies with CYGNSS. *Earth Obs. NASA* **2013**, *25*, 12–21.
18. Gleason, S.; Ruf, C.; Clarizia, M.P.; O'Brien, A. Calibration and Unwrapping of the Normalized Scattering Cross Section for the Cyclone Global Navigation Satellite System (CYGNSS). *IEEE Trans. Geosci. Remote Sens.* **2016**, *54*, 2495–2509. [CrossRef]

19. CYGNSS Data Products. Available online: https://clasp-research.engin.umich.edu/missions/cygnss/data-products.php (accessed on 3 March 2021).
20. Ruf, C. Mission Update. In Proceedings of the CYGNSS Science Team Meeting, Ann Arbor, MI, USA, 6 June 2020.
21. Wang, T.; Zavorotny, V.U.; Johnson, J.; Yi, Y.; Ruf, C.; Gleason, S.; McKague, D.; Hwang, P.; Rogers, E.; Chen, S.; et al. Improvement of CYGNSS Level 1 Calibration Using Modeling and Mesurements of Ocean Surface Mean Square Slope. In Proceedings of the 2020 IEEE International Geoscience and Remote Sensing Symposium (IGARSS), Waikoloa, HI, USA, 26 September–2 October 2020; pp. 5909–5912.
22. Cardellach, E.; Nan, Y.; Li, W.; Padulles, R.; Ribo, S.; Rius, A. Variational Retrievals of High Winds Using Uncalibrated CYGNSS Observables. *Remote Sens.* **2020**, *12*, 3930. [CrossRef]
23. Gleason, S.; O'Brien, A.; Russel, A.; Al-Khaldi, M.M.; Johnson, J.T. Geolocation, Calibration and Surface Resolution of CYGNSS GNSS-R Land Observations. *Remote Sens.* **2020**, *12*, 1317. [CrossRef]
24. *NOAA CYGNSS Level 2 Science Wind Speed 25-km Product*, Version 1.1 ed; PO.DAAC: Pasadena, CA, USA, 2020. Available online: https://doi.org/10.5067/CYGNN-22511 (accessed on 21 March 2021).
25. Crespo, J.A.; Posselt, D.J.; Asharaf, S. CYGNSS Surface Heat Flux Product Development. *Remote Sens.* **2019**, *11*, 2294. [CrossRef]
26. *CYGNSS Level 2 Ocean Surface Heat Flux Climate Data Record*, Version 1.0 ed; PO.DAAC: Pasadena, CA, USA. Available online: https://doi.org/10.5067/CYGNS-C2H10 (accessed on 21 March 2021).
27. Ruf, C.S.; Chew, C.; Lang, T.; Morris, M.G.; Nave, K.; Ridley, A.; Balasubramaniam, R. A New Paradigm in Earth Environmental Monitoring with the CYGNSS Small Satellite Constellation. *Sci. Rep.* **2018**, *8*, 8782. [CrossRef]
28. Rodriguez-Alvarez, N.; Bosch-Lluis, X.; Camps, A.; Vall-Llossera, M.; Valencia, E.; Marchan-Hernandez, J.F.; Ramos-Perez, I. Soil Moisture Retrieval Using GNSS-R Techniques: Experimental Results Over a Bare Soil Field. *IEEE Trans. Geosci. Remote Sens.* **2009**, *47*, 11. [CrossRef]
29. Camps, A.; Park, H.; Pablos, M.; Foti, G.; Gommenginger, C.P.; Liu, P.-W.; Judge, J. Sensitivity of GNSS-R Spaceborne Observations to Soil Moisture and Vegetation. *IEEE J. Sel. Top. Appl. Earth Obs. Remote Sens.* **2016**, *9*, 4730–4742. [CrossRef]
30. Chew, C.; Shah, R.; Zuffada, C.; Hajj, G.; Masters, D.; Mannucci, A.J. Demonstrating Soil Moisture Remote Sensing with Observations from the UK TechDemoSat-1 Satellite Mission. *Geophys. Res. Lett.* **2016**, *43*, 3317–3324. [CrossRef]
31. Eroglu, O.; Kurum, M.; Boyd, D.; Gurbuz, A.C. High Spatio-Temporal Resolution CYGNSS Soil Moisture Estimates Using Artificial Neural Networks. *Remote Sens.* **2019**, *11*, 2272. [CrossRef]
32. Nghiem, S.V.; Zuffada, C.; Shah, R.; Chew, C.; Lowe, S.T.; Mannucci, A.J.; Cardellach, E.; Brakenridge, G.R.; Geller, G.; Rosenqvist, A. Wetland Dynamics Monitoring with Global Navigation Satellite System Reflectometry. *AGU Earth Space Sci.* **2017**, *4*, 16–39. [CrossRef]
33. Zuffada, C.; Chew, C.; Nghiem, S.V. GNSS-R Algorithms for Wetlands Observations. In Proceedings of the 2017 IEEE International Geoscience and Remote Sensing Symposium (IGARSS), Fort Worth, TX, USA, 23–28 July 2017; pp. 1126–1129.
34. Entekhabi, D.; Njoku, E.G.; O'Neill, P.E.; Kellogg, K.H.; Crow, W.T.; Edelstein, W.N.; Entin, J.K.; Goodman, S.D.; Jackson, T.J.; Johnson, J.; et al. The Soil Moisture Active Passive (SMAP) Mission. *Proc. IEEE* **2010**, *98*, 704–716. [CrossRef]
35. Carreno-Luengo, H.; Ruf, C.; Warnock, A.; Brunner, K. Investigating the Impact of Coherent and Incoherent Scattering Terms in GNSS-R Delay Doppler Maps. In Proceedings of the 2020 IEEE International Geoscience and Remote Sensing Symposium (IGARSS), Waikoloa, HI, USA, 26 September–2 October 2020; pp. 6202–6205.
36. Al-Khaldi, M.M.; Johnson, J.T.; Gleason, S.; Loria, E.; O'Brien, A.J.; Yi, Y. An Algorithm for Detecting Coherence in Cyclone Global Navigation Satellite System Mission Level-1 Delay-Doppler Maps. *IEEE Trans. Geosci. Remote Sens.* **2021**, *59*, 4454–4463. [CrossRef]
37. Pekel, J.-F.; Cottam, A.; Gorelick, N.; Belward, A.S. High-resolution mapping of global surface water and its long-term changes. *Nature* **2016**, *540*, 418–422. [CrossRef]
38. Russo, I.M.; di Bisceglie, M.; Galdi, C.; Lavalle, M.; Zuffada, C. Wave Coherence in GNSS Reflectometry: A Signal Processing Point of View. In Proceedings of the 2020 IEEE International Geoscience and Remote Sensing Symposium (IGARSS), Waikoloa, Hawaii, USA, 26 September–2 October 2020; pp. 6214–6217.
39. Santi, E.; Paloscia, S.; Pettinato, S.; Fontanelli, G.; Clarizia, M.P.; Comite, D.; Dente, L.; Guerriero, L.; Pierdicca, N.; Floury, N. Remote Sensing of Forest Biomass Using GNSS Reflectometry. *IEEE J. Sel. Top. Appl. Earth Obs. Remote Sens.* **2020**, *13*, 2351–2368. [CrossRef]
40. Santi, E.; Pettinato, S.; Paloscia, S.; Clarizia, M.P.; Dente, L.; Guerriero, L.; Comite, D.; Pierdicca, N. Soil Moisture and Forest Biomass retrieval on a global scale by using CyGNSS data and Artificial Neural Networks. In Proceedings of the IEEE International Geoscience and Remote Sensing Symposium, Waikoloa, HI, USA, 26 September–2 October 2020; pp. 214–216.
41. Campbell, J.D.; Melebari, A.; Moghaddam, M. Modeling the Effects of Topography on Delay-Doppler Maps. *IEEE J. Sel. Top. Appl. Earth Obs. Remote Sens.* **2020**, *13*, 1740–1751. [CrossRef]
42. Senyurek, V.; Lei, F.; Boyd, D.; Gurbuz, A.C.; Kurum, M.; Moorhead, R. Evaluations of Machine Learning-Based CYGNSS Soil Moisture Estimates against SMAP Observations. *Remote Sens.* **2020**, *12*, 3503. [CrossRef]
43. Yueh, S.H.; Shah, R.; Chaubell, M.J.; Hayashi, A.; Xu, X.; Colliander, A. A Semiempirical Modeling of Soil Moisture, Vegetation, and Surface Roughness Impact on CYGNSS Reflectometry Data. *IEEE Trans. Geosci. Remote Sens.* **2020**. [CrossRef]
44. Lin, L.; Pu, Z. Improving Near-Surface Short-Range Weather Forecasts Using Strongly Coupled Land–Atmosphere Data Assimilation with GSI-EnKF. *Mon. Weather Rev.* **2020**, *148*, 2863–2888. [CrossRef]

45. Lin, L.-F.; Pu, Z. Examining the Impact of SMAP Soil Moisture Retrievals on Short-Range Weather Prediction under Weakly and Strongly Coupled Data Assimilation with WRF-Noah. *Mon. Weather Rev.* **2019**, *147*, 4345–4366. [CrossRef]
46. Chew, C.C.; Small, E.E. Soil Moisture Sensing Using Spaceborne GNSS Reflections: Comparison of CYGNSS Reflectivity to SMAP Soil Moisture. *Geophys. Res. Lett.* **2018**, *45*, 4049–4057. [CrossRef]
47. Park, J.; Johnson, J.T.; O'Brien, A.; Lowe, S.T. An Examination of TDS-1 GNSS-R Returns over Land Surfaces. In Proceedings of the URSI Radio Science Meeting, Pasadena, CA, USA, 6–9 January 2016.
48. Loria, E.; O'Brien, A.; Gupta, I.J. Detection and Separation of Coherent Reflections in GNSS-R Measurements Using CYGNSS Data. In Proceedings of the IEEE International Geoscience and Remote Sensing Symposium, Valencia, Spain, 22–27 July 2018; pp. 3995–3998.
49. Dong, Z.; Jin, S. Evaluation of the Land GNSS-Reflected DDM Coherence on Soil Moisture Estimation from CYGNSS Data. *Remote Sens.* **2021**, *13*, 570. [CrossRef]
50. Balakhder, A.M.; Al-Khaldi, M.M.; Johnson, J.T. On the Coherency of Ocean and Land Surface Specular Scattering for GNSS-R and Signals of Opportunity Systems. *IEEE Trans. Geosci. Remote Sens.* **2019**, *57*, 10426–10436. [CrossRef]
51. Al-Khaldi, M.M.; Johnson, J.T.; Gleason, S.; Chew, C.C.; Gerlein-Safdi, C.; Shah, R.; Zuffada, C. Inland Water Body Mapping Using CYGNSS Coherence Detection. *IEEE Trans. Geosci. Remote Sens.* **2021**. [CrossRef]
52. Gerlein-Safdi, C.; Ruf, C.S. A CYGNSS-Based Algorithm for the Detection of Inland Waterbodies. *Geophys. Res. Lett.* **2019**, *46*, 12065–12072. [CrossRef]
53. Turner, R.; Panciera, R.; Tanase, M.A.; Lowell, K.; Hacker, J.M.; Walker, J.P. Estimation of Soil Surface Roughness of Agricultural Soils Using Airborne LiDAR. *Remote Sens. Environ.* **2014**, *140*, 107–117. [CrossRef]
54. Kim, H.; Lakshmi, V. Use of Cyclone Global Navigation Satellite System (CYGNSS) Observations for Estimation of Soil Moisture. *Geophys. Res. Lett.* **2018**, *45*, 8272–8282. [CrossRef]
55. Al-Khaldi, M.M.; Johnson, J.T.; O'Brien, A.J.; Balenzano, A.; Mattia, F. Time-Series Retrieval of Soil Moisture Using CYGNSS. *IEEE Trans. Geosci. Remote Sens.* **2019**, *57*, 4322–4331. [CrossRef]
56. Clarizia, M.P.; Pierdicca, N.; Costantini, F.; Floury, N. Analysis of CYGNSS Data for Soil Moisture Retrieval. *IEEE J. Sel. Top. Appl. Earth Obs. Remote Sens.* **2019**, *12*, 2227–2235. [CrossRef]
57. Gerlein-Safdi, C.; Ruf, C. CYGNSS Constellation Provides New Insight on Tropical Wetlands Dynamics. In Proceedings of the ESA Annual Meeting, ESA-ESTEC, Noordwijk, The Netherlands, 3–6 August 2020.
58. Ruf, C.; Backhus, R.; Butler, T.; Chen, C.C.; Gleason, S.; Loria, E.; McKague, D.; Miller, R.; O'Brien, A.; van Nieuwstadt, L. Next Generation GNSS-R Instrument. In Proceedings of the 2020 IEEE International Geoscience and Remote Sensing Symposium, Waikoloa, HI, USA, 26 September–2 October 2020; pp. 3353–3356.
59. Morris, M.; Ruf, C.S. Determining Tropical Cyclone Surface Wind Speed Structure and Intensity with the CYGNSS Satellite Constellation. *J. Appl. Meteorol. Clim.* **2017**, *56*, 1847–1865. [CrossRef]
60. Morris, M.; Ruf, C.S. Estimating Tropical Cyclone Integrated Kinetic Energy with the CYGNSS Satellite Constellation. *J. Appl. Meteorol. Clim.* **2017**, *56*, 235–245. [CrossRef]
61. Zhang, S.; Pu, Z.; Posselt, D.J.; Atlas, R. Impact of CYGNSS Ocean Surface Wind Speeds on Numerical Simulations of a Hurricane in Observing System Simulation Experiments. *J. Atmos. Ocean. Technol.* **2017**, *34*, 375–383. [CrossRef]
62. Annane, B.; McNoldy, B.; Leidner, S.M.; Hoffman, R.; Atlas, R.; Majumdar, S.J. A Study of the HWRF Analysis and Forecast Impact of Realistically Simulated CYGNSS Observations Assimilated as Scalar Wind Speeds and as VAM Wind Vectors. *Mon. Weather Rev.* **2018**, *146*, 2221–2236. [CrossRef]
63. Leidner, S.M.; Annane, B.; McNoldy, B.; Hoffman, R.; Atlas, R. Variational Analysis of Simulated Ocean Surface Winds from the Cyclone Global Navigation Satellite System (CYGNSS) and Evaluation Using a Regional OSSE. *J. Atmos. Ocean. Technol.* **2018**, *35*, 1571–1584. [CrossRef]
64. Cui, Z.; Pu, Z.; Tallapragada, V.; Atlas, R.; Ruf, C.S. A Preliminary Impact Study of CYGNSS Ocean Surface Wind Speeds on Numerical Simulations of Hurricanes Harvey and Irma (2017). *Geophys. Res. Lett.* **2019**, *46*, 2984–2992. [CrossRef]
65. Park, J.; Johnson, J.T.; Yi, Y.; O'Brien, A.J. Using "Rapid Revisit" CYGNSS Wind Speed Measurements to Detect Convective Activity. *J. Sel. Top. Appl. Earth Obs. Remote Sens.* **2019**, *12*, 98–106. [CrossRef]
66. Mayers, D.; Ruf, C. Tropical Cyclone Center Fix Using CYGNSS Winds. *J. Appl. Meteorol. Clim.* **2019**, *58*, 1993–2003. [CrossRef]
67. Huang, F.; Garrison, J.L.; Leidner, S.M.; Annane, B.; Hoffman, R.N.; Grieco, G.; Stoffelen, A. A Forward Model for Data Assimilation of GNSS Ocean Reflectometry Delay-Doppler Maps. *IEEE Trans. Geosci. Remote Sens.* **2020**, *59*, 2643–2656. [CrossRef]
68. *CYGNSS Level 3 Storm Centric Grid Science Data Record*, Version 1.0 ed; PO.DAAC: Pasadena, CA, USA. Available online: https://doi.org/10.5067/CYGNS-L3S10 (accessed on 21 March 2021).
69. Lang, T. Comparing Winds Near Tropical Oceanic Precipitation Systems with and without Lightning. *Remote Sens.* **2020**, *12*, 3968. [CrossRef]
70. Bui, H.X.; Maloney, E.D.; Dellaripa, E.M.R.; Singh, B. Wind Speed, Surface Flux, and Intraseasonal Convection Coupling from CYGNSS Data. *Geophys. Res. Lett.* **2020**, *47*, e2020GL090376. [CrossRef]
71. Crespo, J.A.; Naud, C.M.; Posselt, D.J. CYGNSS Observations and Analysis of Low-Latitude Extratropical Cyclones. *J. Appl. Meteorol. Clim.* **2021**, *60*, 527–541. [CrossRef]

*remote sensing*

MDPI

*Technical Note*

# Spatial and Temporal Sampling Properties of a Large GNSS-R Satellite Constellation

**Jack Winkelried [1], Christopher Ruf [1,*] and Scott Gleason [2]**

[1] Department of Climate and Space Sciences and Engineering, University of Michigan, Ann Arbor, MI 48109, USA
[2] Daaxa LLC, Boulder, CO 80305, USA
* Correspondence: cruf@umich.edu

**Abstract:** Using large constellations of smallsats, mission designers can improve sampling density and coverage. We develop performance metrics that characterize key sampling properties for applications in numerical weather prediction and optimize orbit design parameters of the constellation with respect to those metrics. Orbits are defined by a set of Keplerian elements, and the relationship between those elements and the spatial and temporal coverage metrics are examined in order to maximize global and zonal (latitude-dependent) coverage. Additional optimization is performed by dividing a constellation into multiple orbit planes. An iterative method can be applied to this design process to compare the performance of current and previous designs. The main objective of this work is the design of optimized configurations of satellites in low Earth orbiting constellations to maximize the spatial and temporal sampling and coverage provided by its sensors. The key innovations developed are a new cost function which measures the temporal sampling properties of a satellite constellation, and the use of it together with existing cost functions for spatial sampling to design satellite constellations that optimize performance with respect to both performance metrics.

**Keywords:** constellation design; CYGNSS; GNSS reflectometry; SpOCK

**Citation:** Winkelried, J.; Ruf, C.; Gleason, S. Spatial and Temporal Sampling Properties of a Large GNSS-R Satellite Constellation. *Remote Sens.* **2023**, *15*, 333. https://doi.org/10.3390/rs15020333

Academic Editor: Mehrez Zribi

Received: 21 November 2022
Revised: 22 December 2022
Accepted: 28 December 2022
Published: 5 January 2023



## 1. Introduction

The proliferation of smallsats and the small but highly capable scientific sensors on them have ushered in a new era of Earth remote sensing from space. Technology development lifecycles have been shortened, new and improved sensors can reach space more quickly, both launch vehicle and satellite costs have been reduced dramatically, and it is now both possible and affordable to consider flying large constellations of remote sensing instruments [1]. Notably, large constellations of smallsats in low Earth orbit have the potential to significantly improve upon the spatial and temporal sampling densities provided by a single traditional large satellite [2]. The improvement in sampling density with number of spacecraft is generally intuitive—more of them spread out around the globe will tend to sample more places at the same time and to sample the same place more frequently. We consider here the optimization problems of:

1. Achieving a particular sampling density with the minimum number of satellites;
2. Maximizing the sampling density for a given number of satellites; and
3. Spreading out the samples at each location in time to better resolve diurnal variability and support the initialization of numerical weather prediction models.

The optimization is performed with respect to specific orbit parameters of the constellation, namely the number and orientation of orbit planes and the number of satellites in each plane. When quantifying the notion of sampling density, it is helpful to consider a specific type and design of remote sensing instrument. Our optimization study is performed with respect to a Global Navigation Satellite System Reflectometry (GNSS-R) sensor. These sensors are the receiver half of a bistatic radar system in which the transmitters are the existing constellation of navigation satellites, such as GPS and Galileo. Measurements of GNSS signals scattered back into space from the Earth surface contain information about a wide variety of scientifically valuable geophysical surface conditions [3–7]. The relevant instrument design parameters to be considered are the number of simultaneous GNSS surface reflections observed and the gain and field of view of the antenna through which the observations are made. The former design parameter affects the signal processing complexity and power requirement of the instrument, and the latter parameter affects the size and mass of the antenna. GNSS-R sensors are compatible with smallsats [8,9] and have been demonstrated to provide improved sampling density in small constellations [10], so are a natural type of measurement to consider for this study.

The satellite constellation design space being considered here is an extension of the point design used by the NASA CYGNSS mission [11]. Its constellation consists of 8 identical spacecraft spaced roughly equally around a single orbit plane at an altitude of ~525 km and an inclination angle of 35°. Each CYGNSS spacecraft carries a GNSS-R sensor capable of measuring 4 simultaneous surface reflections.

## 2. Discussion of Orbital Parameters, Sampling Process, and SpOCK

### 2.1. Constellation Design Baseline Assumptions

The achievable coverage of a constellation of remote sensing satellites is determined by both the orbit placements of the satellites within the constellation as well as the capability of the instruments carried on each satellite making the observations. Each individual satellite's orbit is defined by the traditional six Keplarian elements: eccentricity, semi-major axis, inclination, right ascension of the ascending node (RAAN), argument of periapsis (AoP) and true anomaly [12]. For purposes of remote sensing efficacy, the key design factors of the instrument itself include antenna configuration, number of parallel observations, and minimum observation signal strength. The principle design considerations for the constellation are the total number of satellites and their orbits relative to each other. Analyzing all the possible combinations of satellites, their orbits, and instrument configurations is not impractical. Therefore, we have made several assumptions and simplifications in order to isolate and evaluate the primary parameters driving the overall measurement coverage performance. This reduces the number of constellation design variables to a manageable and useful subset. The simulations below all include the following orbit and instrument configuration constraints and assumptions:

1. All satellites are in circular orbits (eccentricy = 0, AoP not relevant), all satellites within an orbital plane are equally spaced (true anomalies fixed), all satellites in the constellation are at the same altitude (semi-major axis fixed). The number of satellites, the angular spacing of the orbit planes and the inclinations of the orbit planes are the three remaining variables considered in the constellation design;
2. The instruments on each satellite are constrained as follows: a dual antenna configuration of either (a) the NASA CYGNSS mission antennas or (b) a larger dual antenna design for the higher altitude orbit simulations. The instruments are all capable of tracking up to 16 specular reflection measurements in parallel, based on the current estimated capability of the next generation of GNSS-R instruments [13]. The signal strength required for viable land and ocean observations is based on CYGNSS retrievals vs. range corrected gain (RCG), which considers both antenna gain and path losses for individual surface measurements [14,15].

Using the above assumptions and constraints on the GNSS-R constellation configuration, the design parameters considered have been reduced to a manageable number: number of planes (incl. number of satellites per plane), inclinations of the planes, and minimum observation RCG. Each is explored individually in the sections below. The results and analysis are intended to examine the impacts of the design parameters on sampling and Earth coverage properties for possible future constellation missions.

### 2.2. Overview of Observation Sampling Process and Instrument Considerations

#### 2.2.1. GNSS-R Surface Reflection Global Distribution

The global distribution of GNSS-R surface observations is the result of an asynchronous process between the orbits of the transmitting GNSS satellites (GPS, Galileo, SBAS, etc.) in Medium Earth Orbit (MEO, ~20,200 km altitude) and occasionally Geostationary orbit (~35,786 km altitude) and a constellation of Low Earth Orbit (less than 2000 km altitude) receivers below the transmitters. The GNSS-R observation point is the Earth surface location between the transmitter and receiver that results in a forward specular reflection in the direction of the receiver. An Illustration of three example GNSS-R reflection geometries is shown in Figure 1, with the transmitters above and outside the figure and a single LEO receiver capturing the parallel surface reflections. The specular reflection location point between transmitter and receiver at a single epoch can be calculated using an iterative process as described in [3]. The resulting total of GNSS-R surface reflection points is therefore driven by the pseudo-random alignment of the multiple GNSS satellites and the multiple LEO receivers that result in fortuitous surface reflection locations in the receiver antenna footprint of the GNSS-R instruments. The number of GNSS transmitters can well exceed 100 if multiple constellations are utilized (GPS, Galileo, Beidou, GLONASS) for example, while the optimal number of receivers can range into several dozen in varied orbits resulting in a dense web of overlapping surface observations. Given the large numbers of transmitters and receivers a simulation is used to generate these large data sets of surface observation points over time (nominally one day). The achieved GNSS-R observation coverage was then analyzed with respect to optimizing the spatial and temporal coverage metrics described below.

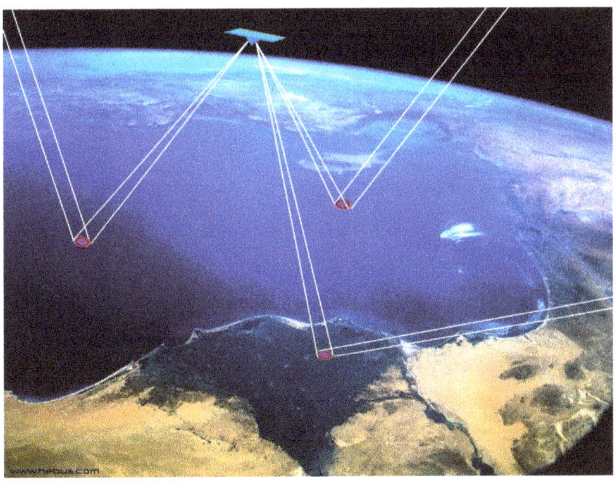


**Figure 1.** General illustration of three GNSS-R reflection specular points captures by a single LEO receiver. Reprinted with permission from ref. [16], Copyright: 2014, IEEE.

#### 2.2.2. GNSS-R Instrument Dependencies

The GNSS-R observation distributions are highly dependent on multiple instrument configuration parameters. The number of viable science observations generated by a given instrument on a single satellite is constrained by two primary instrument capabilities: (a) the processing bandwidth of the instrument and the resulting number of available surface specular points it is capable of processing into science observations and (b) the surface area coverage of the nadir oriented GNSS-R science antennas which are required to capture the reflected signals with sufficient SNR to be usable. Other secondary spacecraft specific system requirements such as satellite down-link capability, attitude knowledge and other aspects are not analyzed here and assumed to be sufficient for the scenarios studied in this analysis.

In this analysis, the processing bandwidth of the GNSS-R instrument considers two cases: the existing on-orbit performance of the CYGNSS GNSS-R instrument which is capable of tracking 4 parallel reflections from GPS only [4], and a next Generation GNSS-R instrument (NGRx) which has shown in initial testing to be capable of tracking up to 16 parallel GNSS reflections from both GPS and Galileo [13]. It is possible that at some point in the future instruments will be capable of tracking more parallel reflections. However, we believe that 16 is a realistic estimate of the state-of-the-art GNSS-R instrument capability and grounds this analysis in existing hardware performance.

The surface coverage of the GNSS-R instrument antennas is also a key element in filtering the actual observations suitable for science applications. As expected, reflections captured in the main high gain lobes of the reflection antennas result in higher SNR observations and generally better retrievals. In this regard, we have extensive data based on the achieved performance of the CYGNSS constellation where it has been observed that ocean wind speed observations can be achieved within the mission error requirements at range corrected gain (RCG) levels of 15 and above [17]. This allows us to assess additional instrument antenna configurations with respect to this threshold and make realistic judgments as to what specular points will result in viable science observations and which specular points to omit due to low quality. In the subsequent analysis, we simulate both the actual CYGNSS antennas as well as theoretical enhanced antennas to assess the performance of different antenna gains and beam widths, as well as at alternative spacecraft altitudes to provide a more complete trade-space with respect to various spacecraft instrument configurations.

### 2.3. SpOCK and Its Operation

Although there are a multitude of functional mission simulation tools available for commercial and research applications, the Spacecraft Orbital Characterization Kit (SpOCK) can better predict the performances of and provide data for mission inputs that operate in a manner similar to CYGNSS. The central capability of SpOCK is a high accuracy numerical propagator of spacecraft orbits and computations of ancillary parameters [18]. The C-based programming of SpOCK allows for data variables, including CYGNSS specular points, to be easily added or subtracted in accordance with user preferences.

When creating constellations for SpOCK simulations, users can select the orbital parameters of individual satellites as well as controlling data processing tools. In text-based input files, users first describe how many satellites will be in the constellation, then provide both the individual sets of orbital parameters of each satellite and the antenna pattern used by all satellites. These inputs are read by SpOCK's C-based simulator, and outputs are provided at a user-specified time delta in the form of multiple data files. Among these outputs is the list of constellation's detected specular points, data that is necessary for CYGNSS's atmospheric measurements and may not be reported by other simulation tools. Using these written output files, various performance metrics can be calculated in other programs.

### 2.4. Performance Metrics

The spatial coverage performance of the various constellation configurations considered is quantified by a Zonal Spatial Coverage (ZSC) metric. The Earth surface is first divided up into a 25 km grid in latitude and longitude. All grid cells in which at least one observation is made within a 24 h period are noted. Then, the percentage of grid cells sampled across all longitudes is computed in each latitude zone and that percentage is reported as a function of latitude.

We quantify the temporal coverage performance of our analyzed constellations using a Zonal Temporal Coverage (ZTC) metric. The analysis uses the same equidistant geo-spatial grid covering the entire surface of the Earth as was used for the ZSC. The simulation calculates a binary decision in every grid cell over a chosen time interval. If there was one or more observations in a given grid cell, the cell is labelled with a one (1). If no observations occurred, it would be labeled with a zero (0). The time interval we use in our analysis is six hours, with four total time intervals in a day. Therefore, the maximum ZTC coverage a location on Earth could achieve per day is 4 (i.e., at least one observation was made in each of the 4 6 h intervals in the day). This 6 h quantization of the 24 h cycle was chosen to support the input data assimilation needs of major weather prediction models with a 6-hourly reporting interval (e.g., the NOAA/NCEP Global Forecast System (GFS) and the ECMWF High Resolution 10-day Forecast (HRES). By matching our ZTC interval to the needs of the models, we address the potential value of GNSS-R observations for use by these major operational models. Naturally, analysis of other intervals with other metrics is possible. Previous studies of the sampling properties of similar satellite constellations have considered the time separation between successive samples in the same grid cell [19]. Our approach expands upon this prior work by considering the ZTC as defined above to provide a more practical and useful quantification of the temporal sampling properties as they relate to the use of the measurements by numerical weather prediction models.

For each of the constellation configurations considered below, sampling performance is derived from a population of sample times and locations generated by an orbit simulation model. The model propagates the orbital locations for each of the science observatories as well as all members of the GPS, Galileo and SBAS constellations of GNSS satellites. At each one-second time step over the course of a 24 h period, the locations of all possible surface reflections are determined for signals propagating from every GNSS satellite transmitter to every science observatory receiver via specular point reflection by the Earth surface. In addition to the time and location of each sample, the value of the receive antenna gain in the direction of the specular point reflection is also noted. This allows the signal-to-noise-ratio quality of the received signal to be determined.

## 3. GNSS-R Coverage Simulations

### 3.1. *Effect of Orbit Inclincation on Constellation Coverage*

One critical design feature of a constellation is the orbit inclination of its satellites. The inclination of a satellite's orbit determines the maximum possible latitude of its surface observations and hence affects the global distribution of its spatial coverage. The inclination will also affect the frequency with which samples are made at different latitudes, which in turn determines the ZTC. These two performance considerations should both be considered in the design optimization process.

For this study, we consider design options in which one third of the total number of satellites is distributed between each of three orbit planes at distinct inclinations. We analyzed three cases, one where all three planes have an inclination of 30°, another where the three planes have inclinations of 30°, 60° and 90°, and a third where all three planes have an inclination of 90°.

The ZSC results of these simulations are illustrated in Figures 2–4. Although these images depict a single orbit of these three constellations over a period of 1 h, 40 min, and 30 s, they can be extrapolated to form the full 24 h spatial coverage maps. The maximum inclination of the constellation can be seen to limit its latitudinal coverage. The coverage will not exceed the inclination by more than 10° north or south. For instance, the constellation with three planes at an inclination of 30° has non-zero ZSC at latitudes slightly lower than 40° north and south. However, after extrapolating the three orbit maps, this truncated coverage will contrast with the almost 100% ZSC performance demonstrated by the 30°-60°-90° and 90°-90°-90° inclination constellations. In these situations where the maximum inclination of the constellation is set to 90°, the constellations are capable of taking measurements over the entire Earth.

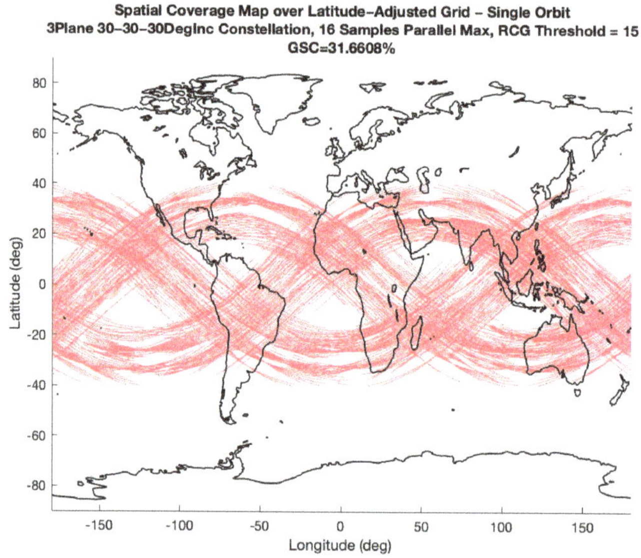


**Figure 2.** Single-Orbit Spatial Coverage for 3-Plane 30°-30°-30° constellation at 800 km, 16 parallel measurements per sampling, and operating with an RCG Threshold of 15. The lower maximum inclination limits spatial coverage to below 40° latitude.

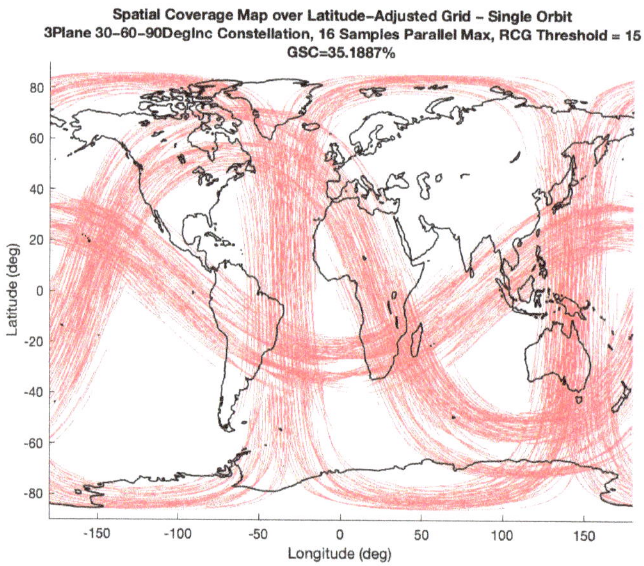


**Figure 3.** Single-Orbit Spatial Coverage for 3-Plane 30°-60°-90° constellation at 800 km, 16 parallel measurements per sampling, and operating with an RCG Threshold of 15. The inclination distribution provides greater coverage at the various orbit peaks, allowing for a greater GSC than the 30°-30°-30° constellation.

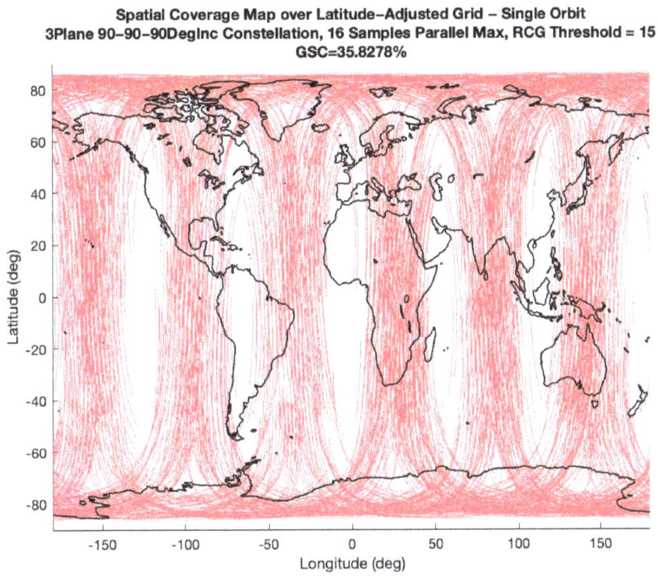


**Figure 4.** Single-Orbit Spatial Coverage for 3-Plane 90°-90°-90° constellation at 800 km, 16 parallel measurements per sampling, and operating with an RCG Threshold of 15. The 90° maximum plane inclination allows the constellation to achieve near 100% GSC after 24 h.

The ZTC results of these simulations are illustrated in Figure 5. When all constellation planes operate at the same inclination, the ZTC peaks around those latitudes. However, when the plane inclinations are more evenly spaced between latitude levels, as shown by the 30°-60°-90° constellations, we get more well-distributed sampling across the Earth at the cost of smaller coverage peaks.

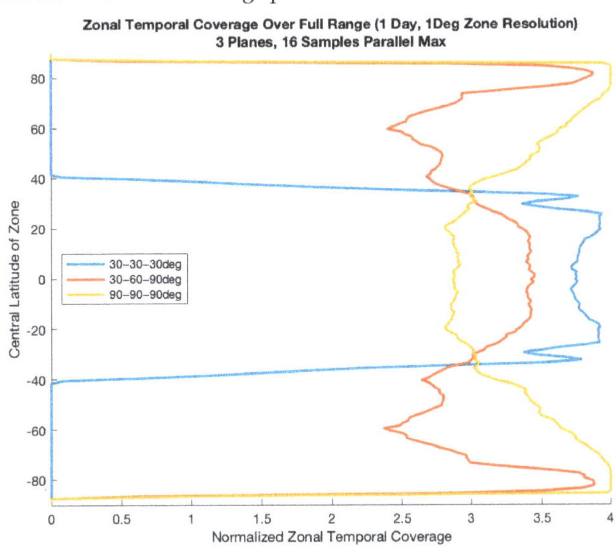


**Figure 5.** 24 h Temporal Coverage measurements at different latitudes for 3-Plane constellations at 800 km, 16 parallel measurements per sampling, and operating with an RCG Threshold of 15. Raising the inclination of individual planes or the entire constellation can alter which latitudes are visited most often.

### 3.2. Effect of Orbit Planes on Constellation Coverage

The second critical design feature of a constellation is how its satellites are distributed between orbit planes. In this series of design options, an orbit plane is defined as a group of satellites orbiting with the same ascending node longitude. With that definition in mind, we find that when a singular plane satellite constellation is divided into a constellation with multiple planes, its ZTC increases. However, when only looking at certain latitude zones on the globe, this trend may not be uniform.

To better understand how the temporal coverage depends on the distribution of planes, two experiments were conducted where the total number of satellites and satellite inclination were held constant. These experiments first simulate the performance of a single plane of 24 evenly distributed satellites all orbiting at the same inclination. Then, the simulation is repeated for a second constellation in which the plane is split into two planes of twelve evenly distributed satellites separated by 180° longitude. Finally, a third constellation is considered in which the satellites are further split into three planes of eight evenly distributed satellites separated by 120° longitude. These results are presented in Figure 6, where the orbit inclination is set to 30°.

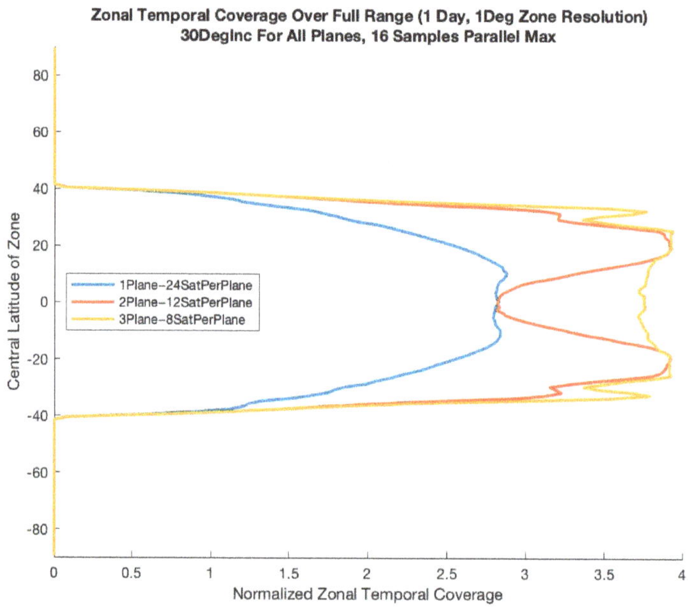


**Figure 6.** 24 h Temporal Coverage measurements at different latitudes for 24-satellite constellations at 800 km, 30° inclination, 16 parallel measurements per sampling, and operating with an RCG Threshold of 15. Increasing the number of constellation planes while keeping the total number of satellites constant will increase the constellation's Zonal and Global Temporal Coverage.

In this scenario, the ZTC in the zone near the orbit peaks grows when the original constellation is divided into more planes. There is also a ZTC increase at latitudes lower than the peaks when the constellation is redistributed from one plane to three planes. However, when observing the two-plane constellation, the ZTC measurements at latitudes near the equator can remain near stagnant.

Using this understanding of orbit planes and inclinations, designers can better tune their constellation orientations so that both spatial and temporal coverage metrics are optimized. For this experiment, we consider a constant number of satellites all at the same inclination and in the same plane. In this case, a 24-satellite constellation at 30° inclination is selected. The performance of this constellation is seen in the blue curve of Figure 7. Note that coverage is lacking near 60°. To improve it, we divide the plane into a constellation of two 12-satellite planes, one of which orbits at 30° and the other at 60°, resulting in the red curves in Figure 7. As these curves demonstrate, adding this second plane at a higher inclination has the desired effect of increasing coverage above 30°. Furthermore, if coverage at even higher latitudes was desired, the constellation could be divided into a third plane with an inclination of 90°. This third plane will also eliminate the coverage drop near the equator present in constellations with two planes.

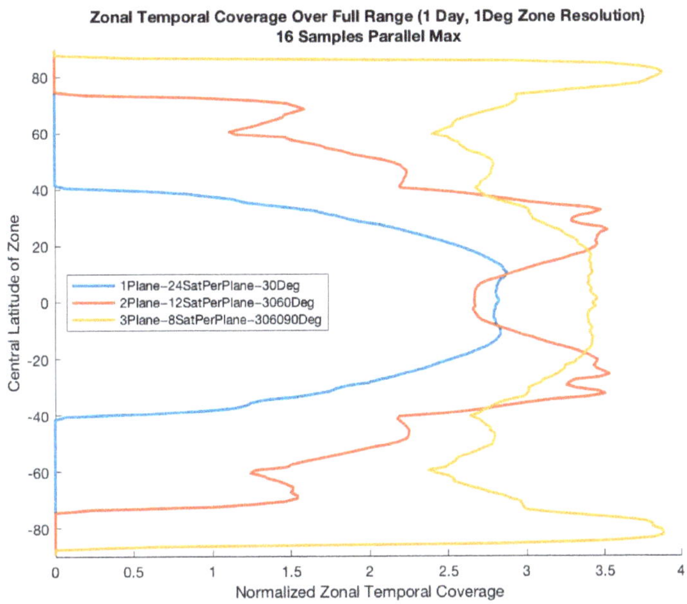


**Figure 7.** 24 h Temporal Coverage measurements at different latitudes for 24-satellite constellations at 800 km, 16 parallel measurements per sampling, and operating with an RCG Threshold of 15. As the initial 30° inclination plane of 24 satellites is divided in smaller planes with raised inclinations allows the constellation's Zonal and Global Temporal Coverage to be shaped and increased.

### 3.3. Effect of Measurement RCG on Constellation Coverage

A third critical design feature of a constellation is the antenna gain of its measurements. The impact of antenna gain on science measurement quality is characterized by the Range Corrected Gain (RCG). A higher RCG represents higher signal-to-noise ratio data and higher quality estimates of geophysical quantities derived from them, such as ocean surface wind speed [17]. Two ways are considered to control the RCG values, changing the design of the receiving antenna assumed on each of the satellites and raising the RCG threshold required for data usage.

The antenna used by a GNSS-R sensor is typically designed to accommodate a particular orbit altitude. A higher altitude will increase the propagation distance and decrease the received signal strength. This decrease can be mitigated by increasing the antenna gain. Alternatively, a decrease in altitude will restrict the field of view of an antenna pattern projected onto the Earth surface. This restriction can be mitigated by widening the antenna pattern. Fortuitously, a wider antenna pattern tends to have a lower antenna gain, so these two considerations can be accommodated jointly by the same adjustment in antenna design as a function of orbit altitude. Two orbit altitudes and corresponding antenna designs are considered here. The first assumes a similar configuration as is used by the CYGNSS constellation—namely a 500 km altitude and a 2 × 3 element phased array antenna. Although CYGNSS operates at an altitude of 525 km, the results at 500 km provides nearly identical performance. The second option is an 800 km orbit and a 3 × 5 element phased array antenna. The advantage of the higher orbit altitude is wider field of view of the antenna, which allows for more available specular point reflections to be sampled. This is demonstrated in the curves of Figure 8. When we compare the RCG values of samples collected from the CYGNSS configuration and the 3 × 5 antenna at 800 km altitude, we can see that the CYGNSS antenna will retrieve significantly fewer parallel measurements for the same RCG threshold as the 3 × 5 patch antenna, demonstrating the advantage of a higher altitude orbit.

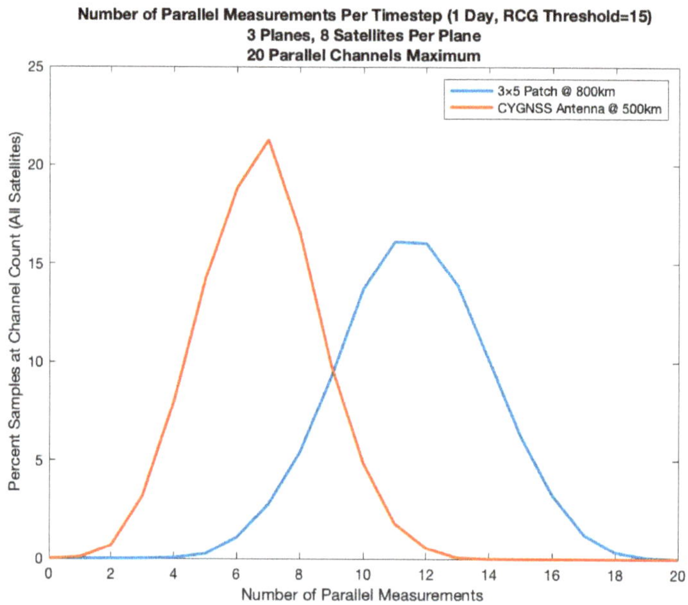


**Figure 8.** Parallel Measurement Comparison between antenna patterns at their corresponding altitudes over 24 h using a 3-Plane constellation with each plane at 30°, 60°, and 90° inclination, 20 Parallel Measurements Maximum per sampling, and an RCG Threshold of 15.

It is important to consider where the minimum RCG threshold is set. A threshold of 15 has been found to be sufficient for providing high-quality science data to most applications. The implications of this threshold on the number of usable samples are illustrated in Figure 9. The figure considers the fraction of all samples retained given different lower-bound RCG thresholds. The reference baseline is usage of all samples (i.e., no threshold). The results show that all three constellations will have 50% of collected data fall below threshold when it is set to RCG > 15.

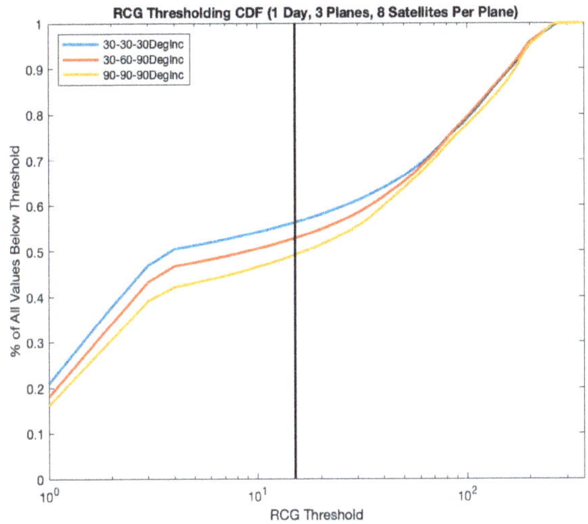


**Figure 9.** RCG Thresholding Comparison for 3-Plane constellations at 800 km, 16 parallel measurements per sampling, and with 8 satellites per plane. For all listed constellations, the fraction of measurements below threshold converges around RCG Value of 15.

## 4. Examples

### 4.1. Application of Design Methodology

By qualifying these three design spaces, they can now be applied towards optimizing a constellation design. One potential optimization task could be to maximize ZTC at all latitudes while maintaining ZSC as close to 100% at those same latitudes.

For this scenario, it can be assumed that 24 satellites are available for use similarly to the examples discussed previously in this report. By doing so, this constellation will have more than enough satellites to guarantee a near 100% ZSC over the course of a 24 h test while allowing for greater manipulability of its ZTC. With 24 satellites provided, a constellation designer would then need to decide how to spatially organize these elements. As described above in Section 3.2, the most practical pattern organization would be to divide the 24 satellites into 3 evenly distributed planes of 8 satellites. The result of doing so while holding other variables constant is further illustrated in Figure 6.

Once the satellites are placed into orbit planes, it is up to the constellation designer to set each plane's orbit inclination. To maintain near 100% GSC while having an evenly distributed ZTC, it would make sense to start with a configuration with planes inclined at 30°, 60°, and 90°. Since there is a plane inclined at 90°, not only is a GSC near 100% as described in Section 3.1, but also a ZTC distribution as illustrated in Figure 5 will be achieved. However, when observing this ZTC distribution over latitude, it can be noted that both the ZTC drops significantly near 60° and the ZTC at other latitudes below 80° are significantly below the maximum coverage value of 4. This behavior can be observed by analyzing the coverage metrics across larger zones, defined as equatorial, mid-latitude, and polar. More specifically, the equatorial zone is bounded by the latitudes 30°S and 30°N, the mid-latitude zone is defined by the two sets of boundaries, 30–60° north and south, and the polar zone is defined by the two sets of boundaries, 60–90° north and south. As illustrated in Table 1, the mid-latitude and polar ZTC measurements for the 30°-60°-90° constellation are significantly lower than the equatorial ZTC. When considering this information along with the ZTC plot in Figure 5, the mid-latitude and polar ZTC drops can be attributed to the smaller quantity of satellites in polar orbits and the placement of the mid-latitude plane at 60° inclination. In order to overcome the lack of uniformity and raise the overall ZTC, it is necessary to adjust each plane's inclination. An alternative approach is to design fitness functions that score constellations based on these coupled global performance metrics [19].

**Table 1.** Constellation Optimization through the Proposed Iterative Method. The constellations all feature 3 evenly distributed orbit planes of 8 evenly spaced satellites per plane at an altitude of 800 km. The coverage metrics described above were evaluated with the 3 × 5 element phased array antenna and a RCG minimum threshold of 15. By making these discrete adjustments in the inclinations of individual planes, we can keep the GSC near 100% while incrementally improving the GTC.

| Constellation (deg-deg-deg) | Global Spatial Coverage (%) | Global Temporal Coverage | ZSC (eq) | ZSC (mid) | ZSC (Polar) | ZTC (eq) | ZTC (mid) | ZTC (Polar) |
|---|---|---|---|---|---|---|---|---|
| 30-30-30 | 62.987 | 2.254 | 99.994 | 35.505 | 0 | 3.788 | 0.983 | 0 |
| 90-90-90 | 99.337 | 3.151 | 99.214 | 99.796 | 98.540 | 2.888 | 3.273 | 3.795 |
| 30-60-90 | 99.688 | 3.069 | 99.949 | 99.756 | 98.238 | 3.283 | 2.716 | 2.881 |
| 35-60-90 | 99.677 | 3.053 | 99.943 | 99.840 | 98.238 | 3.278 | 2.809 | 2.881 |
| 40-60-90 | 99.679 | 3.079 | 99.912 | 99.887 | 98.238 | 3.252 | 2.917 | 2.881 |
| 45-60-90 | 99.676 | 3.102 | 99.886 | 99.916 | 98.238 | 3.214 | 3.032 | 2.881 |
| 45-65-90 | 99.726 | 3.174 | 99.900 | 99.975 | 98.400 | 3.261 | 3.100 | 3.174 |
| 45-70-90 | 99.721 | 3.180 | 99.886 | 99.962 | 98.448 | 3.237 | 3.104 | 3.175 |
| 45-75-90 | 99.814 | 3.181 | 99.870 | 99.951 | 99.227 | 3.207 | 3.096 | 3.313 |
| 50-75-90 | 99.800 | 3.199 | 99.829 | 99.969 | 99.227 | 3.171 | 3.194 | 3.319 |
| 50-75-80 | 99.912 | 3.252 | 99.846 | 99.971 | 99.994 | 3.186 | 3.259 | 3.478 |

There are two possible approaches for correcting the ZTC drop near 60°. First, a designer can either raise the inclination of first plane, which was initially placed at 30°, or lower the inclination of the third plane, which was initially placed at 90°. By adjusting these inclinations, additional coverage can be shifted towards the latitude where ZTC is lacking. However, shifting too much may cause new significant drops to form. Second, a designer can raise or lower the inclination of the second plane, which was initially placed at 60°. Although this approach may not eliminate the ZTC drop, it will reposition it to a latitude where it may become more simple to perform the first method.

Starting from the baseline of a 3-Plane 30°-60°-90° constellation, the progression of this logic and the corresponding changes in GSC, ZSC, GTC, and ZTC are illustrated by the results provided in Table 1. After a few iterations, we find a significant improvement from the original baseline in the 3-Plane 50°-75°-80° constellation. In each step of the process, GSC always remains close to a complete 100%, but most progressions showed growth in the GTC of each prototype constellation. By incrementally raising the first plane inclination, we can slowly increase the GTC. However, once we reach a point where the first two inclinations are starting to get closer, we find that raising the second plane would create a more significant increase in this metric. This is evident by the GTC change between the 45°-60°-90° and the 45°-65°-90° constellations. Additionally, the increase in the inclination of the first plane from 45° to 50° allows for one more small rise in GTC without moving the first plane too close to the second plane while allowing for significant equator coverage. Finally, we can take advantage of the 10° reach from the set inclination to lower the third plane down from 90° to 80° while both improving ZTC at inclinations lower than 80° and maintaining the ZTC between 80° and 90°.

When comparing this 50°-75°-80° constellation to others with no separation in plane inclinations, such as the 30°-30°-30° or 90°-90°-90°, the differences in GTC and ZTC are prominent. Since the 30°-30°-30° constellation's max inclination is equatorial, its mid-latitude ZTC, polar ZTC, and GTC are all much smaller than other constellations considered, making it less than ideal for a globally effective design. On the other hand, the 90°-90°-90° constellation has a relatively large GTC, but its equatorial ZTC is much smaller than other designs considered. Although the 50°-75°-80° constellation does not have the largest ZTC of these three constellations in any of the regions, its zonal and global metrics are still relatively good. Considering the goal of this design study is to maximize ZTC at all latitudes, the 50°-75°-80° constellation is a good design candidate, for its zonal metrics are large at all latitudes and are further characterized by a large GTC.

### 4.2. Sampling of a Landfalling Hurricane

To illustrate the scientific implications of the sampling properties of an optimized constellation of GNSS-R satellites, a specific example is considered. Hurricane Ida made landfall along the U.S. Louisiana coast on 29 August 2021 at 1655 UTC. The measurements that would have been made on that day by two different GNSS-R constellations have been simulated and overlaid with GOES imagery of the storm. One constellation consists of 8 satellites arranged in a common orbit plane at an inclination of 30°, similar to the CYGNSS configuration. The other constellation consists of 8 satellites in each of 3 orbit planes with RAANs of 0°, 120° and 240° and a common 30° inclination. The global coverage with this configuration was shown in Figure 2. All spacecraft in both constellations are assumed to be capable of 16 simultaneous specular reflection measurements. Results of the simulation are shown in Figure 10. In the figure, four 6-hourly intervals of time are considered, consistent with the initialization time increment between operational numerical hurricane forecasts. All samples made during each 6 h interval are shown, together with the GOES image of the storm taken at the center of the four time intervals, namely at 0600, 1200, 1800 and 2400 UTC on 29 August.

In Figure 10, the right column corresponds to measurements made by the 24 spacecraft distributed over 3 orbit planes. In each 6 h interval, the storm is well sampled both in its inner core region and across the surrounding wind field in all four principle quadrants. Inner core measurements support the determination of maximum sustained winds and hurricane intensity, while coverage of the wider wind field supports the determination of storm size and duration during landfall. The left-hand column in Figure 10 corresponds to measurements made by the 8 spacecraft constellation. The storm is quite well sampled throughout the inner core and surrounding wind field during only one of the four 6 h intervals (1500–2100 UTC). In two of the other intervals, samples are made over only two of the four principle quadrants of the storm, with little of the inner core sampled in either case. The fourth 6 h interval has no samples whatsoever of the storm.

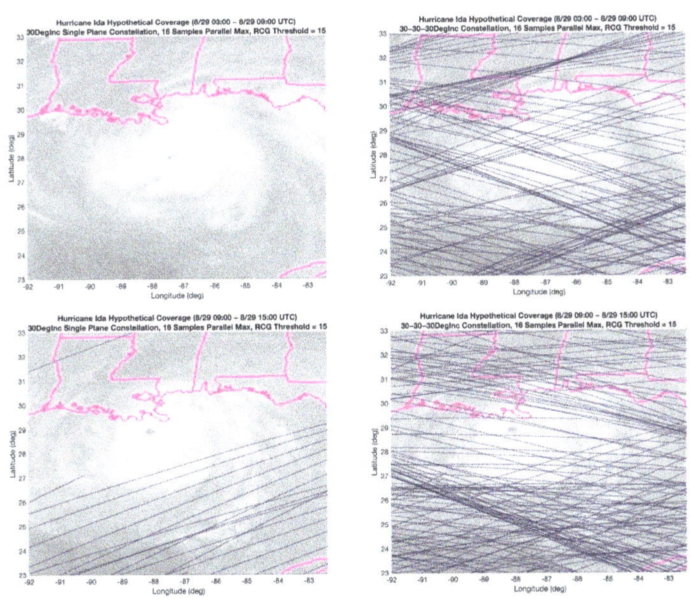


**Figure 10.** *Cont.*

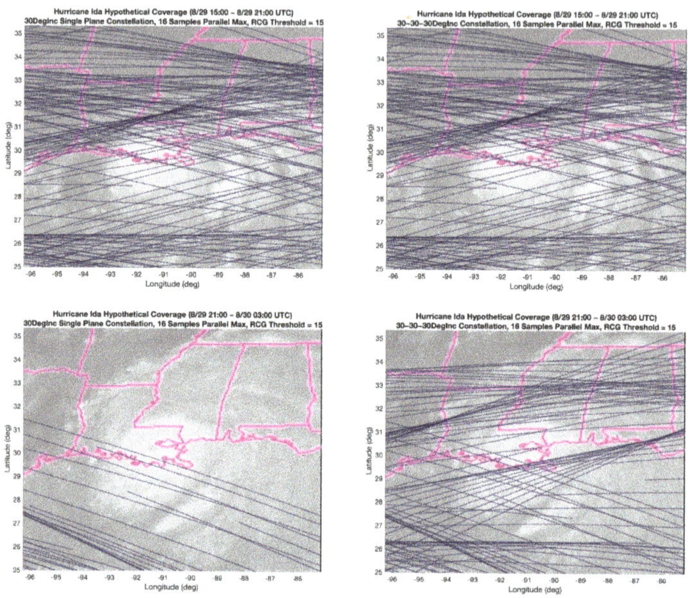


**Figure 10.** Simulated measurements by two GNSS-R constellations of Hurricane Ida during landfall on 29 August 2021. The left column corresponds to an 8-satellite constellation and the right to one with 24 satellites, all at 30° inclination. Rows correspond to 6 h time intervals centered on 0600, 1200, 1800 and 2400 UTC. GNSS-R tracks are overlaid on GOES-16 Band 8 (6.17 μm) images at the center times. Landfall occurred at 1655 UTC.

## 5. Conclusions

Large constellations of smallsats in low Earth orbit can provide spatial and temporal coverages with greater sampling density relative to constellations with fewer large satellites. This increase in coverage is primarily characterized by three factors and quantified by zonal and global measurements. The selection of satellite inclinations allows designers to choose which latitude zones are receiving more coverage while also optimizing global visibility. Meanwhile, dividing a constellation into multiple orbit planes allows designers to further improve global and zonal temporal coverage figures. Additionally, the quality and quantity of the permissible sensed data can be controlled by selecting an appropriate RCG threshold in accordance with constellation's onboard GNSS-R instruments. Ultimately, these parameters allow for designers to iteratively evaluate constellation prototypes and maximize spatial and temporal coverage both globally and in regions of interest.

When designing a constellation, each mission will have its own set of objectives where it wants to maximize its coverage. However, the most generic definition of a successful constellation design would be maximizing global coverage metrics while keeping zonal coverage metrics evenly distributed. Through an iterative process of adjusting the satellite inclinations of 3-plane constellations, it has been found that a constellation with planes of 8 satellites orbiting at inclinations of 50°, 75° and 80° meet these design goals. Although it may be possible to adjust the definition of success to meet specific mission requirements, this constellation has been found to provide quite good overall coverage both globally and zonally.

**Author Contributions:** Conceptualization, C.R. and S.G.; methodology, J.W., C.R. and S.G.; software, J.W.; validation, J.W., C.R. and S.G.; formal analysis, J.W., C.R. and S.G.; investigation, J.W., C.R. and S.G.; resources, C.R.; data curation, C.R.; writing—original draft preparation, J.W., C.R. and S.G.; writing—review and editing, J.W., C.R. and S.G.; visualization, J.W.; supervision, C.R. and

S.G.; project administration, C.R.; funding acquisition, C.R. All authors have read and agreed to the published version of the manuscript.

**Funding:** This research and the APC were funded by NASA contract number 80LARC21DA003.

**Data Availability Statement:** No new data were created in performance of the work presented here.

**Conflicts of Interest:** The authors declare no conflict of interest.

## References


1. NASEM. *Leveraging Commercial Space for Earth and Ocean Remote Sensing*; The National Academies Press: Washington, DC, USA, 2022. [CrossRef]
2. Bussy-Virat, C.D.; Ruf, C.S.; Ridley, A.J. Relationship Between Temporal and Spatial Resolution for a Constellation of GNSS-R Satellites. *IEEE J. Sel. Top. Appl. Earth Obs. Remote Sens.* **2018**, *12*, 16–25. [CrossRef]
3. Gleason, S.; Gebre-Egziabher, D. *GNSS Applications and Methods*, 1st ed.; Artech House: Norwood, MA, USA, 2009; p. 508.
4. Katzberg, S.J.; Dunion, J.; Ganoe, G.G. The use of reflected GPS signals to retrieve ocean surface wind speeds in tropical cyclones. *Radio Sci.* **2013**, *48*, 371–387. [CrossRef]
5. Unwin, M.; Jales, P.; Tye, J.; Gommenginger, C.; Foti, G.; Rosello, J. Spaceborne GNSS-Reflectometry on TechDemoSat-1: Early Mission Operations and Exploitation. *IEEE J. Sel. Top. Appl. Earth Obs. Remote Sens.* **2016**, *9*, 4525–4539. [CrossRef]
6. Carreno-Luengo, H.; Lowe, S.; Zuffada, C.; Esterhuizen, S.; Oveisgharan, S. Spaceborne GNSS-R from the SMAP Mission: First Assessment of Polarimetric Scatterometry over Land and Cryosphere. *Remote Sens.* **2017**, *9*, 362. [CrossRef]
7. Pierdicca, N.; Comite, D.; Camps, A.; Carreno-Luengo, H.; Cenci, L.; Clarizia, M.P.; Costantini, F.; Dente, L.; Guerriero, L.; Mollfulleda, A.; et al. The Potential of Spaceborne GNSS Reflectometry for Soil Moisture, Biomass, and Freeze–Thaw Monitoring: Summary of a European Space Agency-funded study. *IEEE Geosci. Remote Sens. Mag.* **2021**, *10*, 8–38. [CrossRef]
8. Ruf, C.S.; Atlas, R.; Chang, P.S.; Clarizia, M.P.; Garrison, J.L.; Gleason, S.; Katzberg, S.J.; Jelenak, Z.; Johnson, J.T.; Majumdar, S.J.; et al. New Ocean Winds Satellite Mission to Probe Hurricanes and Tropical Convection. *Bull. Am. Meteorol. Soc.* **2016**, *97*, 385–395. [CrossRef]
9. Jales, P.; Esterhuizen, S.; Masters, D.; Nguyen, V.; Nogués-Correig, O.; Yuasa, T.; Cartwright, J. The new Spire GNSS-R satellite missions and products. In Proceedings of the Image and Signal Processing for Remote Sensing XXVI 2020, Online, 20 September 2020; Volume 11533, p. 1153316. [CrossRef]
10. Ruf, C.; Asharaf, S.; Balasubramaniam, R.; Gleason, S.; Lang, T.; McKague, D.; Twigg, D.; Waliser, D. In-Orbit Performance of the Constellation of CYGNSS Hurricane Satellites. *Bull. Am. Meteorol. Soc.* **2019**, *100*, 2009–2023. [CrossRef]
11. Rose, R.; Ruf, C.; Rose, D.; Brummitt, M.; Ridley, A. The CYGNSS flight segment; A major NASA science mission enabled by micro-satellite technology. In Proceedings of the 2013 IEEE Aerospace Conference, Big Sky, MT, USA, 2–9 March 2013; pp. 1–13. [CrossRef]
12. Roddy, D. *Satellite Communications*, 4th ed.; McGraw-Hill: New York, NY, USA, 2006; pp. 32–36.
13. Ruf, C.; Backhus, R.; Butler, T.; Chen, C.-C.; Gleason, S.; Loria, E.; McKague, D.; Miller, R.; O'Brien, A.; van Nieuwstadt, L. Next Generation GNSS-R Instrument. In Proceedings of the IGARSS 2020-2020 IEEE International Geoscience and Remote Sensing Symposium, Waikoloa, HI, USA, 26 September–2 October 2020; pp. 3353–3356. [CrossRef]
14. Gleason, S.; Ruf, C.S.; O'Brien, A.J.; McKague, D.S. The CYGNSS Level 1 Calibration Algorithm and Error Analysis Based on On-Orbit Measurements. *IEEE J. Sel. Top. Appl. Earth Obs. Remote Sens.* **2018**, *12*, 37–49. [CrossRef]
15. Ruf, C.; Chang, P.; Clarizia, M.P.; Gleason, S.; Jelenak, Z.; Murray, J.; Morris, M.; Musko, S.; Posselt, D.; Provost, D.; et al. *CYGNSS Handbook*; Michigan Pub.: Ann Arbor, MI, USA, 2016; p. 154. ISBN 978-1-60785-380-0.
16. Zavorotny, V.U.; Gleason, S.; Cardellach, E.; Camps, A. Tutorial on Remote Sensing Using GNSS Bistatic Radar of Opportunity. *IEEE Geosci. Remote Sens. Mag.* **2015**, *2*, 8–45. [CrossRef]
17. Ruf, C.; Gleason, S.; McKague, D.S. Assessment of CYGNSS Wind Speed Retrieval Uncertainty. *IEEE J. Sel. Topics Appl. Earth Obs. Remote Sens.* **2018**, *12*, 87–97. [CrossRef]
18. Bussy-Virat, C.; Getchius, J.; Ridley, A. The Spacecraft Orbital Characterization Kit and its Applications to the CYGNSS Mission. In Proceedings of the 2018 Space Flight Mechanics Meeting, Kissimmee, FL, USA, 8–12 January 2018. [CrossRef]
19. Tan, C.; Xu, Y.; Luo, R.; Li, Y.; Yuan, C. Low Earth orbit constellation design using a multi-objective genetic algorithm for GNSS reflectometry missions. *Adv. Space Res.* **2022**, in press. [CrossRef]

*remote sensing*

MDPI

*Article*

# Spaceborne GNSS-R Wind Speed Retrieval Using Machine Learning Methods

**Changyang Wang [1,2], Kegen Yu [1,2,*], Fangyu Qu [3], Jinwei Bu [1,2], Shuai Han [1,2] and Kefei Zhang [1,2]**

[1] MNR Key Laboratory of Land Environment and Disaster Monitoring, China University of Mining and Technology, Xuzhou 221116, China; tb20160008b1@cumt.edu.cn (C.W.); b_jinwei@cumt.edu.cn (J.B.); ts20160051a31ld@cumt.edu.cn (S.H.); kefei.zhang@cumt.edu.cn (K.Z.)

[2] School of Environment Science and Spatial Informatics, China University of Mining and Technology, Xuzhou 221116, China

[3] College of Computer Science, Nankai University, Tianjin 300073, China; qufangyu@mail.nankai.edu.cn

* Correspondence: kegen.yu@cumt.edu.cn

**Abstract:** This paper focuses on sea surface wind speed estimation using L1B level v3.1 data of reflected GNSS signals from the Cyclone GNSS (CYGNSS) mission and European Centre for Medium-range Weather Forecast Reanalysis (ECMWF) wind speed data. Seven machine learning methods are applied for wind speed retrieval, i.e., Regression trees (Binary Tree (BT), Ensembles of Trees (ET), XGBoost (XGB), LightGBM (LGBM)), ANN (Artificial neural network), Stepwise Linear Regression (SLR), and Gaussian Support Vector Machine (GSVM), and a comparison of their performance is made. The wind speed is divided into two different ranges to study the suitability of the different algorithms. A total of 10 observation variables are considered as input parameters to study the importance of individual variables or combinations thereof. The results show that the LGBM model performs the best with an RMSE of 1.419 and a correlation coefficient of 0.849 in the low wind speed interval (0–15 m/s), while the ET model performs the best with an RMSE of 1.100 and a correlation coefficient of 0.767 in the high wind speed interval (15–30 m/s). The effects of the variables used in wind speed retrieval models are investigated using the XGBoost importance metric, showing that a number of variables play a very significant role in wind speed retrieval. It is expected that these results will provide a useful reference for the development of advanced wind speed retrieval algorithms in the future.

**Keywords:** wind speed; Cyclone Global Navigation Satellite System (CYGNSS); regression model; machine learning

**Citation:** Wang, C.; Yu, K.; Qu, F.; Bu, J.; Han, S.; Zhang, K. Spaceborne GNSS-R Wind Speed Retrieval Using Machine Learning Methods. *Remote Sens.* **2022**, *14*, 3507. https://doi.org/10.3390/rs14143507

Academic Editors: Hugo Carreno-Luengo, Dallas Masters and Chun-Liang Lin

Received: 3 June 2022
Accepted: 20 July 2022
Published: 21 July 2022



## 1. Introduction

With the continuous development of global navigation satellite systems (GNSSs), spaceborne GNSS reflectometry (GNSS-R) technology has become a hot research direction in the field of remote sensing. In 1993, Martín-Neira proposed the concept of the Passive Reflectometry and Interferometry System (PARIS) and the use of GNSS-R for ocean altimetry [1]. Since then, GNSS-R has been utilized for a range of ocean and land applications, including sea surface altimetry [2], sea surface wind speed measurements [3], sea ice detection [4], and soil moisture measurements [5]. Over the past few decades, a number of ground-based GNSS-R experiments have been conducted. Many airborne experiments have also been conducted to investigate this new remote sensing technology. Notwithstanding some technological challenges, satellite-based GNSS-R technology has the advantages of low cost and great coverage in some applications [6]. Currently, there are more than 14 satellites in operation carrying a GNSS-R payload.

UK-DMC (United Kingdom—Disaster Monitoring Constellation), the first satellite carrying a GNSS-R receiver, was launched on 27 September 2003; data from this system have been used to sense ocean roughness. UK TDS-1 (TechDemoSat-1), the second GNSS-R satellite, was launched on the 8 July 2014. On the 15 December 2016, NASA launched eight microsatellites to form the cyclone GNSS (CYGNSS) constellation with the initial objective of monitoring hurricane intensity [7,8]. Both TDS-1 and CYGNSS have generated a large amount of data which can be downloaded for scientific research [9]. On the 5 June 2019, the BuFeng-1 A/B twin satellites, developed by CASTC (China Aviation Smart Technology Co., Shenzhen, China), were launched from the Yellow Sea. One focus of the satellite mission is on the sensing of sea surface wind velocities, and especially typhoons, using GNSS-R [10].

Sea surface wind speed is an important and commonly used ocean geophysical parameter [11]. The stability of the wind field plays an important role in ocean circulation and global climate [12,13]. Traditional sea surface wind field monitoring methods generally use buoys or coastal meteorological stations, but these methods can only cover small areas with low spatial resolution and expensive equipment [14]. Microwave scatter meters and synthetic aperture radars can also monitor the global sea surface wind field [15,16]. Compared with these traditional wind measurement methods, spaceborne GNSS-R has several advantages, such as rich signal sources and all-weather, all-day, low cost, and large coverage [17,18].

GNSS-R technology is basically mature in retrieving sea surface wind speeds. Zavorotny and Voronovich proposed the scattering model theory in 2000 [19], which can simulate different waveforms of GNSS reflection signals, thus inverting sea surface wind speeds by delayed waveform matching methods [20]. Since then, observations extracted from DDMs (Delay Doppler Maps) have been widely used. DDM is the basic observation data of airborne and spaceborne GNSS-R receivers [21]. Some DDM observations, such as DDM average (DDMA), are directly related to sea surface roughness [21]. Other DDM observations can be used as variables for retrieving sea surface parameters. The normalized bistatic radar cross-section (NBRCS), leading edge slope (LES) and signal-to-noise ratio (SNR) have good correlations with the mean square slope (MSS) of the sea surface. Generally, the MSS is mainly affected by the sea surface wind speed [22].

In recent years, many spaceborne GNSS-R wind speed retrieval models have been developed. Jing et al. demonstrated the effectiveness of NBRCS by proposing some geophysical model functions (GMFs) related thereto [10]. Bu et al. proposed double- and triple-parameter GMFs with higher retrieval accuracy [14]. Machine learning methods have also been used to improve the performance of spaceborne GNSS-R wind speed retrieval. Liu Y. et al. proposed a machine learning algorithm based on a multi-hidden layer neural network. The accuracy of their models was significantly higher than that of GMFs [23]. Many subsequent studies have adopted similar algorithms and obtained results with RMSE of about 1.5–2.0 [24–26]. However, most of the above studies observed that it is difficult to use their algorithms to accurately retrieve high sea surface wind speeds [27,28]. A few studies have tried to enhance the ability of GNSS-R to retrieve high wind speeds. For instance, Zhang et al. developed machine learning-based models to retrieve wind speeds (20–30 m/s) with an RMSE of 2.64 and a correlation coefficient of 0.25 [29].

With high wind speed intervals, the Spaceborne GNSS-R data present different distributions and physical characteristics compared to when low wind speed intervals are applied, which leads to the inconsistent performance of different machine learning models. Therefore, this study analyzes the performance of various machine learning models in different wind speed intervals using the following methods: Regression trees (Binary Tree (BT), Ensembles of Trees (ET), XGBoost (XGB), LightGBM (LGBM) ), ANN (Artificial neural network), Stepwise Linear Regression (SLR), and Gaussian Support Vector Machine (GSVM). In this research, the selection of the input parameters for machine learning methods was significant. In this article, a range of variables are considered and evaluated, which are directly or indirectly relevant to sea surface wind speed. The main contributions of the article are as follows:

(1) Seven machine learning methods are used to retrieve sea surface wind speed, and their performance is evaluated under two different wind speed ranges.

(2) A ranking of the effects of 10 variables on wind speed retrieval is obtained by comparing the performance of different combinations of the variables. This provides a useful guide for variable selection when considering both complexity and accuracy.
(3) A filtering algorithm is proposed to process DDM data, achieving both low complexity and good performance.
(4) The effects of the number of neurons and activation functions on the performance of ANN wind speed retrieval are analyzed.

The rest of the paper is organized as follows. Section 2 introduces the GNSS-R variables and then describes the basic principles of the machine learning methods used in this study. Section 3 provides details of the applied data preprocessing strategies, the data filtering algorithm, and the construction of the machine learning-based model; the experimental results are also presented. Section 4 discusses the effects of the variables on wind speed retrieval. Section 5 presents the conclusions.

## 2. Methods

### 2.1. The CYGNSS Variables

#### 2.1.1. Variables Calculated with DDM

The DDM glistening zone (the area from which scattered signals are observed) depends on the sea state, and the DDM volume has a significant correlation with wind speed [30,31]. Five variables (LES, DDMA, Noise Floor, SNR and NBRCS), extracted from DDM, can better reflect the sea state than the simple DDM volume [10,31]. Meanwhile, these variables are calibrated in the CYGNSS Level 1B product, which is commonly used for the retrieval of wind speed [32]. The LES of the integrated delay waveform, such as that generated with the delay waveforms of five different Doppler shifts, is strongly correlated with wind speed [33]. DDMA is the average of scattered power computed from the center 5 Doppler $\times$ 3 delay bin box [34], which is also significantly affected by wind speed. Noise Floor is the average power of DDM pixels which only contain noise. The signal-to-noise ratio (SNR) is defined as 10log(Smax/Noise Floor), where Smax is the maximum value in DDM, which has a strong correlation with sea surface roughness [35]. NBRCS is one of the two observables that were used to produce the global tropical cyclone product of CYGNSS [26], which is effective for wind speed retrieval [32].

#### 2.1.2. Other Variables

In addition to the five variables derived from DDM data, five other variables were considered, representing the signal status, so that they can be used to enhance the performance of the models [8]. Instrument gain is the black body noise count divided by the sum of the black body power and the instrument noise power, which is an important parameter to calculate the DDM values. Scattering Area is the area of the central part of the DDM; generally, the larger this area, the rougher the reflective surface. Sp_inc_angle and sp_az_body are the incidence angle and azimuth angle of a given specular point, respectively. By taking sp_inc_angle and sp_az_body into account, the models can better reflect the situation of the received reflected signal [26]. Additionally, GNSS-R wind retrievals are affected by the ocean state [33]. Ocean swells are waves which travel from a long distance. The significant wave height of a swell (SWH_swell) will affect the reflection of the GNSS signals, which is a form of interference which can be used as a variable [27]. Table 1 lists all the variables used in this study.

**Table 1.** List of input variables used in wind speed retrieval.

| Input Variables | Long Name | Unit |
|---|---|---|
| NBRCS | Normalized bistatic radar cross-section | no unit |
| LES | Leading edge slope | no unit |
| SNR | DDM signal-to-noise ratio | dB |
| DDMA | DDM average | no unit |
| Noise Floor | DDM noise floor | no unit |
| sp_inc_angle | Specular point incidence angle | degree |
| sp_az_body | Specular point azimuth angle | degree |
| Instrument Gain | Instrument gain | no unit |
| Scatter Area | Scattering area of NBRCS and LES | square meter |
| SWH_swell | Significant wave height of ocean swell | meter |

### 2.2. *Regression Trees*

Four out of the aforementioned seven machine learning algorithms comprise regression trees, which are briefly described in this subsection.

#### 2.2.1. Binary Tree

A binary tree (BT) is easy to interpret, fast for fitting and prediction, and low on memory usage. It consists of nodes and directed edges. There are two types of nodes: internal and leaf. In this paper, the internal nodes represent the variables of CYGNSS data and the leaf nodes represent the wind speed value. Each step in a prediction involves checking the value of one predictor variable. Figure 1 shows a simple sample BT composed of 100 CYGNSS-ERA5 matchups. In the experiments described in Section 3, the BT models are much more complex than this example, and the retrieval accuracy is much improved, because the amount of data used to build BT models is much larger.

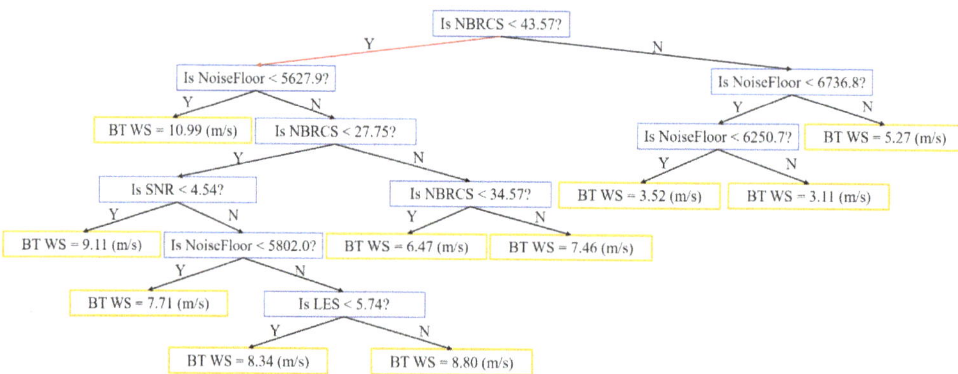


**Figure 1.** An example of a BT model structure.

When BT is used for regression tasks, variables of the sample are tested from the root node, and the sample is assigned to its child node according to the test results. In this way, the samples are tested and allocated recursively until they reach the leaf node, and each leaf node corresponds to a wind speed value. The criteria of splitting nodes are defined to balance predictive power and parsimony [36]. It is necessary to specify the minimum number of training samples used to calculate the response of each leaf node. When growing a regression tree, its simplicity and predictive power need to be considered at the same time. A very leafy tree tends to overfit, and its validation accuracy is often far lower than its training (or resubstitution) accuracy. In contrast, a coarse tree with fewer large leaves does not attain high training accuracy. However, a coarse tree can be more robust in that its training accuracy can be near that of a representative test set. In this paper, the minimum leaf size is set at 4.

### 2.2.2. Ensembles of Trees

Ensembles of Trees (ET) is one of the most popular techniques for building regression models [37,38]. Ensemble models combine results from many weak learners into one high-quality ensemble model. This approach has been applied frequently in fields such as remote sensing and statistics [39,40]. The function used to predict values is as follows:

$$\hat{y}_i = \sum_{j=1}^{K} f_j(x_i),\ f_j \in F \tag{1}$$

where $\hat{y}_i$ is the predicted value of the $i$-th sample, $K$ is the number of trees, $x_i$ is the $i$-th sample vector, $f_j$ denote the structure of the $j$-th independent tree and $F$ is the ensemble space of trees.

In this paper, a bagging tree is applied to build the ET. It draws its training set from the original sample set. In each round, n training samples are drawn from the original sample set using Bootstraping (some samples may be drawn multiple times in the training set, while some samples may not be drawn at all) [41]. A total of k rounds of extraction are performed to obtain k training sets, which means that k models will be built. The k training sets are independent of each other [42]. In this paper, k = 30 and the minimum leaf size is 8. Therefore, if several similar datasets are created by resampling with replacement and regression trees are grown without pruning, the variance component of the output error is reduced [41].

### 2.2.3. XGBoost

XGBoost (XGB) is a scalable, end-to-end tree boosting system which has been widely used in classification, regression and other machine learning tasks [43]. Based on Equation (1), XGBoost improves the running speed of model by using the regularized learning objective, which consists of two parts: the training loss term and regularization term, as given by:

$$Obj = \sum_{i}^{N} l(\hat{y}_i, y_i) + \sum \Omega(f_k) \tag{2}$$

where $l(\hat{y}_i, y_i)$ is the loss function which represents the deviation of $\hat{y}_i$ (predicted value) from $y_i$ (true value); $\Omega(f_k)$ represents the complexity of the model as a regularization term, which helps to control the complexity of the model and avoid overfitting; and $N$ is the number of samples. In order to minimize the regularized learning objective as much as possible, Equation (2) will be minimized for multiple rounds. In each round, $f_t$ is added to Equation (2). The regularized learning objective of $t$-th round can be written as follows:

$$Obj^{(t)} = \sum_{i=1}^{N} l\left(y_i, \hat{y}_i^{(t-1)} + f_t(x_i)\right) + \Omega(f_k) \tag{3}$$

The regularized learning objective can be approximated using the Taylor formula expansion:

$$Obj^{(t)} \cong \sum_{i=1}^{N} \left( l\left(y_i, \hat{y}_i^{(t-1)}\right) + g_i f_t(x_i) + \frac{1}{2} h_i f_t^2(x_i) \right) + \Omega(f_k) \tag{4}$$

where $g_i = \partial_{\hat{y}^{(t-1)}} l\left(y_i, \hat{y}_i^{(t-1)}\right)$ is the first gradient statistics on the loss function, $h_i = \partial^2_{\hat{y}^{(t-1)}} l\left(y_i, \hat{y}_i^{(t-1)}\right)$ is the second gradient statistics on the loss function. The regularized learning objective of the $t$-th round is as follows:

$$Obj^{(t)} = \sum_{j=1}^{T}\left(G_j w_j + \frac{1}{2}(H_j + \lambda)w_j^2\right) + \gamma T \tag{5}$$

where $G_j = \sum_{i \in I_j} g_i$ and $H_j = \sum_{i \in I_j} h_i$ are the accumulation of $g_i$ and $h_i$, and $I_j$ denote the instance of $j$-th leaf. $T$ is the number of leaves in the tree. The optimal weight $w_j$ of the $j$-th leaf node can be determined as:

$$w_j = -\frac{G_j}{H_j + \lambda} \tag{6}$$

and the corresponding optimal value of the objective function $Obj^{(t)}$ is given by:

$$Obj^{(t)} = -\frac{1}{2}\sum_{j=1}^{T}\frac{G_j^2}{H_j + \lambda} + \gamma T \tag{7}$$

The parameter settings of XGBoost are shown in Table 2.

**Table 2.** Parameter settings of XGBoost.

| Parameter | Meaning | Value |
|---|---|---|
| n_estimators | Number of gradient boosted trees; equivalent to the number of boosting rounds. | 100 |
| importance_type | The type of variable importance | gain |

#### 2.2.4. LightGBM

LightGBM (LGBM) is an efficient gradient boosting decision tree, which serves to enhance the efficiency of the model when the variable dimension of the data sample is high and the data scale is large [44]. Compared with Xgboost, LightGBM is faster to compute and consumes less memory. LightGBM uses an Exclusive Feature Bundling (EFB) strategy to bundle mutually exclusive variables in order to reduce the number of variables and achieve the purpose of dimensionality reduction. Finding the optimal binding variable has been proven to be an NP-hard problem, as the enumeration method cannot be applied. In actual operation, EFB uses the greedy algorithm to approximate the optimal solution, i.e., which reduces the number of variables without affecting the accuracy of split nodes. Table 3 shows the parameter settings of LightGBM.

**Table 3.** Parameter settings of LightGBM.

| Parameter | Meaning | Value |
|---|---|---|
| n_estimators | Number of boosted trees to fit | 100 |
| num_leaves | Maximum tree leaves for base learners | 31 |
| learning_rate | Boosting learning rate | 0.1 |

### 2.3. *Artificial Neural Network*

Artificial neural networks (ANNs) are relatively new computational tools that have been used extensively to solve many complex real-world problems [45]. In order to avoid the effects of dimension and order of magnitude, before using an ANN to process data, the CYGNSS variables need to be normalized:

$$X_i' = \frac{X_i - X_{min}}{X_{max} - X_{min}} \tag{8}$$

where $X_{min}$ and $X_{max}$ are the minimum and maximum values of the CYGNSS variables and $\{X_i'\}$ are the normalized CYGNSS variables. The Full Connection Network (FCN) is often used in regression problems. Regarding GNSS-R wind speed retrieval, many researchers have demonstrated a significant improvement compared with traditional methods [23,25]. Figure 2 shows the ANN structure adopted in this paper, including input layers, hidden layers and the result of wind speed retrieval. Input layers are the 10 CYGNSS variables used in this paper. Three hidden layers are adopted; their neurons are N, 2N and N respectively. Figure 2 shows the structure of ANN when N = 5.

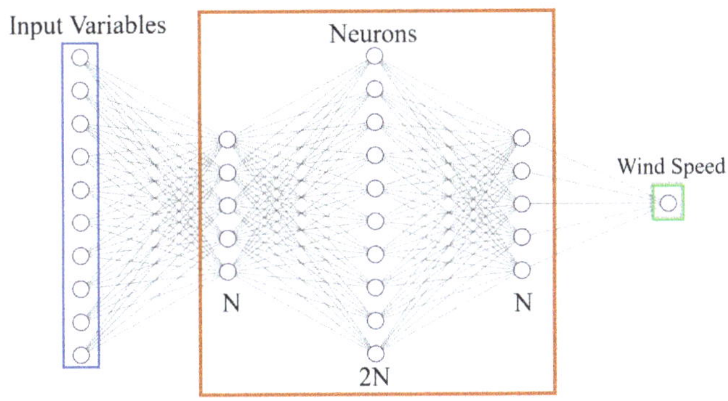


**Figure 2.** ANN structure adopted in this paper when N = 5. Small circles represent neurons in the model.

The number of neurons in an ANN affects the retrieval results, so the size of N is an important parameter when setting up a network. Herein, we analyze the impact of different activation functions on the performance of FCN, which makes connections between neurons. Generally, the accuracy of linear models is low, so activation functions improve the performance of ANN models by adding nonlinear factors. Determining the optimal activation function in an artificial neural network is an important task, because it is directly linked with the network performance. However, unfortunately, it is hard to determine this function analytically; rather, the optimal function is generally determined by trial and error or by tuning [46]. Three activation functions are analyzed in this paper, i.e., ReLu, Tanh and Sigmoid:

$$f_{ReLu} = \max(0, v) = \begin{cases} 0 \ (v < 0) \\ v \ (v \geq 0) \end{cases} \tag{9}$$

$$f_{Tanh} = \frac{e^v - e^{-v}}{e^v + e^{-v}} \tag{10}$$

$$f_{Sigmoid} = \sigma(v) = \frac{1}{1 + e^{-v}} \tag{11}$$

where $v$ is the input value of the previous neuron. The advantages of ReLU include the fast convergence speed of the network being trained, low computational complexity, and the absence of saturation and vanishing of gradient problems when $v > 0$. The ReLU activation and combinations of multiple instances are non-linear. The Tanh function provides stronger non-linearity but is plagued from with saturating and vanishing gradient problems. The advantage of Tanh and Sigmoid is their stability.

### 2.4. Stepwise Linear Regression

Stepwise linear regression (SLR) is able to establish the optimal multi-variable linear regression equation. First, linear regression model $SLR^0$ is constructed with all variables $\{v_1, v_2, \ldots, v_p\}$:

$$f_{SLR}^0 = \beta_0 + \sum_{i=1}^{p} \beta_i v_i \tag{12}$$

where $\{\beta_0, \beta_1, \beta_2, \ldots, \beta_p\}$ are constant parameters. The model is then used to estimate an unknown parameter such as wind speed for $n$ times, where $n$ is the number of observation datasets. The root mean square error (RMSE) of m estimations of model $SLR^0$ is calculated and denoted as $RMSE_0$. Next, the first variable, $v_1$, is removed, and estimation is performed m times again. Finally, the RMSE may be calculated and denoted as $RMSE_1$. If $RMSE_1$ is smaller than $RMSE_0$, $v_1$ may be removed; otherwise, it should be retained. This process is repeated until all variables are tested. Then, the variable with the smallest RMSE is selected. Therefore, this method is efficient for seeking localized variables [47]. SLR has good predictive ability and lower computational complexity than other methods [48].

### 2.5. Gaussian Support Vector Machine

Support Vector Machines (SVMs) are based on statistical learning theory, which contains polynomial classifiers, neural networks and radial basis function (RBF) networks in special cases. The SVM is thus not only theoretically well-founded but also superior in practical applications [49]. It is also commonly used to construct regression models. The function used to estimate the unknown parameter vector (such as the wind speed estimate vector) is given by:

$$f_{SVM} = \sum_{i=1}^{m}\sum_{j=1}^{m} (l_i' - l_i)\left(l_j' - l_j\right) x_i^T x_j + b \tag{13}$$

where $m$ is the number of samples and $l_i'$ and $l_i$ are the Lagrange multipliers. In this paper, $x$ is observation metric, which is composed of 10 variable rows and $m$ sample columns; $x_i$ and $x_j$ are the $i$-th column and $j$-th column, respectively; and $b$ is the threshold. By introducing the kernel functions replacing $x_i^T x_j$ with $K(x_i, x_j)$, where $K(x_i, x_j)$ is a transformation that maps $x_i$ to a high-dimensional space, the performance of the model can be improved. The choice of kernel function and parameters directly affects the performance of SVM [50]. The following are the commonly used positive semidefinite kernel functions, which are named as Linear function, Polynomial function and Gaussian function:

$$K(x_i, x_j) = x_i^T x_j \tag{14}$$

$$K(x_i, x_j) = \left(1 + x_i^T x_j\right)^p \tag{15}$$

$$K(x_i, x_j) = e^{-\|x_i - x_j\|^2} \tag{16}$$

After testing these kernel functions, it was found that the Gaussian function had the best effect in this study. Thus, Gaussian SVM (GSVM) is considered in this paper. In this study, the Box Constraint is 0.9762, the Epsilon is 0.09762 and the Kernel Scale is set at 3.7.

## 3. Experiments and Results

### 3.1. Data Processing Flow

This study makes use of the CYGNSS Level 1B (L1B) product, which contains Delay Doppler Maps (DDM), together with other engineering and science measurement parameters. CYGNSS data are in the range of 40°S to 40°N and work with a spatial resolution of ~25 km. The sampling rate of the data used in this study is 2 Hz. Different from most previous studies on wind speed estimation, this study adopts the latest CYGNSS v3.1 data instead of CYGNSS v2.1 data. Several data fields have been empirically corrected in the v2.1 L1 calibration algorithm. Therefore, they need to be carefully examined before modeling. Additionally, time-dependent variations have been observed in v2.1 data due to the variability of the transmitter and receiver. All these problems have been addressed in v3.1 data. The data are encapsulated by NASA in the netCDF file format and can be downloaded from https://podaac.jpl.nasa.gov/dataset/CYGNSS_L1_V3.1 (accessed on 26 March 2022) [29,30].

ECMWF reanalysis data (i.e. ERA-5) were used as the ground-truth data. ECMWF obtains hourly ERA-5 reanalysis datasets by assimilating meteorological data from different sources. The current sea surface wind speed product of ECMWF can be used as the ground-truth data in CYGNSS sea surface wind speed retrieval [25]. In this study, we use two ERA-5 parameters: the 10 m (above sea surface) u-component of neutral wind speed $WS_{u10}$ and the 10 m v-component of wind speed $WS_{v10}$, i.e., the eastward component and the northward component of the 10 m wind speed. The horizontal wind speed of 10 m above sea surface $WS_{10}$ can be readily obtained as the root square of the sum of the squares of these two parameters. However, CYGNSS data are sampled at an interval of half second and therefore need to be matched temporally with ERA-5 data. The spatial resolution of ERA-5 is $0.5^\circ \times 0.5^\circ$, which is rather different from that of CYGNSS, so spatial matching is also required.

In order to analyze the performance of the machine learning methods in different wind speed intervals, two datasets are constructed according to the wind speed distribution. They are a low wind speed dataset with wind speeds within 0–15 m/s and a high wind speed dataset with wind speeds within 15–30 m/s. To ensure the data is representative and generalizable, and to improve the generalization ability of the models, this study mainly uses randomly selected data from 2019 to 2021. Figure 3 shows the spatial distribution of all data used in this paper. Red points represent low wind speed data and green points represent high wind speed data. Most high wind speed data generally appear in high latitudes, while low wind speed data appear in all latitudes. It should be noted that the sea surface roughness near the coast may be affected by land [6], which leads to performance degradation of GNSS-R technology in terms of retrieving sea surface wind speeds and other parameters [26].

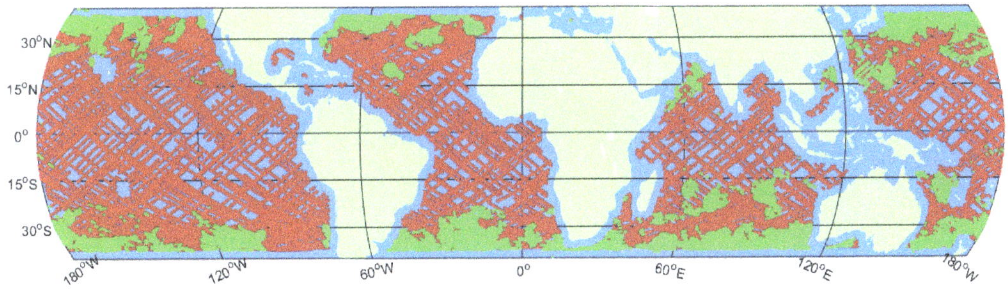


**Figure 3.** The spatial distribution of all data used in this paper.

The process of wind speed retrieval can be briefly summarized as containing four steps:

(1) Selecting the datasets used in this study and dividing them into a training set and a testing set in a proportion of about 3:1;
(2) Filtering the data;
(3) Training the processed data with the machine learning methods described in Section 2. It should be noted that five folders cross validation is adopted when training the model. By dividing the dataset into several folders and estimating the accuracy of each fold, the cross validation prevents over fitting.
(4) Evaluating the performance of different models by using test data.

Figure 4 shows a flow chart of the proposed model construction and evaluation methods. Figure 5 shows the histogram of wind speed distribution. High wind speed data are more difficult to obtain than low wind speed data, and a great deal of the former are concentrated in the range of 15–20 m/s. Next, in order to evaluate the performance of the models and the effect of variables, three metrics are chosen, i.e., the root mean square error (RMSE), the correlation coefficient (R) and mean difference (MD), defined as:

$$RMSE = \sqrt{\frac{1}{n}\sum_{i=1}^{n}(X_i - Y_i)^2} \tag{17}$$

$$R = \frac{\sum_{i=1}^{n}(X_i - \overline{X})(Y_i - \overline{Y})}{\sqrt{\sum_{i=1}^{n}(X_i - \overline{X})^2 \sum_{i=1}^{n}(Y_i - \overline{Y})^2}} \tag{18}$$

$$MD = \frac{1}{n}\sum_{i=1}^{n}(Y_i - X_i) \tag{19}$$

where $n$ is the number of total data samples, $\{X_i\}$ are the wind speed estimates, $\{Y_i\}$ are the wind speed data of ERA5, $\overline{X}$ is the mean of $\{X_i\}$ and $\overline{Y}$ is the mean of $\{Y_i\}$.

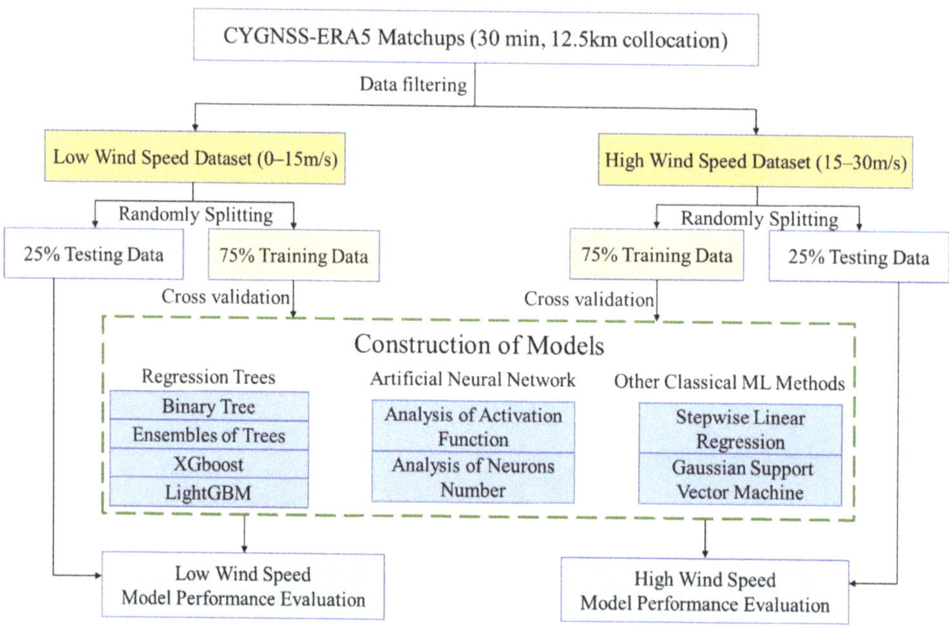


**Figure 4.** Model construction process and evaluation methods.

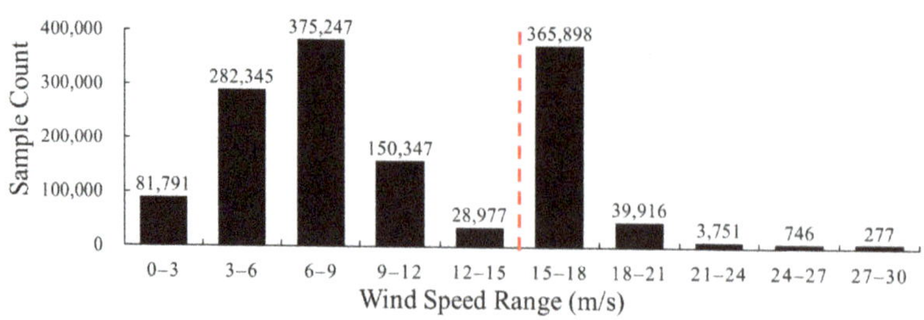


**Figure 5.** Wind speed distribution histogram. The red dotted line divides the dataset into the low wind speed dataset and the high wind speed dataset.

### 3.2. Data Filtering

In this study, after discarding all abnormal values of observations (marked with NaN and negative numbers) and using quality control (QC) flags, a filtering algorithm based on DDM images is proposed. CYGNSS DDM is composed of 11 Doppler rows and 17 delay columns. When the signal condition is poor, the DDMs obtained by GYGNSS will not have an obvious horseshoe shape [14]. Such DDMs are unable to represent the MSS of the reflected surface effectively, and therefore, cannot be used for sea surface wind speed retrieval. In order to analyze the shapes of DDMs more easily, all DDMs are normalized according to:

$$nDDM(\tau, f) = \frac{DDM(\tau, f)}{DDM_{max}} \tag{20}$$

where $nDDM(\tau, f)$ represents the measured power of the reflected signal when the time delay and frequency shift are $\tau$ and $f$ in the normalized DDM. $DDM_{max}$ represents the maximum power in the original DDM. CYGNSS compresses the DDM from a $128 \times 20$ matrix to a $17 \times 11$ matrix [6]. The red solid box in Figure 6a indicates the selected area of the noise floor part where the signal is absent. All the data whose noise floor maximum powers exceed the threshold value of 0.4 are excluded. This step screens out most of the DDMs influenced by noise without involving much computation. Some remaining DDMs may still be influenced by noise, so it is necessary to verify whether a basic horseshoe-shaped emerges. In order to reduce computation, this paper proposes a parameter called $EdgeA$, i.e., the difference between the mean value of the Edge Box and the mean of the noise floor. The orange and red boxes in Figure 6a indicate the trailing edge part and the floor noise part of the DDMs, respectively. The mean value of the noise floor is derived from Equation (21) [30], and $EdgeA$ is derived from Equation (22).

$$Noise_{floor} = \frac{1}{N_1} \sum_{i=1}^{2} \sum_{j=1}^{11} nDDM(\tau_i, f_j) \tag{21}$$

$$EdgeA = \frac{1}{N_2} \sum_{i=\tau_{max}}^{2} \sum_{j=1}^{11} nDDM(\tau_i, f_j) - Noise_{floor} \tag{22}$$

where $N_1$ and $N_2$ are the number of all power values in the noise box and edge box. $\tau_{max}$ is the column number when the power of nDDM is maximum. In this study, $EdgeA$ must be greater than 0.1 to ensure that all DDMs have a basic horseshoe shape.

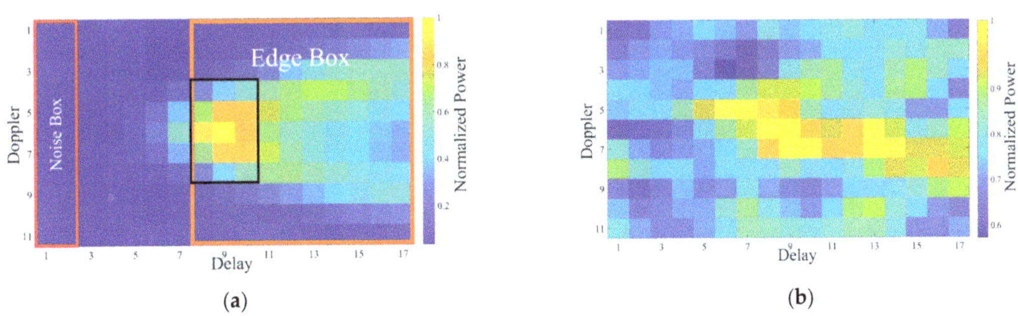


**Figure 6.** (**a**) DDM with a distinct horseshoe shape, (**b**) DDM without a distinct horseshoe shape.

### 3.3. The Results of Regression Trees

This section analyzes the effects of the four regression trees modeling methods (i.e., BT, ET, XGB and LGBM) that were described in Section 2.3. Figure 7 shows the scatter plots of the true and estimated wind speeds. In the figure, the color (from cool to warm) indicates the density of the points. Table 4 shows the retrieval performance of each regression tree model. The bold font represents the best results. It may be seen that many high wind speed data are concentrated in the range of 15–20 m/s, causing elevated inversion accuracy in this range. In order to avoid the influence of data distribution on the analysis of the result, the performance of high wind speed models was analyzed in three data intervals: (1) overall (15–30 m/s), (2) 15–20 m/s and (3) 20–30 m/s.

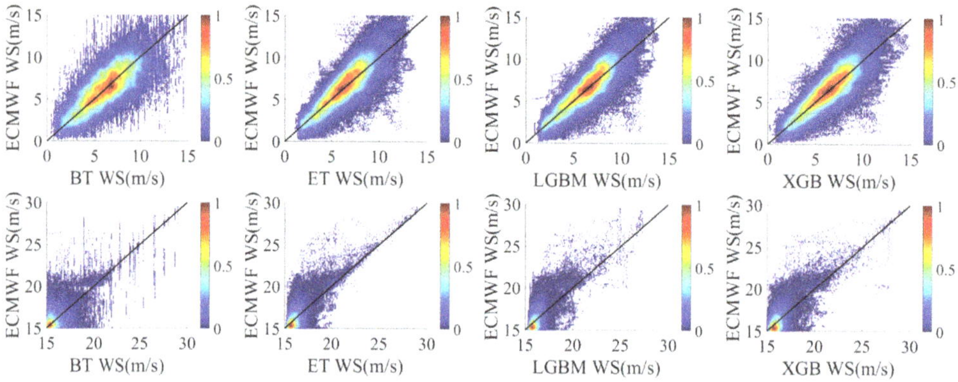

**Figure 7.** Results of wind speed retrievals based on regression trees methods. The subgraphs in the first row represent the retrieval results in low wind speed, while those in the second row represent the retrieval results in high wind speed. The black line shows the 1:1 performance line.

**Table 4.** The retrieval performance of each regression tree models.

| Methods | 0–15 (m/s) | | | 15–30 (m/s) | | | 15–20 (m/s) | | | 20–30 (m/s) | | |
|---|---|---|---|---|---|---|---|---|---|---|---|---|
| | RMSE | R | MD | RMSE | R | MD | RMSE | R | MD | RMSE | R | MD |
| BT | 1.970 | 0.724 | 0.089 | 1.440 | 0.627 | 0.150 | 1.320 | 0.300 | 0.070 | 2.577 | 0.567 | 1.255 |
| ET | 1.496 | 0.831 | 0.097 | **1.100** | **0.767** | **0.145** | 0.971 | 0.497 | 0.047 | 2.204 | 0.625 | 1.487 |
| XGB | 1.483 | 0.835 | 0.085 | 1.145 | 0.744 | 0.165 | 1.005 | 0.470 | 0.214 | 2.336 | 0.611 | 1.559 |
| LGBM | **1.419** | **0.849** | **0.066** | 1.148 | 0.746 | 0.162 | 0.971 | 0.489 | 0.210 | 2.542 | 0.614 | 1.961 |

As shown in Figure 7, all four regression tree-based modeling methods have the ability to retrieve wind speeds in different intervals. As the simplest regression tree modeling method, BT demonstrated the worst retrieval results, i.e., the greatest dispersion, as shown in Figure 7. Further analysis showed that the performance of the other three methods was superior to that of BT. LGBM had the best performance in the low wind speed interval; the RMSE and R of LGBM were improved by 27.97% and 17.27% compared with BT. In the high wind speed interval, the performance of ET was the best. For instance, the RMSE and R of ET were improved by 23.61% and 22.33% compared with BT. It should be noted that the RMSEs of high wind speed models are basically smaller than those of low wind speed models, which does not mean that the former have better performance in general. In fact, this situation is mainly affected by the wind speed distribution of the dataset used in this paper. The performance of all regression trees modeling methods was better in low wind speed interval, which is consistent with the conclusions of many previous studies [26–29]. From the calculated MD, a slight underestimation of true wind speed in both figures was observed. Besides, more obvious underestimations at high winds were shown by both models. This result is similar to that of [28]. Most of the research results demonstrate that GNSS-R data are more suitable for retrieving low wind speeds, while significant performance degradation occurs when retrieving high wind speeds [27–29]. This might be due to the reduced sensitivity of an ocean scattering cross-section to the high wind speed and the increased random error in the DDM signal [14].

### 3.4. The Results of ANNs

As shown in Figure 2, we adopted the three-layer neural network structure; the number distribution of three-layer neurons was N-2N-N. In this section, the influence of the value of N and activation function is analyzed. N is set at 5, 10, 20, 30, 40, 50 and 60, respectively. Tables 5–7 show the RMSEs, Rs and MDs of the wind speed retrieval using ANN models with different activation functions and N values, respectively. As in Section 3.3, the bold font represents the best result, and the performance of the high wind speed models was analyzed in three data intervals. Figure 8 shows RMSEs and Rs of the wind speed retrieval models in a more intuitive form, i.e., in the form of line chart.

**Table 5.** RMSEs of the ANN models.

| N | 0–15 (m/s) | | | 15–30 (m/s) | | | 15–20 (m/s) | | | 20–30 (m/s) | | |
|---|---|---|---|---|---|---|---|---|---|---|---|---|
| | Relu | Tanh | Sigmoid | Relu | Tanh | Sigmoid | Relu | Tanh | Sigmoid | Relu | Tanh | Sigmoid |
| 5 | 1.466 | 1.474 | 1.473 | 1.291 | 1.245 | 1.283 | 1.016 | 0.960 | 0.956 | 3.226 | 3.402 | 3.204 |
| 10 | 1.501 | 1.473 | **1.460** | 1.390 | 1.232 | 1.286 | 1.022 | 0.980 | 0.970 | 3.755 | 3.342 | 3.072 |
| 20 | 1.541 | 1.472 | 1.512 | 1.289 | 1.320 | 1.248 | 1.055 | 1.004 | 0.988 | 3.028 | 3.019 | 3.503 |
| 30 | 1.556 | 1.524 | 1.503 | 1.914 | **1.211** | 1.223 | 1.794 | 0.996 | 1.007 | 3.129 | 2.902 | 2.766 |
| 40 | 1.616 | 1.522 | 1.544 | 3.876 | 1.239 | 1.248 | 3.861 | 0.999 | 1.013 | 4.074 | 3.039 | 2.917 |
| 50 | 1.611 | 1.554 | 1.523 | 2.468 | 1.219 | 1.209 | 2.380 | 0.999 | 0.977 | 3.455 | 2.800 | 2.963 |
| 60 | 1.663 | 1.563 | 1.577 | 2.022 | 1.224 | 1.243 | 1.863 | 0.982 | 0.992 | 3.542 | 3.086 | 2.926 |

**Table 6.** Rs of the ANN models.

| N | 0–15 (m/s) | | | 15–30 (m/s) | | | 15–20 (m/s) | | | 20–30 (m/s) | | |
|---|---|---|---|---|---|---|---|---|---|---|---|---|
| | Relu | Tanh | Sigmoid | Relu | Tanh | Sigmoid | Relu | Tanh | Sigmoid | Relu | Tanh | Sigmoid |
| 5 | 0.838 | 0.836 | 0.836 | 0.659 | 0.695 | 0.671 | 0.434 | 0.498 | 0.501 | 0.518 | 0.540 | 0.543 |
| 10 | 0.829 | 0.836 | **0.839** | 0.635 | 0.699 | 0.665 | 0.452 | 0.473 | 0.487 | 0.413 | 0.491 | 0.547 |
| 20 | 0.821 | 0.837 | 0.827 | 0.662 | 0.667 | 0.685 | 0.440 | 0.471 | 0.482 | 0.550 | 0.547 | 0.418 |
| 30 | 0.818 | 0.825 | 0.829 | 0.453 | **0.708** | 0.703 | 0.228 | 0.468 | 0.464 | 0.484 | 0.524 | 0.578 |
| 40 | 0.803 | 0.826 | 0.820 | 0.264 | 0.693 | 0.687 | 0.112 | 0.470 | 0.457 | 0.372 | 0.527 | 0.543 |
| 50 | 0.804 | 0.817 | 0.825 | 0.385 | 0.703 | 0.708 | 0.187 | 0.475 | 0.495 | 0.426 | 0.562 | 0.512 |
| 60 | 0.794 | 0.816 | 0.811 | 0.455 | 0.702 | 0.690 | 0.219 | 0.483 | 0.481 | 0.442 | 0.563 | 0.536 |

**Table 7.** MDs of the ANN models.

| N | 0–15 (m/s) | | | 15–30 (m/s) | | | 15–20 (m/s) | | | 20–30 (m/s) | | |
|---|---|---|---|---|---|---|---|---|---|---|---|---|
| | Relu | Tanh | Sigmoid | Relu | Tanh | Sigmoid | Relu | Tanh | Sigmoid | Relu | Tanh | Sigmoid |
| 5 | 0.047 | 0.038 | 0.036 | 0.192 | 0.183 | 0.208 | 0.022 | 0.005 | 0.017 | 2.533 | 2.633 | 2.833 |
| 10 | 0.050 | 0.028 | **0.027** | 0.117 | 0.183 | 0.192 | 0.016 | 0.021 | 0.009 | 1.509 | 2.423 | 2.703 |
| 20 | 0.062 | 0.040 | 0.053 | 0.160 | 0.137 | 0.152 | 0.009 | 0.045 | 0.007 | 2.233 | 1.400 | 2.148 |
| 30 | 0.071 | 0.028 | 0.050 | 0.172 | **0.100** | 0.182 | 0.036 | 0.011 | 0.061 | 2.043 | 1.321 | 1.849 |
| 40 | 0.084 | 0.080 | 0.087 | 0.165 | 0.191 | 0.175 | 0.040 | 0.046 | 0.037 | 1.877 | 2.198 | 2.075 |
| 50 | 0.077 | 0.064 | 0.083 | 0.122 | 0.165 | 0.144 | 0.012 | 0.017 | 0.021 | 1.630 | 2.194 | 1.843 |
| 60 | 0.056 | 0.064 | 0.095 | 0.113 | 0.136 | 0.184 | 0.004 | 0.022 | 0.029 | 1.619 | 1.688 | 2.315 |

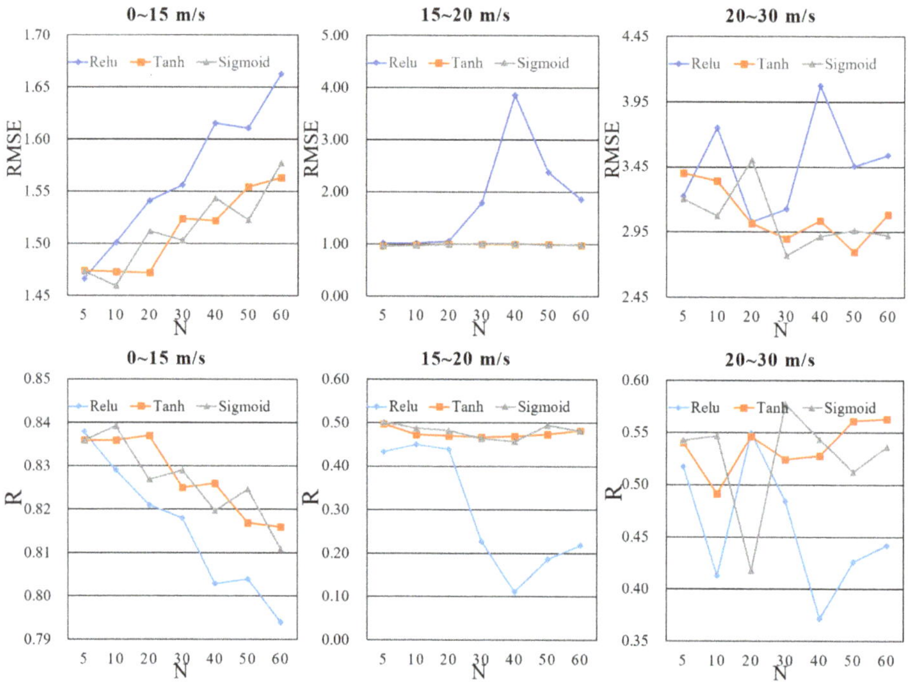


**Figure 8.** RMSEs and Rs of the wind speed retrieval models using ANN models with different activation functions and different numbers of neurons.

In the low wind speed interval, it is obvious that the choice of activation function hardly affected the ANN models, as shown in Figure 8. However, the increase of the number of neurons significantly reduced the accuracy of the models, although the accuracy variation was not very significant. In the high wind speed interval, the increase in the number of neurons had little effect on Sigmoid and Tanh, but it had an obvious effect on ReLu. As shown in Table 6, for a low wind speed interval, when the activation function was Sigmoid and N was 10, the performance of ANN was the best. For a high wind speed interval, when the activation function was Tanh and N was 30, the performance of ANN was the best. Overall, although the underestimation of ANNs at high winds was smaller than that of regression trees, the retrieval performance of ANNs was slightly worse than that of the regression tree modeling methods. In order to facilitate a comparison with other methods, scatter plots of low wind speeds retrieved by the Sigmoid function and of high wind speeds retrieved by the Tanh function are presented as examples, as shown in Figure 9.

### *3.5. The Results of SLR and SVM*

This section analyzes the effects of two other classical machine learning methods (i.e., SLR and SVM), as described in Sections 2.4 and 2.5. Figure 10 shows scatter plots of the true and estimated wind speeds. Table 8 shows the retrieval performance of each regression trees model. Similarly, the bold font represents the best result, and the performance of high wind speed models was analyzed in three data intervals. It is obvious that the retrieval results of SVM had less dispersion than those of SLR, which means that the performance of SVM was better. However, the retrieval performance of the models described in the previous two sections was better than that of the models presented in this section.

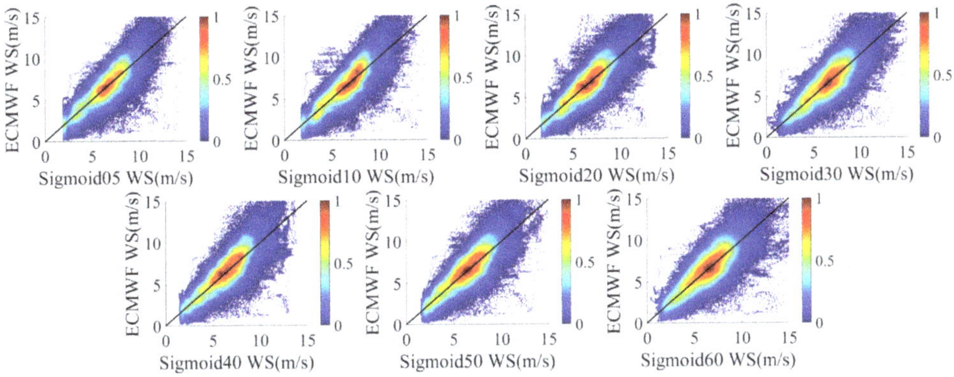


(**a**)

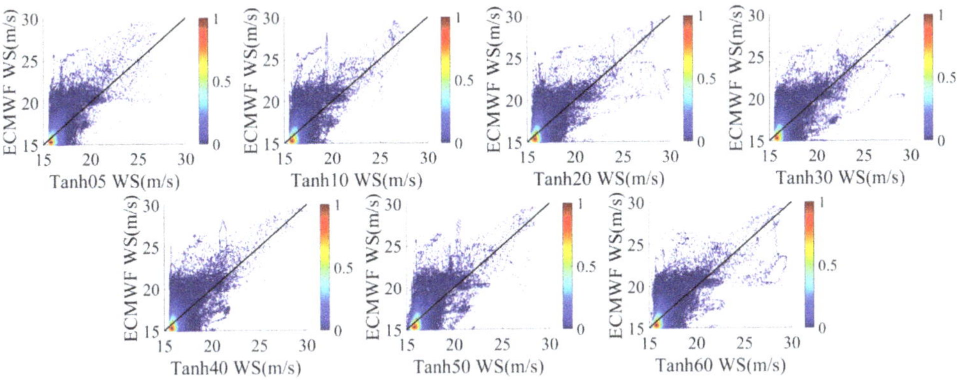


(**b**)

**Figure 9.** (**a**) ANN scatter plot with the activation function Sigmoid in the low wind speed interval. (**b**) ANN scatter plot with the activation function Tanh in the high wind speed interval.

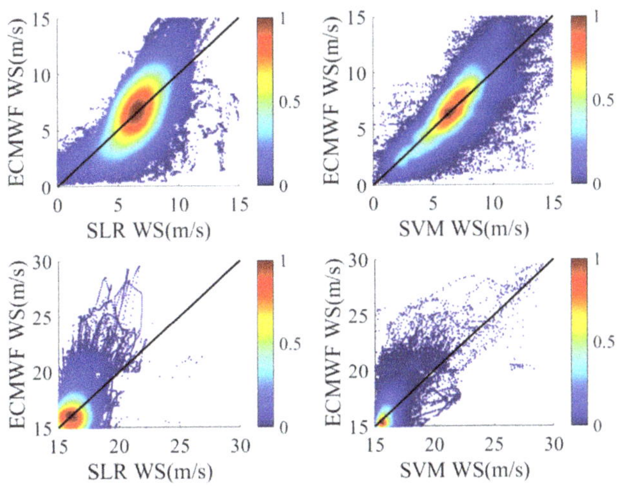


**Figure 10.** Results of wind speed retrievals based on SLR and SVM.

**Table 8.** Retrieval performance of SLR and SVM.

| Methods | 0–15 (m/s) | | | 15–30 (m/s) | | | 15–20 (m/s) | | | 20–30 (m/s) | | |
|---|---|---|---|---|---|---|---|---|---|---|---|---|
| | RMSE | R | MD | RMSE | R | MD | RMSE | R | MD | RMSE | R | MD |
| SLR | 1.929 | 0.697 | 0.055 | 1.541 | 0.475 | 0.364 | 1.213 | 0.280 | 0.199 | 3.844 | 0.393 | 3.456 |
| SVM | **1.577** | **0.810** | **−0.021** | **1.290** | **0.684** | **0.304** | 1.007 | 0.488 | 0.075 | 3.254 | 0.556 | 2.623 |

### 3.6. Summary

The preceding subsections presented and analyzed the retrieval performance of several machine learning methods in different wind speed intervals. This subsection summarizes and analyzes their performance gaps. Figure 11 shows the Rs and RMSEs of the models using machine learning methods. It is obvious from Figure 11 that the RMSE of LGBM is smaller than those of other models in a low wind speed interval, while the RMSE of ET is smaller than those of other models in a high wind speed interval. The R values are usually larger when the RMSE values are smaller. The performance of LGBM, ET, ANN and XGB are significantly better than that of SVM, BT and SLR, which means that they are more suitable for wind speed retrieval.

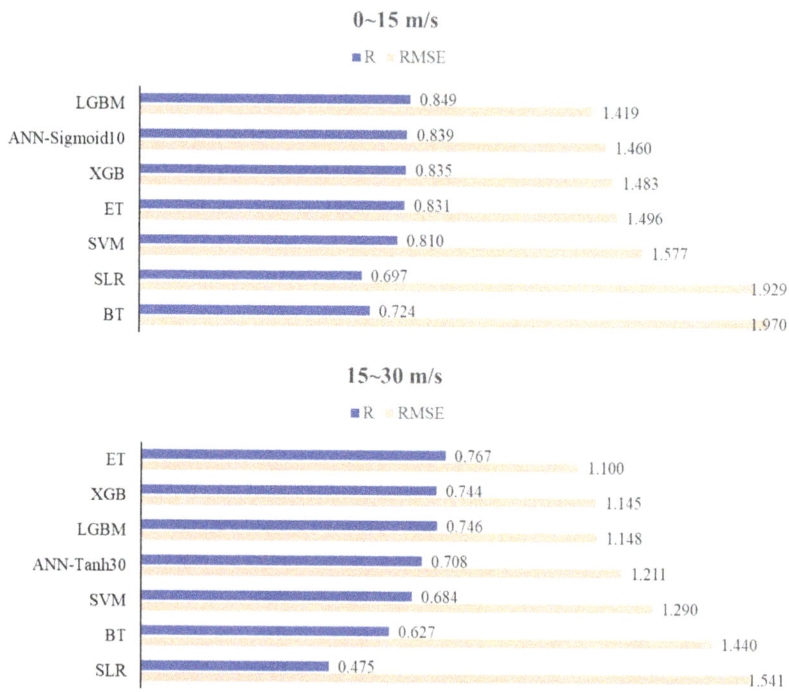


**Figure 11.** Rs and RMSEs of machine learning models.

## 4. Discussion

By analyzing the performance of all seven models, it can be concluded that LGBM performed best in the low wind speed interval, while ET performed best in the high wind speed interval. However, the above experimental results do not prove that all the variables in Table 1 can be used to optimize the performance of the model. On the contrary, some variables may reduce the accuracy of the model. Therefore, it is very important to analyze the effects of different variables. It should be noted that in the high wind speed interval, the data of spaceborne GNSS-R also present different data distributions and characteristics from those in the low wind speed interval, and the roles of the variables were not always consistent. Here, we use the characteristics of XGBoost as the basis for evaluating the effect of each variable. XGBoost uses the average gain (AG) of data splits across all trees to measure the effects of variables [51]. After model training, by analyzing the XGBoost model structure, the AG related to each variable is defined as:

$$AG_{v_i} = \frac{\sum Gain_{v_i}}{S_{v_i}} \tag{23}$$

where $v_i$ is a variable used in the XGBoost model, $S_{v_i}$ is the number of times that $v_i$ is used to split the data across all trees and $Gain_{v_i}$ is the gain value of each tree after splitting with $v_i$. Table 9 shows the AG of each variable in the low and high wind speed intervals, respectively.

**Table 9.** Rankings of the effects of variables.

| 0–15 (m/s) | | | 15–30 (m/s) | | |
|---|---|---|---|---|---|
| **Rank** | **AG** | **Variables** | **Rank** | **AG** | **Variables** |
| 1 | 9452.12 | NBRCS | 1 | 364.99 | SWH_swell |
| 2 | 2649.02 | LES | 2 | 100.85 | NoiseFloor |
| 3 | 1887.14 | SNR | 3 | 90.88 | NBRCS |
| 4 | 1602.55 | SWH_swell | 4 | 88.56 | ScatterArea |
| 5 | 443.67 | InstGain | 5 | 84.76 | InstGain |
| 6 | 360.00 | NoiseFloor | 6 | 76.91 | AzBody |
| 7 | 337.60 | DDMA15 | 7 | 73.86 | DDMA15 |
| 8 | 320.35 | ScatterArea | 8 | 71.09 | IncAngle |
| 9 | 284.02 | AzBody | 9 | 64.71 | SNR |
| 10 | 253.56 | IncAngle | 10 | 44.97 | LES |

Although AG helps to verify the effectiveness of feature selection, it cannot be used as a direct basis thereof. As such, the rationale of Table 9 needs to be demonstrated through experimental results. In order to analyze the influences of different variables more intuitively, this study constructed 60 models based on ET, XGB and LGBM with different variables. Line charts were used to help in analyzing the influence of these variables. The $x$-axis in Figure 12 indicates the number of variables, which is consistent with the ranking of the effects of variables in Table 9. For example, in the low wind speed interval, if the number of variables was set at 4, NBRCS, LES, SNR and SWH_swell were used in the modeling; in the high wind speed interval, if the number of variables was set at 3, SWH_swell, NoiseFloor and NBRCS were used in the modeling.

In Figure 12, the relationship between variables and models can be analyzed clearly. It is obvious that Figure 12 and Table 9 are highly consistent. In the low wind speed interval, the AG of NBRCS is much larger than that of other variables, which means that NBRCS is the most important variable in the low wind speed models. In the two subgraphs of the first column of Figure 12, it is obvious that LES, SNR and SWH_swell improved the performance of the model greatly, as also confirmed in Table 9. In Table 9, the AGs of LES, SNR and SWH_swell are significantly greater than those of the other variables. These variables effectively reduced the RMSE of the model and increased the correlation coefficient between the wind speed estimates and the true values of wind speed. In the high wind speed interval, the models were mostly affected by SWH_swell; this may have been due to the degradation of the performance of spaceborne GNSS-R technology in a high wind speed. This result also indicates that, especially in the high wind speed interval, spaceborne GNSS-R technology needs to fuse more reliable auxiliary information to achieve better retrieval results. The contributions of other variables to the model are basically similar. Different from the results of the low wind speed interval, the effects of NoiseFloor and ScatterArea were significantly greater, while the effects of SNR and LES were lower. In the high wind speed interval, the quality of DDM became lower, decreasing the correlation coefficients between sea surface MSS and the variables SNR and LES.

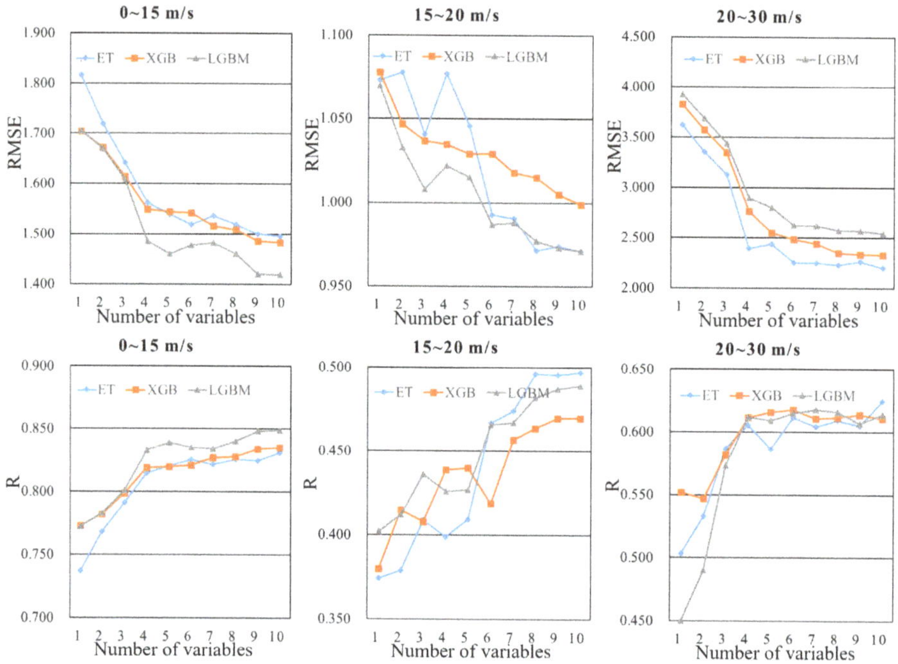


**Figure 12.** RMSEs and Rs of the wind speed retrieval models using different numbers of variables.

In general, from the above analysis, it is obvious that the results of the models with all variables are the best in both high and low wind speed intervals. In most cases, the accuracy of the model is directly proportional to the number of variables. Additionally, for different modeling methods, the influence of the number of variables was different; for different wind speed intervals, the rankings of the effects of variables were different. The above conclusions may be helpful for the future research of spaceborne GNSS-R sea surface wind speed retrieval.

## 5. Conclusions

By using machine learning methods, this study investigated wind speed retrieval in different wind speed intervals. Through extensive processing of experimental data, it was observed that different machine learning methods have different properties in different wind speed intervals. In particular, a range of multi-variable models was developed and evaluated. The results showed that the LGBM model performs best with an RMSE of 1.419 m/s and a correlation coefficient of 0.849 in the low wind speed interval (0–15 m/s), while the ET model performs best with an RMSE of 1.100 and a correlation coefficient of 0.767 in the high wind speed interval (15–30 m/s). In addition, through experiments, some characteristics of ANN models were found in wind speed retrieval. In the low wind speed interval, the choice of activation function hardly affects the ANN models, while the increase of the number of neurons significantly reduces the accuracy of the model. In the high wind speed interval, the increase in the number of neurons has little effect on Sigmoid and Tanh, but it has an obvious effect on ReLu.

The effects of the variables used in the wind speed retrieval models described in this paper were analyzed. Through processing experimental data, it was observed that the models with all variables (i.e. NBRCS, LES, SNR, DDMA, Noise Floor, sp_inc_angle, sp_az_body, Instrument Gain, Scatter Area, and SWH_swell) achieved the highest accuracy. In the low wind speed interval, NBRCS, LES, SNR and SWH_swell were the most important variables. In the high wind speed interval, the models were mostly affected by SWH_swell, and the ranking of the effects of variables was very different from that in the low wind speed interval.

Future studies will focus on further performance enhancements of the models developed in this paper. For instance, the accuracy of the model would decrease in the presence of large wind speed and high SWH_swell. It would thus be useful to develop techniques to handle the retrieval of high wind speeds with minimal performance degradation.

**Author Contributions:** All authors have made significant contributions to this manuscript. C.W. constructed a part of machine learning models of this paper, analyzed the data, wrote the initial version of paper and validated all the models; K.Y. conceived the improved method, wrote the revised version of the paper and provided supervision; F.Q. constructed some of the machine learning models used in this paper; J.B., S.H. and K.Z. checked and revised this paper. All authors have read and agreed to the published version of the manuscript.

**Funding:** This work was supported in part by the National Natural Science Foundation of China under Grants 42174022 and in part by the Programme of Introducing Talents of Discipline to Universities, Plan 111, Grant No. B20046.

**Data Availability Statement:** Not applicable.

**Acknowledgments:** We would like to thank the NASA and European Center for Medium-Range Weather Forecasts (ECMWF) for providing the data. The results (figures and tables) presented in this article are mainly generated by MATLAB software (https://ww2.mathworks.cn/ (accessed on 26 March 2022)). The authors thank the anonymous reviewers for their in-depth reviews and helpful suggestions that have largely contributed to improving this paper.

**Conflicts of Interest:** The authors declare no conflict of interest.

## References

1. Martin-Neira, M. A Passive Reflectometry and Interferometry System (PARIS): Application to ocean altimetry. *ESA J.* **1993**, *17*, 331–355.
2. Hu, C.J.; Benson, C.R.; Qiao, L.; Rizos, C. The validation of the weight function in the leading-edge-derivative path delay estimator for space-based GNSS-R altimetry. *IEEE Trans. Geosci. Remote Sens.* **2020**, *58*, 6243–6254. [CrossRef]
3. Clarizia, M.P.; Ruf, C.S. Wind speed retrieval algorithm for the Cyclone Global Navigation Satellite System (CYGNSS) mission. *IEEE Trans. Geosci. Remote Sens.* **2016**, *54*, 4419–4432. [CrossRef]
4. Alonso-Arroyo, A.; Zavorotny, V.U.; Camps, A. Sea ice detection using UK TDS-1 GNSS-R data. *IEEE Trans. Geosci. Remote Sens.* **2017**, *55*, 4989–5001. [CrossRef]
5. Arroyo, A.A.; Camps, A.; Aguasca, A.; Forte, G.F.; Monerris, A.; Rudiger, C.; Walker, J.P.; Park, H.; Pascual, D.; Onrubia, R. Dual-polarization GNSS-R interference pattern technique for soil moisture mapping. *IEEE J. Sel. Top. Appl. Earth Obs. Remote Sens.* **2014**, *7*, 1533–1544. [CrossRef]
6. Peng, Q.; Jin, S.G. Significant wave height estimation from space-borne cyclone-GNSS reflectometry. *Remote Sens.* **2019**, *11*, 584. [CrossRef]
7. Ruf, C.; Lyons, A.; Unwin, M.; Dickinson, J.; Rose, R.; Rose, D.; Vincent, M. CYGNSS: Enabling the future of hurricane prediction. *IEEE Geosci. Remote Sens. Mag.* **2013**, *1*, 52–67. [CrossRef]
8. Gleason, S. *CYGNSS Algorithm Theoretical Basis Documents, Level 1A and 1B*; University of Michigan: Ann Arbor, MI, USA, 2018; pp. 4–19.
9. Yu, K. *Theory and Practice of GNSS Reflectometry*; Springer Nature: Berlin/Heidelberg, Germany, 2021; pp. 1–376. [CrossRef]
10. Jing, C.; Niu, X.L.; Duan, C.D.; Lu, F.; Di, G.D.; Yang, X.F. Sea surface wind speed retrieval from the first Chinese GNSS-R mission: Technique and preliminary results. *Remote Sens.* **2019**, *11*, 3013. [CrossRef]
11. Huang, B.Y.; Stone, P.H.; Sokolov, A.P.; Kamenkovich, I.V. Ocean heat uptake in transient climate change: Mechanisms and uncertainty due to subgrid-scale eddy mixing. *J. Clim.* **2003**, *16*, 3344–3356. [CrossRef]
12. Barthelmie, R.J. The effects of atmospheric stability on coastal wind climates. *Meteorol. Appl.* **1999**, *6*, 39–47. [CrossRef]
13. Kirincich, A. Remote sensing of the surface wind field over the coastal ocean via direct calibration of HF radar backscatter power. *J. Atmos. Oceanic Technol.* **2016**, *33*, 1377–1392. [CrossRef]
14. Bu, J.W.; Yu, K.G.; Zhu, Y.C.; Qian, N.J.; Chang, J. Developing and testing models for sea surface wind speed estimation with GNSS-R delay doppler maps and delay waveforms. *Remote Sens.* **2020**, *12*, 3760. [CrossRef]

15. Jacobson, M.D.; Emery, W.J.; Westwater, E.R. Oceanic wind vector determination using a dual-frequency microwave airborne radiometer theory and experiment. In Proceedings of the IGARSS'96 1996 International Geoscience and Remote Sensing Symposium, Lincoln, NE, USA, 31 May 1996; pp. 1138–1140.
16. Monaldo, F.M.; Thompson, D.R.; Beal, R.C.; Pichel, W.G.; Clemente-Colon, P. Comparison of SAR-derived wind speed with model predictions and ocean buoy measurements. *IEEE Trans. Geosci. Remote Sens.* **2001**, *39*, 2587–2600. [CrossRef]
17. Liu, X.X.; Bai, W.H.; Xia, J.M.; Huang, F.X.; Yin, C.; Sun, Y.Q.; Du, Q.F.; Meng, X.G.; Liu, C.L.; Hu, P.; et al. FA-RDN: A hybrid neural network on GNSS-R sea surface wind speed retrieval. *Remote Sens.* **2021**, *13*, 4820. [CrossRef]
18. Bu, J.W.; Yu, K.G. Sea surface rainfall detection and intensity retrieval based on GNSS-reflectometry data from the CYGNSS mission. *IEEE Trans. Geosci. Remote Sens.* **2022**, *60*, 1–15. [CrossRef]
19. Zavorotny, V.U.; Voronovich, A.G. Scattering of GPS signals from the ocean with wind remote sensing application. *IEEE Trans. Geosci. Remote Sens.* **2000**, *38*, 951–964. [CrossRef]
20. Komjathy, A.; Zavorotny, V.U.; Axelrad, P.; Born, G.H.; Garrison, J.L. GPS Signal scattering from sea surface: Wind speed retrieval using experimental data and theoretical model. *Remote Sens. Environ.* **2000**, *73*, 162–174. [CrossRef]
21. Clarizia, M.P.; Ruf, C.S.; Jales, P.; Gommenginger, C. Spaceborne GNSS-R minimum variance wind speed estimator. *IEEE Trans. Geosci. Remote Sens.* **2014**, *52*, 6829–6843. [CrossRef]
22. Ruffini, G.; Soulat, F.; Caparrini, M.; Germain, O.; Martin-Neira, M. The Eddy Experiment: Accurate GNSS-R ocean altimetry from low altitude aircraft. *Geophys. Res. Lett.* **2004**, *31*, 1–4. [CrossRef]
23. Liu, Y.X.; Collett, I.; Morton, Y.J. Application of neural network to GNSS-R wind speed retrieval. *IEEE Trans. Geosci. Remote Sens.* **2019**, *57*, 9756–9766. [CrossRef]
24. Asgarimehr, M.; Wickert, J.; Reich, S. Evaluating impact of rain attenuation on space-borne GNSS reflectometry wind speeds. *Remote Sens.* **2019**, *11*, 1048. [CrossRef]
25. Asgarimehr, M.; Zhelavskaya, I.; Foti, G.; Reich, S.; Wickert, J. A GNSS-R Geophysical model function: Machine Learning for wind speed retrievals. *IEEE Geosci. Remote Sens. Lett.* **2020**, *17*, 1333–1337. [CrossRef]
26. Li, X.H.; Yang, D.K.; Yang, J.S.; Zheng, G.; Han, G.Q.; Nan, Y.; Li, W.Q. Analysis of coastal wind speed retrieval from CYGNSS mission using artificial neural network. *Remote Sens. Environ.* **2021**, *260*, 112454. [CrossRef]
27. Guo, W.F.; Du, H.; Guo, C.; Southwell, B.J.; Cheong, J.W.; Dempster, A.G. Information fusion for GNSS-R wind speed retrieval using statistically modified convolutional neural network. *Remote Sens. Environ.* **2022**, *272*, 112934. [CrossRef]
28. Asgarimehr, M.; Arnold, C.; Weigel, T.; Ruf, C.; Wickert, J. GNSS Reflectometry global ocean wind speed using deep learning: Development and assessment of CyGNSSnet. *Remote Sens. Environ.* **2022**, *269*, 112801. [CrossRef]
29. Zhang, Y.; Yin, J.W.; Yang, S.H.; Meng, W.T.; Han, Y.L.; Yan, Z.Y. High wind speed inversion model of CYGNSS sea surface data based on machine learning. *Remote Sens.* **2021**, *13*, 3324. [CrossRef]
30. Zhu, Y.C.; Yu, K.G.; Zou, J.G.; Wickert, J. Sea ice detection based on differential delay-doppler maps from UK TechDemoSat-1. *Sensors* **2017**, *17*, 1614. [CrossRef]
31. Marchan-Hernandez, J.F.; Valencia, E.; Rodriguez-Alvarez, N.; Ramos-Perez, I.; Bosch-Lluis, X.; Camps, A.; Eugenio, F.; Marcello, J. Sea-state determination using GNSS-R data. *IEEE Geosci. Remote Sens. Lett.* **2010**, *7*, 621–625. [CrossRef]
32. Li, B.W.; Yang, L.; Zhang, B.; Yang, D.K.; Wu, D. Modeling and simulation of GNSS-R observables with effects of swell. *IEEE J. Sel. Top. Appl. Earth Obs. Remote Sens.* **2020**, *13*, 1833–1841. [CrossRef]
33. Lin, W.M.; Portabella, M.; Foti, G.; Stoffelen, A.; Gommenginger, C.; He, Y.J. Toward the generation of a wind geophysical model function for spaceborne GNSS-R. *IEEE Trans. Geosci. Remote Sens.* **2019**, *57*, 655–666. [CrossRef]
34. Bu, J.W.; Yu, K.G.; Han, S.; Qian, N.J.; Lin, Y.R.; Wang, J. Retrieval of sea surface rainfall intensity using spaceborne GNSS-R data. *IEEE Trans. Geosci. Remote Sens.* **2022**, *60*, 5803116. [CrossRef]
35. Chen-Zhang, D.D.; Ruf, C.S.; Ardhuin, F.; Park, J. GNSS-R nonlocal sea state dependencies: Model and empirical verification. *J. Geophys. Res.-Ocean.* **2016**, *121*, 8379–8394. [CrossRef]
36. White, D.; Sifneos, J.C. Regression tree cartography. *J. Comput. Graphical Stat.* **2002**, *11*, 600–614. [CrossRef]
37. Prasad, A.M.; Iverson, L.R.; Liaw, A. Newer classification and regression tree techniques: Bagging and random forests for ecological prediction. *Ecosystems* **2006**, *9*, 181–199. [CrossRef]
38. Chen, W.; Shahabi, H.; Zhang, S.; Khosravi, K.; Shirzadi, A.; Chapi, K.; Pham, B.T.; Zhang, T.Y.; Zhang, L.Y.; Chai, H.C.; et al. Landslide susceptibility modeling based on GIS and novel bagging-based kernel logistic regression. *Appl. Sci.* **2018**, *8*, 2540. [CrossRef]
39. Hothorn, T.; Lausen, B.; Benner, A.; Radespiel-Troger, M. Bagging survival tree. *Stat. Med.* **2004**, *23*, 77–91. [CrossRef]
40. Chan, J.C.W.; Huang, C.Q.; DeFries, R. Enhanced algorithm performance for land cover classification from remotely sensed data using bagging and boosting. *IEEE Trans. Geosci. Remote Sens.* **2001**, *39*, 693–695.
41. Breiman, L. Bagging predictors. *Mach. Learn.* **1996**, *24*, 123–140. [CrossRef]
42. Bauer, E.; Kohavi, R. An empirical comparison of voting classification algorithms: Bagging, boosting, and variants. *Mach. Learn.* **1999**, *36*, 105–139. [CrossRef]
43. Chen, T.Q.; Guestrin, C.; Assoc Comp, M. XGBoost: A scalable tree boosting system. In Proceedings of the 22nd ACM SIGKDD International Conference on Knowledge Discovery and Data Mining (KDD), San Francisco, CA, USA, 13–17 August 2016; pp. 785–794.

44. Ke, G.L.; Meng, Q.; Finley, T.; Wang, T.F.; Chen, W.; Ma, W.D.; Ye, Q.W.; Liu, T.Y. LightGBM: A highly efficient gradient boosting decision tree. In Proceedings of the 31st Annual Conference on Neural Information Processing Systems (NIPS), Long Beach, CA, USA, 4–9 December 2017.
45. Basheer, I.A.; Hajmeer, M. Artificial neural networks: Fundamentals, computing, design, and application. *J. Microbiol. Methods* **2000**, *43*, 3–31. [CrossRef]
46. Ertugrul, O.F. A novel type of activation function in artificial neural networks: Trained activation function. *Neural Netw.* **2018**, *99*, 148–157. [CrossRef] [PubMed]
47. Siddiqi, M.H.; Ali, R.; Khan, A.M.; Park, Y.T.; Lee, S. Human facial expression recognition using stepwise linear discriminant analysis and hidden conditional random fields. *IEEE Trans. Image Process.* **2015**, *24*, 1386–1398. [CrossRef] [PubMed]
48. Krusienski, D.J.; Sellers, E.W.; McFarland, D.J.; Vaughan, T.M.; Wolpaw, J.R. Toward enhanced P300 speller performance. *J. Neurosci. Methods* **2008**, *167*, 15–21. [CrossRef] [PubMed]
49. Scholkopf, B.; Sung, K.K.; Burges, C.J.C.; Girosi, F.; Niyogi, P.; Poggio, T.; Vapnik, V. Comparing support vector machines with Gaussian kernels to radial basis function classifiers. *IEEE Trans. Signal Process.* **1997**, *45*, 2758–2765. [CrossRef]
50. Burges, C.J.C. A Tutorial on support vector machines for pattern recognition. *Data Min. Knowl. Discov.* **1998**, *2*, 121–167. [CrossRef]
51. Zhang, A.W.; Dong, Z.; Kang, X.Y. Feature selection algorithms of airborne LiDAR combined with hyperspectral images based on XGBoost. *Chin. J. Lasers-Zhongguo Jiguang* **2019**, *46*, 0404003. [CrossRef]

*remote sensing*

MDPI

*Article*

# GloWS-Net: A Deep Learning Framework for Retrieving Global Sea Surface Wind Speed Using Spaceborne GNSS-R Data

**Jinwei Bu [1,2], Kegen Yu [2,*], Xiaoqing Zuo [1], Jun Ni [3], Yongfa Li [1] and Weimin Huang [4]**

1 Faculty of Land Resources Engineering, Kunming University of Science and Technology, Kunming 650093, China
2 School of Environment Science and Spatial Informatics, China University of Mining and Technology, Xuzhou 221116, China
3 School of Information Science and Engineering, Yunnan University, Kunming 650500, China
4 Department of Electrical and Computer Engineering, Memorial University of Newfoundland, St. John's, NL A1B 3X5, Canada
* Correspondence: kegen.yu@cumt.edu.cn

**Abstract:** Spaceborne Global Navigation Satellite System Reflectometry (GNSS-R) is a new remote sensing technology that uses GNSS signals reflected from the Earth's surface to estimate geophysical parameters. Because of its unique advantages such as high temporal and spatial resolutions, low observation cost, wide coverage and all-weather operation, it has been widely used in land and ocean remote sensing fields. Ocean wind monitoring is the main objective of the recently launched Cyclone GNSS (CYGNSS). In previous studies, wind speed was usually retrieved using features extracted from delay-Doppler maps (DDMs) and empirical geophysical model functions (GMFs). However, it is a challenge to employ the GMF method if using multiple sea state parameters as model input. Therefore, in this article, we propose an improved deep learning network framework to retrieve global sea surface wind speed using spaceborne GNSS-R data, named GloWS-Net. GloWS-Net considers the fusion of auxiliary information including ocean swell significant wave height (SWH), sea surface rainfall and wave direction to build an end-to-end wind speed retrieval model. In order to verify the improvement of the proposed model, ERA5 and Cross-Calibrated Multi-Platform (CCMP) wind data were used as reference for extensive testing to evaluate the wind speed retrieval performance of the GloWS-Net model and previous models (i.e., GMF, fully connected network (FCN) and convolutional neural network (CNN)). The results show that, when using ERA5 winds as ground truth, the root mean square error (RMSE) of the proposed GloWS-Net model is 23.98% better than that of the MVE method. Although the GloWS-Net model and the FCN model have similar RMSE (1.92 m/s), the mean absolute percentage error (MAPE) of the former is improved by 16.56%; when using CCMP winds as ground truth, the RMSE of the proposed GloWS-Net model is 2.16 m/s, which is 20.27% better than the MVE method. Compared with the FCN model, the MAPE is improved by 17.75%. Meanwhile, the GloWS-Net outperforms the FCN, traditional CNN, modified CNN (MCNN) and CyGNSSnet models in global wind speed retrieval especially at high wind speeds.

**Keywords:** Cyclone Global Navigation Satellite System (CYGNSS); delay-Doppler maps (DDMs); ocean wind speed; geophysical model functions (GMFs); deep learning

**Citation:** Bu, J.; Yu, K.; Zuo, X.; Ni, J.; Li, Y.; Huang, W. GloWS-Net: A Deep Learning Framework for Retrieving Global Sea Surface Wind Speed Using Spaceborne GNSS-R Data. *Remote Sens.* **2023**, *15*, 590. https://doi.org/10.3390/rs15030590

Academic Editors: Hugo Carreno-Luengo, Dallas Masters and Chun-Liang Lin

Received: 16 November 2022
Revised: 13 January 2023
Accepted: 17 January 2023
Published: 18 January 2023



## 1. Introduction

Ocean wind speed is of great significance for the numerical prediction of the marine environment, monitoring of marine disasters, air sea interaction, meteorological forecasting, climate research, etc. In particular, tropical cyclones, typhoons or storm surges have seriously damaged infrastructure and endangered people's lives. For these reasons, it is very important to monitor the sea wind speed to study and predict some complex weather (such as tropical cyclone, typhoon or storm surge warning) [1–3]. Long-term measurements of ocean wind speed can be made using conventional observational methods such as buoys and ships. However, due to less spatial sampling and high cost, they cannot meet the needs of global observation. At present, there are two main ways for observing the sea surface wind field through satellite. One is to measure the sea wind through satellite cloud images, but it only obtains the cloud top wind, which is different from the sea surface wind field; second, the satellite load, such as the scatterometer, actively transmits microwave signals to the sea surface to measure the sea breeze. However, the satellite weight is large, and it can only measure the sea breeze within 50 m/s. If it is applied to the construction of constellations for quasi real-time monitoring of the sea wind field, the cost is very high. Although many monitoring and early warning systems have been established worldwide, it is also important to use GNSS to achieve an effective and cost-effective approach to disaster monitoring. Therefore, the spaceborne GNSS-R technology can provide a more effective alternative for ocean wind speed retrieval because of its remarkable advantages such as rich signal sources, high resolution, all-weather, and global fast coverage, etc. At present, many disaster monitoring systems using GNSS signals have been developed. For example, on 8 July 2014, the United Kingdom launched the TechDemoSat-1 (TDS-1) satellite [4]. In December 2016, National Aeronautics and Space Administration (NASA) successfully deployed the Cyclone Global Navigation Satellite System (CYGNSS) composed of eight small satellites, bringing greater opportunities for retrieving sea surface wind speed using GNSS-R technology [5]. More fortunately, BuFeng-1 A/B dual satellites were launched on 5 June 2019, which is China's first GNSS-R satellite mission dedicated to GNSS-R applications. China has become the fourth country after the United Kingdom, the United States and Japan to install GNSS-R satellites in the Earth's orbit. At the same time, BF-1 A/B is the only two GNSS-R constellations in the world. The mission also acquired the world's first on-board BDS DDM [6]. In addition to BF-1 A/B satellites, China also launched FengYun-3E meteorological satellite on 5 July 2021. The GNSS-R data observed by the satellite have been used for the retrieval of sea surface wind speed [7]. In addition, the satellite mission is also committed to GNSS radio occultation (GNSS-RO)-related applications. It is worth mentioning that although the main objective of the above spaceborne GNSS-R tasks is to monitor ocean wind speed, these satellite tasks can also be used for sea surface height retrieval [8,9], sea ice detection [10,11], sea ice concentration (SIC) retrieval [12,13], sea ice thickness retrieval [14], tsunami detection [15], storm surge monitoring [16,17], wave height retrieval [18–24], rainfall detection [25,26], rainfall intensity retrieval [27,28], soil moisture retrieval [29] and biomass retrieval [30], etc.

For the spaceborne platform, Clarizia et al. proposed to use the DDM average (DDMA), DDM variance (DDMV), DDM Allan variance (DDMAV), leading edge slope (LES) and trailing edge slope (TES) observables to retrieve wind speed [31]. Although the use of multiple observables has enriched the data information, it increases the computation load of the algorithm. Ruf et al. developed parametric models for high and low wind speeds using the normalized bistatic radar cross section (NBRCS) of the sea surface and the LES. Through a thorough evaluation of wind speed retrieval performance of geophysical model functions (GMF) by the NBRCS and LES methods, the overall root mean square (RMS) uncertainty of wind speed retrieved by CYGNSS has been found to be 1.4 m/s [32]. At high wind speed, the increased retrieval error is mainly due to the decrease in the sensitivity of the ocean scattering section to wind speed change. Clarizia and Ruf proposed a retrieval algorithm for the level 2 sea surface wind speed data product of CYGNSS mission [33]. This algorithm combines the DDMA and LES wind speed estimates into an optimal weighted estimator based on the method described by Clarizia et al. in 2014. Due to the limitation of cumulative time and resolution, this algorithm is only applicable to two specific observables (DDMA and LES) [31]. In [34], the generalized observation (defined as the linear combination of DDM samples) was optimized by using three different methods (i.e., minimum wind speed variance, maximum signal to noise ratio and principal component analysis (PCA)). The results show that PCA has the best performance. Guo et al. proposed a sea surface wind speed retrieval method for spaceborne GNSS-R based on the particle swarm optimization (PSO) algorithm. The experimental results show that the retrieval accuracy of the PSO method is superior to the classical MVE method. Hammond et al. used the spaceborne Galileo and BDS GNSS-R data collected by the UK TechDemoSat-1 mission to assess the geophysical sensitivity for the first time [35]. Unfortunately, due to the limited amount of reflective BDS data currently available, it is impossible to conduct a comprehensive analysis of the geophysical sensitivity of BDS signals. This research shows the possibility of wind speed retrieval and sea ice detection using spaceborne GNSS-R to acquire different GNSS reflection signals (such as GPS, Galileo and BDS) in the future. Liu et al. proposed a machine learning (ML) method for the retrieval of sea surface wind speed based on a multi hidden layer neural network (MHL-NN). The results show that MHL-NN method is superior to other methods in terms of root mean square error (RMSE) and mean absolute percentage error (MAPE). The method mainly takes DDMA and LES observables as input variables [36]. With the rapid development of artificial intelligence, the machine learning algorithm has been applied to spaceborne GNSS-R wind speed retrieval [36–43], and results with high accuracy have been obtained. In recent years, some researchers have proposed and improved the calculation method of NBRCS [44–46]. Combined with power calibration, NBRCS can be calculated more accurately, to obtain better retrieval results, so it can be consistent with the entire sea surface wind speed range. Based on the DDM data obtained by Bufeng-1 satellite, Jing et al. developed an empirical geophysical function model using the NBRCS observable. Preliminary tests showed that the RMSE between the results obtained by the model and the ECMWF reanalysis wind speed was 2.63 m/s, and the determination coefficient was 0.55. Bu et al. proposed a combined wind speed retrieval model using both the NBRCS and LES observables [47]. The RMSE and determination coefficient of the combined method is 2.1 m/s and 0.906, respectively. Compared with the single-observable model based on NBRCS or LES, it achieves considerable performance improvement.

Table 1 summarizes the comparison of methods for spaceborne GNSS-R sea surface wind speed (WS) retrieval, mainly from the aspects of retrieval methods, GNSS-R observables, retrieval models and retrieval accuracy. Only representative literature is listed in the table, and "-" means not involved. Among them, MHL-NN represents a multi hidden layer neural network, ANN means an artificial neural network, HMDL denotes a heterogeneous multi-mode deep learning network, CNN means a convolutional neural network, MCNN indicates a statistically modified CNN and FA-RDN represents a cyclic neural network of feature attention mechanisms. It can be seen from Table 1, for the GMF method, some features (e.g., LES, TES and normalized bistatic radar cross section (NBRCS)) related to sea surface wind speed are extracted from DDM first. Next, by fitting an empirical function that links DDM observables with wind speed, and linear regression is applied to build GMF, then the minimum variance estimator (MVE) is used to combine the retrieval results of single observables and eliminate the residuals [31]. Although, the MVE method can achieve an accuracy of about 2 m/s for wind speed in the range of 0–30 m/s. However, the accuracy improvement of the MVE method over a single observable based method is still limited. Further research shows that the GNSS-R wind retrieval can be affected by incidence angle, reflection geometric parameters (such as longitude and latitude of specular point) [42,48] and sea conditions (such as significant wave height). Ocean swell caused by non-local winds may also have an impact on sea surface roughness [41,49,50], especially at very low wind speeds. It is difficult to develop a joint model through an analytical approach while considering all the factors. Therefore, in the traditional algorithm development, only a limited set of observations and features are used, which limits the accuracy of wind speed retrieval. However, it is encouraging that the GNSS-R wind speed retrieval algorithm has been significantly improved in recent years, opening a window for retrieving global wind speed from spaceborne GNSS-R data. In particular, the algorithm based on deep learning has been used to measure global ocean wind speed and achieved satisfactory results from spaceborne GNSS-R data [40–42]. Deep learning provides the ability to mitigate the impact caused by interfering data. However, due to the uneven distribution of wind speed samples, the retrieval performance of the deep learning model for high wind speeds is poor. Therefore, samples in a wider wind speed range should be selected for a more generally applicable spaceborne GNSS-R wind speed retrieval model.

**Table 1.** Comparison of spaceborne GNSS-R sea surface wind speed retrieval methods.

| Methods | Literature | GNSS-R Observables | Retrieval Models | Retrieval Accuracy |
|---|---|---|---|---|
| Waveform matching method | [51] | Normalized power value of DDM | - | 1–2 m/s for WS < 20 m/s |
| Empirical model method | [31,33,34,47, 52–54] | DDMA, DDMAV, ADDMV, LES, TES, NBRCS, Generalized Linear Observables | Polynomial fitting, exponential function, power function, combination of multiple observables. | ~2 m/s for WS < 20 m/s |
| Intelligent optimization algorithm | [31,55] | NBRCS、LES | Combination model based on MVE and PSO. | PSO method is better than MVE method. The RMSE is better than 1.95 m/s for WS < 20 m/s. |
| Machine/Deep learning method | [36,38–43] | DDM, DDMA, LES, incidence angle, NBRCS, GPS satellite number, RCG, longitude and latitude of specular point, etc. | MHL-NN, ANN, HMDL, CNN, MCNN, FA-RDN | The retrieval accuracy of CNN, MCNN and FA-RDN models is the best. When the WS is less than 20-30 m/s, they are better than 1.36 m/s, 1.53 m/s and 1.45 m/s, respectively. |

To solve the problem that the accuracy of wind speed retrieval of the empirical combination model is limited, the fully connected network (FCN) is applied to develop a multi feature wind speed retrieval model [37–43]. However, the improvement is mainly for the wind speed of 5–10 m/s because most data used are in this range. Due to overfitting, the estimated RMSE is much larger when it is above 15 m/s. Recently, Chu et al. used the one-month long TDS-1 data set to develop a more advanced architecture and proposed a multimodal deep learning model that fuses 34 auxiliary parameters with the DDM features extracted by CNN. The results show that the proposed model is substantially improved compared with FCN [39]. Asgarimehr and Guo et al. also applied multimodal CNN to the CYGNSS data and achieved decent performance [40,41]. Although these models are still affected by the probability density distribution function of wind speed, they open up a new prospect for GNSS-R wind speed inversion through a heterogeneous multimodal deep learning (HMDL) method.

In this paper, we propose an improved deep learning framework (i.e., GloWS-Net) for global sea surface wind speed retrieval from spaceborne GNSS-R data. GloWS-Net considers the fusion of auxiliary information (ocean swell SWH, sea surface rainfall and wave direction) to build an end-to-end wind speed retrieval model. Compared with previous methods, the proposed model has the following advantages:

(1) GloWS-Net includes a convolution layer for extracting effective features from the combination of BRCS DDM and corresponding effective scattering area, and a full connection layer for processing auxiliary parameters and higher-level input parameters. The best architecture is obtained based on the validation set and evaluated on the testing set to verify the generality of the model.
(2) When training GloWS-Net, some traditional machine learning strategies are used to prevent overfitting, including batch normalization and early stop.
(3) Contribution from different inputs is exploited by the GloWS-Net deep learning model. Meanwhile, the influence of swell and rainfall on wind speed retrieval performance is considered and corrected.
(4) The GloWS-Net model architecture has significantly reduced the underestimation phenomenon for high wind speeds. It outperforms the CyGNSSnet and MCNN architectures.

The rest of this paper is arranged as follows. Section 2 describes the data sets used and the basic modeling methods. Section 3 focuses on the design, training and validation of the GloWS-Net model. Section 4 compares the wind speed retrieval performance of different modeling methods, and discusses the results in detail. Section 5 provides the conclusion of this paper.

## 2. Data Set and Method Description

### 2.1. Data Set

To develop and test the retrieval model of the spaceborne GNSS-R sea surface wind speed, in this paper, six different data sets are used, including the CYGNSS GNSS-R L1 level data, ERA5 wind speed data, ERA5 swell significant wave height (SWH), ERA5 wave direction data, IMERG rainfall data, and the Cross-Calibrated Multi-Platform (CCMP) analysis wind (v02.1 near real time) products produced by Remote Sensing Systems (RSS). These data cover all the days of 2021.

(1) CYGNSS GNSS-R L1 level data

At present, the National Aeronautics and Space Administration (NASA) mainly provides users with three levels of CYGNSS GNSS-R data for free. The L1 level data product of version V3.1 is used in this article. This version mainly includes GNSS-R data from August 1, 2018 (DOY 213) to December 20, 2022. We can access the GNSS-R data from the website (https://podaac.jpl.nasa.gov/dataset/CYGNSS_L1_V3.1 (accessed on 19 December 2022)). This type of data records the DDM, specular point coordinates, transmitter positions and other relevant information of eight CYGNSS satellites (i.e., CY01, CY02, CY03, CY04, CY05, CY06, CY07 and CY08). In addition, CYGNSS L1 V3.1 also provides multiple DDM products (such as brcs, raw_counts, and power_analog). This article uses brcs and power_analog together with the effective scanning area corresponding to brcs.

(2) ERA5 reanalysis data sets

The Copernicus climate change service (C3S) climate database provides users with ECMWF reanalysis data products. ERA5 is the fifth generation atmospheric reanalysis of global climate by ECMWF. The reanalysis data combine the model data with the observation data from all over the world to form a global complete and consistent data set. ERA5 replaces its predecessor ERA Interim reanalysis. ERA5 contains multiple types of data sets. In this paper, we use ERA5 hourly data on a single level from 1959 to present (https://cds.climate.copernicus.eu/cdsapp#!/dataset/reanalysis-era5-single-levels (accessed on 1 July 2022)), which includes the data of 2 m temperature, 2 m dewpoint temperature, total precipitation, average sea level pressure, surface pressure, 10 m u-component of wind, 10 m v-component of wind, boundary layer height and sea surface wave height. We used ERA5 swell height data, ERA5 wave direction data, ERA5 10 m u-component of wind and 10 m v-component of wind data. Among them, the ERA5 swell height data represent the significant height of total swell. The 10 m u-component of wind and 10 m v-component of wind data are matched with other data through bilinear interpolation in space and linear interpolation in time. It should be noted that both the cubic and bilinear interpolation methods can be applied for matching data in a spatial domain, the bilinear method is usually used in the GNSS-R field. Since the results difference between them is negligible, the bilinear method is chosen here. Then, wind synthesis is conducted, and the resultant wind speed is used as the reference value of the real sea surface wind speed at 10 m above the sea surface. It is worth mentioning that the spatial resolution of ERA5 swell height data and ERA5 wave direction data is $0.5^\circ \times 0.5^\circ$ and temporal resolution is 1 h. However, the spatial resolution of 10 m u-component of wind and 10 m v-component of wind data is $0.25^\circ \times 0.25^\circ$ and temporal resolution is 1 h.

(3) Integrated Multi-satellite Retrievals for Global Precipitation Measurement (GPM IMERG) precipitation data sets

The Integrated Multi-satellite Retrievals for GPM (IMERG) algorithm combines information from the GPM satellite constellation to estimate precipitation over the majority of the Earth's surface. This algorithm is particularly valuable over the majority of the Earth's surface that lacks precipitation measuring instruments on the ground. Now in the latest Version 06 release of IMERG, the algorithm fuses the early precipitation estimates collected during the operation of the TRMM satellite (2000–2015) with more recent precipitation estimates collected during the operation of the GPM satellite (2014–present). GPM-IMERG has three types of products, namely IMERG Early Run (IMERG-E) (real-time products with a delay of about 4 h), IMERG Late Run (IMERG-L) (near real-time processing with a delay of about 14 h) and IMERG Final Run (IMERG-F). The spatial resolution of three types of products is $0.1^{\circ} \times 0.1^{\circ}$, they all have 1 month, 1 day and 30 min temporal resolutions. This article uses an IMERG-F product with a temporal resolution of 30 min, which can be downloaded from the GPM website for free (https://gpm.nasa.gov/data/directory (accessed on 1 July 2022)).

(4) RSS CCMP wind analysis product

The CCMP wind product was also used to evaluate the performance of the proposed model in retrieving wind speed. The CCMP is a combination of ocean surface (10 m) wind speed retrievals from multiple types of satellite microwave sensors and a background field from reanalysis. That is, CCMP simultaneously integrates multi-satellite sea surface winds of scatterometers from QuikSCAT and METOP-A/ASCAT, radiometers from SSM/I, SSMIS, AMSR, TMI, WindSat and GMI as well as in situ observations from NDBC, TAO, TRITON, RAMA, PIRATA and Canadian [56,57]. This data set has its own unique characteristics. CCMP wind data including u- and v-components were downloaded from Remote Sensing Systems (https://data.remss.com/ccmp/ (accessed on 19 December 2022)). The spatial resolution of its 10 m u-component of wind and 10 m v-component of wind data is $0.25^{\circ} \times 0.25^{\circ}$ and temporal resolution is 6 h. The coverage range is (0, 360°) for longitude and (−78.375°, −78.375°) for latitude. Wind speed at 10 m (U10) was calculated after respective interpolation of u and v wind components, which was used as the ground truth wind speeds.

Please note that in consideration of the spatial and temporal resolution difference between the CYGNSS GNSS-R data and ERA5 data set, CCMP wind data and IMERG rainfall data set, these data must be matched with the real sea surface ERA5 wind speed data at 10 m above the sea surface. Linear interpolation in time and bilinear interpolation in space are implemented to obtain the co-located sea surface wind speed.

### 2.2. Method Description

The data processing flow and model building process of the spaceborne GNSS-R sea surface wind speed retrieval method are shown in Figure 1. First, the CYGNSS DDM data are down sampled and spatially matched with the ERA5 wind speed, ERA5 wave direction, ERA5 swell SWH and IMERG rainfall data to produce the matched data set. To ensure the quality of the data set, the sampling data need to be quality controlled and filtered according to the standards listed in [24,28,42], so as to obtain the filtered high-quality data set. Then, the data set is randomly divided into training set, verification set and test set. They account for 30%, 15% and 55% of the filtered data set, respectively. The training set is used to train the model and the verification set is used to evaluate the performance of the network configuration. The content of network configuration is described in detail in Section 3.2. In the model training stage, it is necessary to specify the number of hidden layers and the number of neurons in each layer, and select the configuration with the best performance for the ultimate retrieval of global ocean wind speed.

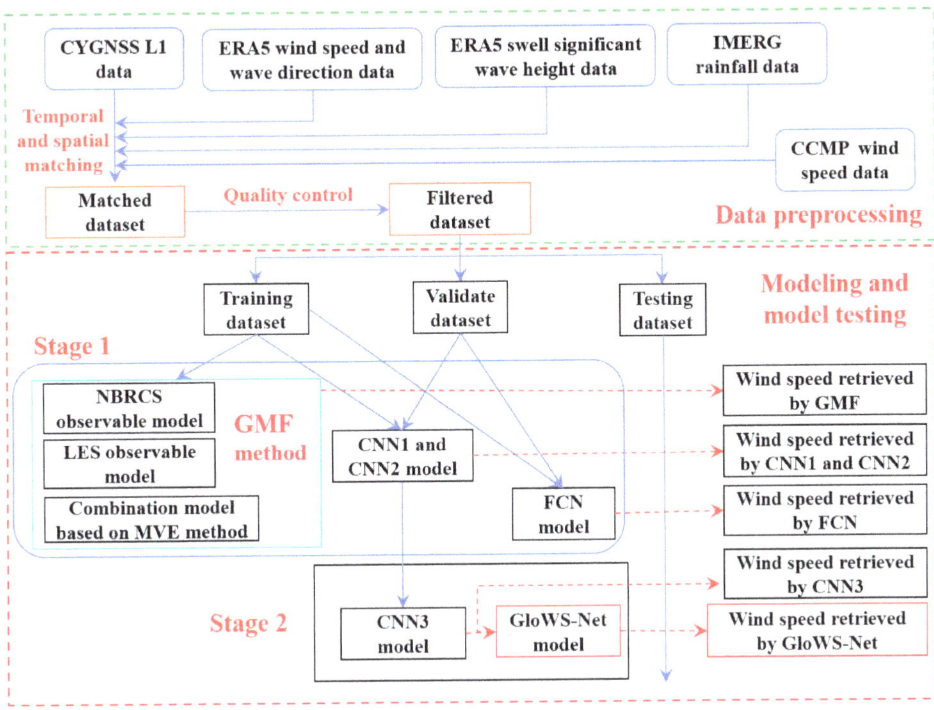


**Figure 1.** Data processing and model development process of spaceborne GNSS-R sea surface wind speed retrieval.

In order to confirm the improvement of the proposed model, the training model is further divided into two stages, namely the first stage and the second stage indicated by the red dotted box in Figure 1. In the first stage, the training set is used to establish the GMF, FCN and CNN models, and the latter two use the early stop method through the verification set to avoid overfitting. It should be noted that the ERA5 swell SWH, IMERG rainfall data and ERA5 wave direction data are not used to retrieve wind speed based on the GMF method. In this paper, the FCN models are divided into FCN1, FCN2, FCN3 and FCN4 models according to the number of input parameters in the input layer. The CNN models only have DDM images as input. According to the different types of DDM images, the CNN models are divided into CNN1 and CNN2 models (see Table 3 below). In the second stage, the CNN3 model is established by adding effective scattering area and additional auxiliary parameters corresponding to BRCS DDM on the basis of the CNN1 model. Please note that the CNN1, CNN2, CNN3 and GloWS-Net models are independent, and they are directly trained by the training set and supervised by the verification set. The CNN3 model is only designed to analyze the impact of introducing additional auxiliary parameters. For convenience, the proposed improved CNN model is named GloWS-Net model, and its retrieved wind speed is hereinafter referred to as GloWS-Net wind speed. Finally, the performance of these four models (GMF, CNN1, CNN2 and FCN) in the first stage and two models (CNN3 and GloWS-Net) in the second stage are tested on the test set. The reference wind speed data in the test set include the ERA5 and CCMP wind speed data, among which, the CCMP wind speed data are not used for model training, so this data set can better evaluate the robustness and generality of the proposed model. Details of each stage are provided later.

## 3. Development of the GloWS-Net Deep Learning Model

### *3.1. CYGNSS Observations and Other Auxiliary Observations*

DDM is the basic data recorded by a GNSS-R receiver and also the most useful data. For the retrieval of sea surface wind speed using spaceborne GNSS-R technology, the peak value of bistatic reflected power will change, and the shape of the delayed Doppler maps will also change due to the variation in sea surface roughness. In the case of bistatic scattering, the relationship between wind speed and the peak DDM power is different from that of the backscattering geometry. When wind speed is low, the strong forward reflection will cause the received signal to be very strong. With the increase in wind speed, the peak power decreases. This is consistent with the situation observed in the DDM diagram under different wind speed conditions shown in Figure 2. It can be seen that the DDM of the GNSS signal reflected from the sea surface is horseshoe shaped, and the power value is usually the largest at the specular point. These differences provide a valuable opportunity for retrieving the sea surface wind speed based on spaceborne GNSS-R. Therefore, four variables (NBRCS, LES, TES and SNR) extracted from DDM have strong correlation with wind speed [58,59]. These variables are calibrated in the CYGNSS Level 1B product, which is usually used for wind speed retrieval [33]. The signal to noise ratio (SNR) is also strongly correlated with sea surface roughness [34,35]. NBRCS and LES are the two most important observables in CYGNSS products, which are very effective for wind speed retrieval. TES is the slope of the trailing edge of the integrated delay waveform, which is also useful for sea surface wind speed [31]. Please note that the TES observation value in this paper is calculated from the power DDM. In addition, this paper also uses two main measurement products, BRCS DDM and its corresponding effective scattering area. BRCS DDM has a size of 17 × 11, which is different from power DDM.

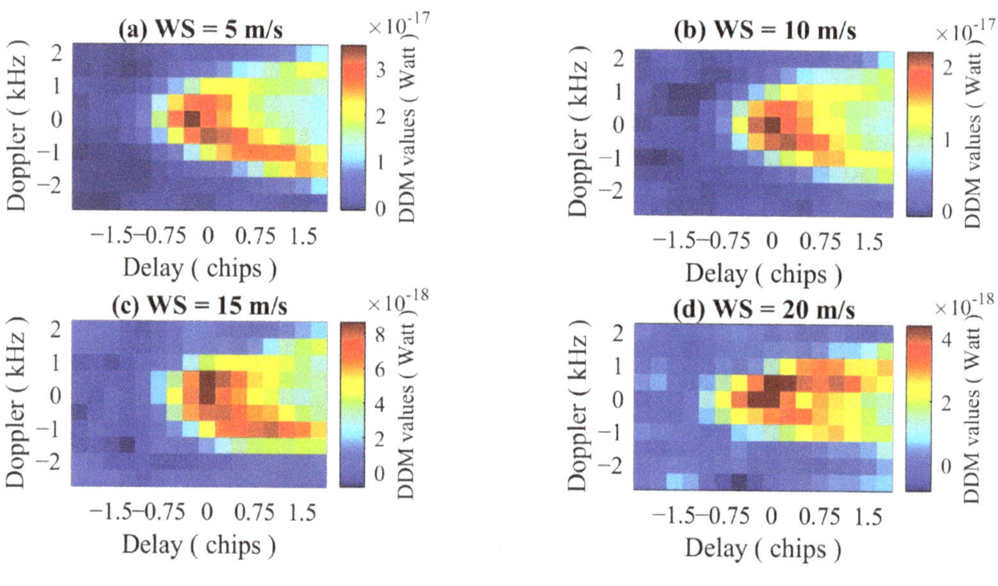


**Figure 2.** DDMs under different wind speed (WS) conditions.

In addition to the above observables, several other variables related to receiver, satellite geometry, sea state factors and weather conditions are also considered to improve the performance of the model. The receiver antenna gain reflects the state of signal reception, and the incident angle and specular reflection point describe the satellite geometry. Sea state factors mainly include swell and wave development degree will increase the uncertainty of spaceborne GNSS-R wind speed retrieval. For example, the wind speed retrieval in CYGNSS L2 is based on two independent GMFs (i.e., one for fully developed sea and one for developing sea), taking into account the development of the ocean [54]. The GMF used by NOAA for CYGNSS wind speed retrieval includes SWH as the input parameter [60,61]. Li and Guo et al. also used swell SWH as input parameter to establish their wind speed retrieval model [41,42]. The swell is a long wave. The SWH of the swell will affect the GNSS signal reflection, so it is used as one input parameter. Similarly, this paper also uses SWH as a prior knowledge to improve the performance of the wind speed retrieval model. Secondly, because the sea wave information has an important contribution to wind speed retrieval based on the deep learning network, it is also helpful to use more sea wave related parameters as the input of the neural network training. However, considering that the structure complexity of the network and the computational load of the network training will increase significantly, which will affect the convergence of the training, the SWH of the swell and the wave direction parameters are mainly considered in the GloWS-Net model. In addition, rainfall will also affect the retrieval of spaceborne GNSS-R wind speed, especially at low wind speeds (<10 m/s), rainfall will lead to wind speed retrieval deviation [25,61,62], and wind speed deviation will increase with the increase in rainfall [61]. In order to consider the effect of rain, rainfall intensity is also added as an input parameter. For clarity, Table 2 lists the auxiliary parameters used in the GloWS-Net model.

**Table 2.** Specific description of input auxiliary parameters information of GloWS-Net model.

| | Input Variables | Description |
|---|---|---|
| Related to DDM | NBRCS | Normalized bistatic radar cross section |
| | LES | Leading edge slope |
| | TES | Trailing edge slope |
| | SNR | DDM signal to noise ratio |
| Related to receiver | sp_rx_gain | Range corrected gain |
| Related to satellite geometry | sp_lon | Specular point longitude |
| | sp_lat | Specular point latitude |
| | sp_inc_angle | Specular point incidence angle |
| Other parameters | Swell SWH | ERA5 swell SWH |
| | Wave direction | ERA5 wave direction |
| | Rainfall intensity | IMERG rainfall intensity |

### 3.2. GloWS-Net Model Design

In this section, we describe the general architecture of GloWS-Net (as shown in Figure 3) and the scheme for optimizing the network.

Please note that the GloWS-Net architecture proposed in this paper is different from the network architecture of Asgarimehr and Guo et al. [40,41]. In Figure 3, the GloWS-Net architecture has three input ports, the first for BRCS DDM, the second for effective scattering area and the third for auxiliary parameters. In particular, the third port inputs the longitude and latitude, incidence angle, NBRCS, LES, TES, rainfall information and other auxiliary variables of the specular reflection point. It can be seen from Figure 3 that the size of the characteristic map gradually decreases through the processing of different layers, and the results (i.e., wind speed) are finally output through the full connecting layer.

To compare with the GloWS-Net architecture proposed here, we provide the results of CNN1 architecture with only BRCS DDM input and CNN2 architecture with only power DDM input, and the results of FCN1, FCN2, FCN3 and FCN4 architectures with only auxiliary parameters but no convolution layer. These architectures are different in terms of auxiliary parameters and input image types. Information about architectures and auxiliary parameters is shown in Table 3.

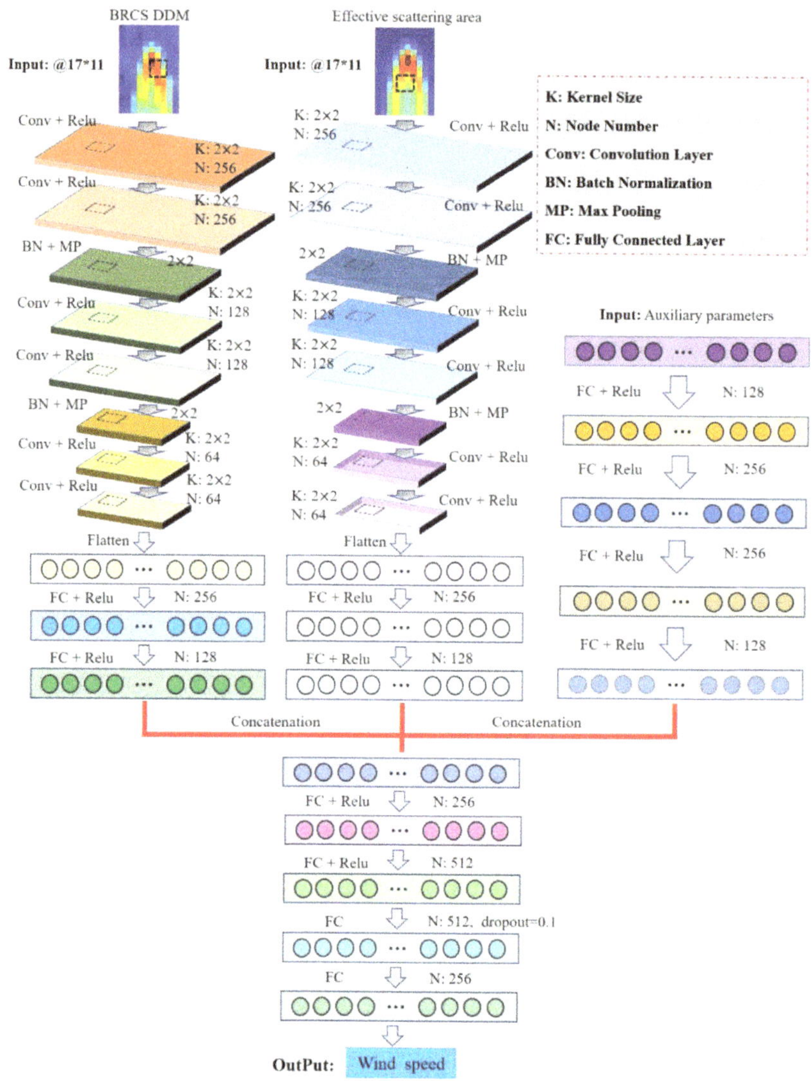


**Figure 3.** Architecture of GloWS-Net.

**Table 3.** Input parameters of FCN, CNN and GloWS-Net models.

| Architecture (Model) | Input Image | Input Auxiliary Parameters (Variables) |
|---|---|---|
| FCN1 | No | NBRCS, LES |
| FCN2 | No | NBRCS, LES, TES, SNR, incidence angle, receiver antenna gain |
| FCN3 | No | NBRCS, LES, TES, SNR, incidence angle, receiver antenna gain, longitude and latitude of specular point |
| FCN4 | No | NBRCS, LES, TES, SNR, incidence angle, receiver antenna gain, longitude and latitude of specular point, swell SWH, wave direction, rainfall intensity |
| CNN1 | BRCS DDM | No |
| CNN2 | power_analog | No |

**Table 3.** *Cont.*

| Architecture (Model) | Input Image | Input Auxiliary Parameters (Variables) |
|---|---|---|
| CNN3 | BRCS DDM, effective scattering area | NBRCS, LES, TES, SNR, incidence angle, receiver antenna gain, longitude and latitude of specular point |
| GloWS-Net | BRCS DDM, effective scattering area | NBRCS, LES, TES, SNR, incidence angle, receiver antenna gain, longitude and latitude of specular point, swell SWH, wave direction, rainfall intensity |

Determining the optimal activation function in neural networks is an important task because it is directly related to network performance. Unfortunately, however, it is difficult to determine this function analytically. On the contrary, an optimal activation function is usually determined through repeated trials or adjustments [63]. This paper considers five activation functions, namely Relu, Tanh, Elu, Sigmoid and softplus:

$$f_{\text{Relu}}(x) = \max(0, x) = \begin{cases} 0, \ x < 0 \\ x, \ x >= 0 \end{cases} \tag{1}$$

$$f_{\text{Tanh}}(x) = \frac{2}{1 + e^{-2x}} - 1 \tag{2}$$

$$f_{\text{Elu}}(x) = \begin{cases} x, \ x > 0 \\ \alpha(\mathrm{e}^{x} - 1), \ x \leqslant 0 \end{cases} \tag{3}$$

$$f_{\text{Sigmoid}}(x) = \frac{1}{1 + e^{-x}} \tag{4}$$

$$f_{\text{Softplus}}(x) = \ln(1 + \exp(x)) \tag{5}$$

In the above formula, $x$ is the input value of the previous layer neuron.

The advantage of the Sigmoid activation function is that the gradient is smooth, avoiding jumping output values. However, the gradient tends to be zero, so the exponential calculation is slow. The Tanh function has some advantages over the Sigmoid function. However, when the input is large or small, the output is almost smooth and the gradient is small, which is not conducive to weight update. The Relu function is a popular activation function in deep learning. Compared with the Sigmoid function and Tanh function, the Relu function does not have the gradient saturation problem, and because the Relu function only has a linear operation function, its calculation speed is faster than the Sigmoid function and Tanh function. However, the Relu function also has disadvantages. For example, when the input is negative, in the back propagation process, the gradient will be completely zero.The Elu function solves the problem of Relu. Compared with Relu, Elu has a negative value, which will make the average value of activation close to zero. The mean activation close to zero can make learning faster because it makes the gradient closer to the natural gradient. Please note that although the Elu function is better than the Relu function in theory, there is no sufficient evidence to show that Elu is always better than Relu in practice. The Softplus function is similar to the Relu function, but it is relatively smooth. As with Relu, it has unilateral suppression and has a wide acceptance range (i.e., (0, +inf)).

In order to determine the activation function for optimal performance, we analyzed and discussed the influence of the number of neurons in different hidden layers and different activation functions on the accuracy of wind speed retrieval using the FCN architecture. Generally, the accuracy of linear models is low, so the activation function improves the performance of the FCN model by adding nonlinear factors. As stated in [38], the number of hidden layers used in FCN is usually between 1 and 3, and 4 layers are rarely used because more hidden layers will cause a higher computing cost [38]. Generally, the FCN architecture with three hidden layers can obtain the best results in wind speed retrieval [50]. Therefore, we also adopted the FCN architecture with three hidden layers, and set the number of neurons in each layer to 4N, 2N and N, respectively, with N being 4, 8, 12, 24, 32, 48, 64 or 128.

Figure 4 shows the RMSE comparison of wind speed retrieval performance for the different number of neurons and different activation functions in the three hidden layers. It can be seen from the figure that different activation functions have a great impact on the accuracy of wind speed retrieval, so it is very important to select an appropriate activation function. When N = 128, the Relu activation function achieves the best performance in RMSE, followed by the Tanh function. Therefore, this paper uses the Relu function as the best activation function for all neural network architectures, and the hidden layers of FCN1, FCN2, FCN3 and FCN4 models are set as 512, 256 and 128, respectively.

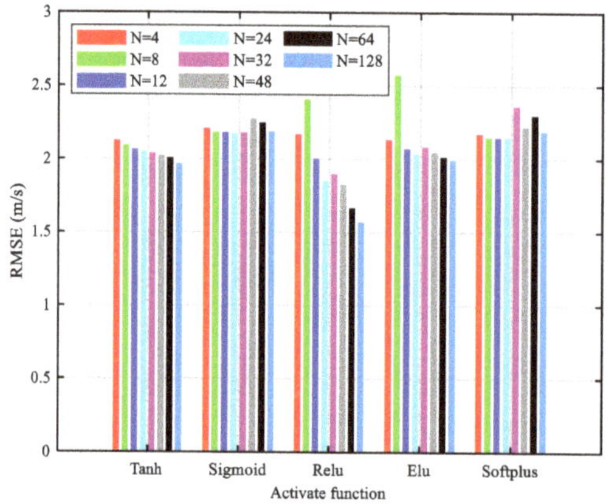


**Figure 4.** Comparison of different activation functions.

In this paper, the mean square error between the predicted wind speed and the reference wind speed is used as a loss function:

$$\text{Loss} = \frac{1}{N}\sum_{i=1}^{N}(\hat{u}_i - u_i)^2 \tag{6}$$

where $N$ is the number of samples, $\hat{u}_i$ is the predicted wind speed, and $u_i$ is the reference wind speed.

The optimizer is another important configuration of the training model. After comparison, we chose Adam optimization algorithm to train small batches randomly disordered.

### 3.3. GloWS-Net Model Training and Verification

The related network is implemented under the Keras architecture and TensorFlow backend. Table 4 shows the basic configuration of the experimental platform. These networks can be trained on the GeForce GTX 1050Ti GPU. The preprocessed data are normalized to zero mean and unit variance according to the characteristics. The verification set is used to avoid overfitting. We adopt the early stop condition with six epochs. The learning rate is set to 0.0005. To further prevent the overfitting problem, the random deactivation (dropout) is set to 0.1.

The training set and verification set are separated. The training set is used to train the network model, and the verification set is used to supervise the model training. The batch size in this paper is set to 32. When the network updates the weight coefficient each time, only samples from one batch are used to train the model. To ensure good performance, the number of epoch is set to 100.

**Table 4.** The configuration of experiment platform.

| | |
|---|---|
| Programming Language | Python 3.7 |
| Deep learning API | Tensorflow 2.3 and Keras 2.4 |
| CPU | Intel core i7-7700HQ |
| Running memory | 16 GB |
| GPU | GeForce GTX 1050Ti |

## 4. Performance Analysis and Discussion of Wind Speed Retrieval

### *4.1. Comparison with ERA5 Data*

In this section, the performance of the GloWS-Net model architecture is compared with the traditional MVE method and some existing network architectures (such as FCN and CNN models). The wind speed retrieval performance test is conducted on the test data set, which is not used during training and is separated from the training set and verification set. Therefore, it is very suitable for evaluating the generalization ability of corresponding network models on blind data sets. However, this training method also has a disadvantage that any systematic errors will be invisible to the algorithm. In the following, we first compare and analyze the wind speed retrieval performance of different retrieval models, then evaluate the performance of different retrieval models in retrieving global wind speed, and finally discuss the results.

As shown in Figure 1 above, wind speed retrieval can be divided into two stages. The first stage involves the GMF method, CNN1, CNN2, FCN1, FCN2, FCN3 and FCN4 models. The second stage includes wind speed retrieval by CNN3 model and GloWS-Net model. Among them, the CNN3 model fuses the DDM image and auxiliary parameters excluding SWH, wave direction and rainfall information. The GloWS-Net model includes SWH of swell, wave direction and rainfall information as input parameters. Table 5 shows the retrieval accuracy statistics of different retrieval models for wind speed less than 10 m/s, 10–15 m/s, 15 m/s and 0–30 m/s. In the table, RMSE, mean absolute error (MAE), correlation coefficient (CC) and MAPE represent root mean square error, mean absolute error, correlation coefficient and mean absolute percentage error, respectively. The following conclusions can be drawn from the table:

**Table 5.** Accuracy of wind speed retrieved from different models using ERA5 data as reference.

| Models | <10 m/s | 10–15 m/s | >15 m/s | 0–30 m/s | | | |
|---|---|---|---|---|---|---|---|
| | RMSE (m/s) | RMSE (m/s) | RMSE (m/s) | RMSE (m/s) | MAE (m/s) | CC | MAPE (%) |
| NBRCS | 2.05 | 2.46 | 4.48 | 2.42 | 1.85 | 0.83 | 29.93 |
| LES | 2.26 | 2.54 | 4.65 | 2.71 | 1.98 | 0.70 | 32.41 |
| MVE | 2.23 | 2.42 | 4.29 | 2.53 | 1.98 | 0.81 | 30.64 |
| FCN1 | 2.48 | 2.52 | 4.08 | 2.66 | 2.08 | 0.80 | 31.69 |
| FCN2 | 1.78 | 2.02 | 3.33 | 2.03 | 1.57 | 0.89 | 23.40 |
| FCN3 | 1.61 | 2.13 | 3.42 | 1.99 | 1.52 | 0.89 | 22.63 |
| FCN4 | 1.76 | 1.89 | 2.89 | 1.92 | 1.48 | 0.90 | 24.78 |
| CNN1 | 1.71 | 2.59 | 4.68 | 2.37 | 1.81 | 0.85 | 25.72 |
| CNN2 | 2.05 | 2.46 | 4.54 | 2.46 | 1.90 | 0.82 | 28.64 |
| CNN3 | 1.81 | 2.18 | 2.81 | 2.02 | 1.53 | 0.89 | 21.91 |
| GloWS-Net | 1.69 | 2.10 | 2.68 | 1.92 | 1.43 | 0.90 | 20.67 |

(1) When the wind speed is within the range of 0-30 m/s, the RMSE of the FCN4 and the proposed GloWS-Net model architectures are the lowest (1.92 m/s). However, it can be observed from the table that the proposed GloWS-Net model architecture is superior to the FCN4 model in terms of MAE and MAPE, especially for MAPE. Generally, MAE and MAPE are less affected by extreme values. However, RMSE uses the square of error and it is more sensitive to outliers. The FCN3 and CNN3 model architectures with the

same number of auxiliary parameter inputs also show a similar situation, which indicates that after adding DDM images, the GloWS-Net model architecture can obtain the same RMSE as the FCN model, but MAPE has been significantly improved by 16.56%. In addition, in terms of the four indicators (MRSE, MAE, CC and MAPE), the performance of the proposed GloWS-Net model architecture is significantly better than the model combing the results from NBRCS and LES based on the MVE method, with an increase of 23.98%, 27.95%, 11.02% and 32.52%, respectively. Compared with the NBRCS method, it increased by 20.67%, 22.78%, 8.95% and 30.92%, respectively. Compared with the LES method, it has increased by 29.16%, 27.85%, 29.19% and 36.21%, respectively.

(2) In the case of strong wind (>15 m/s), the RMSE of all architectures and MVE methods increases. However, adding auxiliary parameters, as done in FCN2, FCN3, FCN4, CNN3, and GloWS-Net model architectures, will reduce RMSE. Compared with other architectures, CNN1 and CNN2 that only handle BRCS DDM or power DDM have poor performance. This shows that even when the wind speed is lower than 9 m/s, it is necessary to add auxiliary parameters to obtain satisfactory results. It is encouraging to note that the GloWS-Net model architecture has the best performance in retrieving wind speed under strong wind conditions and has obtained the minimum RMSE.

(3) In the case of moderate wind speed (10–15 m/s), the retrieval performance of FCN4 and GloWS-Net model architecture is better than that of the GMF method. However, the performance of GloWS-Net is slightly inferior to that of FCN4. Under low wind speed (<10 m/s), the performance of GloWS-Net is slightly better than that of FCN4. In this case, although the RMSE of the MVE method is small, it is still worse than FCN (such as FCN2, FCN3 and FCN4), CNN (CNN1, CNN2 and CNN3) and GloWS-Net. In addition, it is worth noting that the accuracy of the FCN1 model with only NBRCS and LES observables as auxiliary parameters is much lower than that of the traditional MVE method, which indicates that more auxiliary parameters need to be added to the FCN1 model for wind speed retrieval to obtain satisfactory results (such as FCN2, FCN3 and FCN4). Another important aspect we can notice is that the CNN1 model with BRCS DDM as the input only achieves the same performance as the FCN4 and GloWS-Net model architectures under low wind speed. This shows that the CNN1 wind model for low wind speed can benefit from the accretion layer.

In sum, in terms of overall RMSE, except for FCN1, CNN1 and CNN2, other model architectures have better wind speed retrieval performance than the NBECS, LES and MVE methods. For the accuracy obtained from the wind speed range of 0–30 m/s, although the RMSEs of GloWS-Net and FCN4 are equivalent, the performance of the proposed GloWS-Net is better than that of FCN4 in terms of MAE and MAPE, especially for MAPE. In addition, FCN4 performs worse than GloWS-Net under low and high wind speeds. It should be emphasized that only a verification using in situ measurements provided by weather stations or a campaign with boats on the investigated area will fully validate the proposed approach.

In order to evaluate the performance of different models, Figure 5 shows the wind speed scatter density plots of different models and ERA5. The color bar in the figure represents the data density, the red dotted line represents the 1:1 reference line, the magenta solid line represents the linear fitting result between the retrieved and ERA5 wind speeds, and the CC represents Pearson correlations between retrieved wind speed of the model and ERA5 wind speed are also given. The following observations can be seen from the figure:

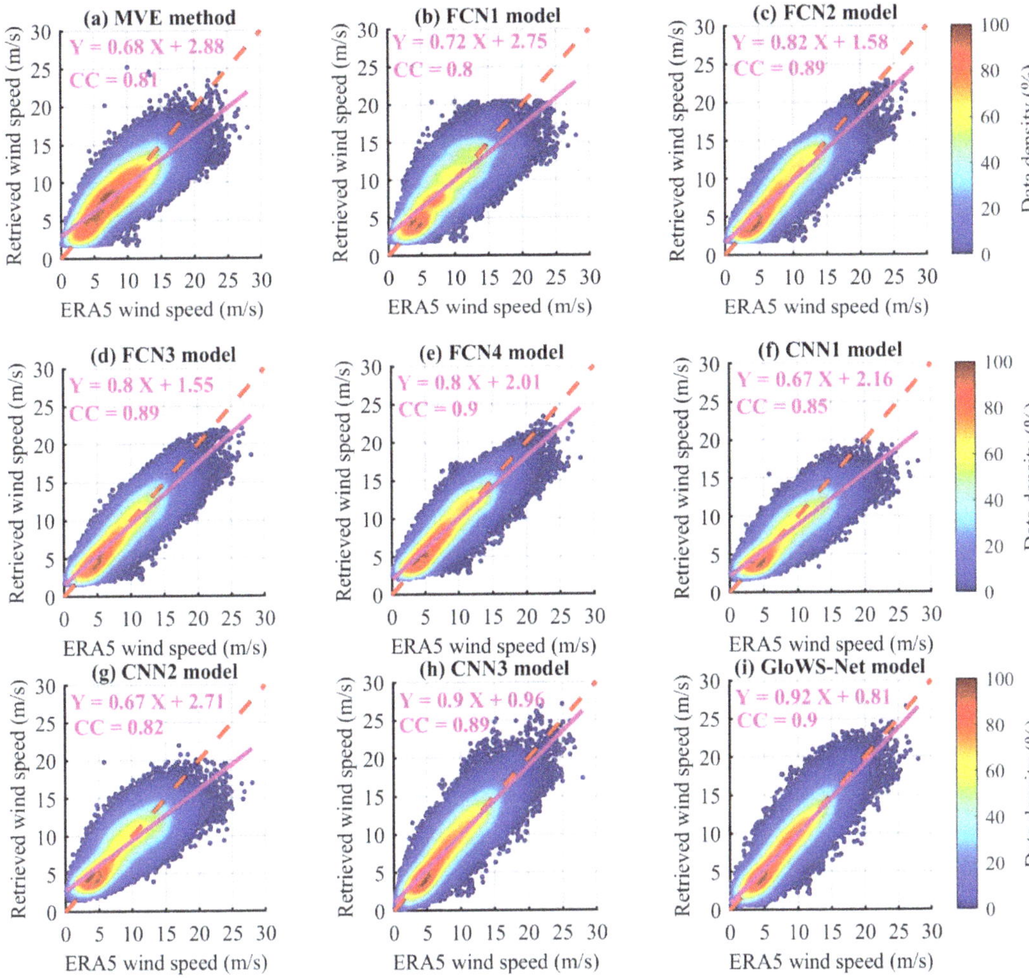


**Figure 5.** Scatter density plots of retrieved wind speed by different models and ERA5 wind speed.

(1) Although the FCN4 model analyzed above and the GloWS-Net model produce the same RMSE, the proposed GloWS-Net model and CNN3 model show better performance in terms of correlation between retrieved wind speed and ERA5 data. Obviously, there are more data points distributed symmetrically along the y = x line, and fewer data points scattered around the line.

(2) The MVE method obviously overestimated the wind speed of 5 to 12 m/s, which may be caused by insufficient power calibration of CYGNSS data and high noise [48,54]. Within this range, a large number of samples and some measured power values may be underestimated. This problem has been improved by FCN, CNN and GloWS-Net. In particular, the results retrieved by CNN3 and GloWS-Net both are more concentrated along the 1:1 reference line when wind speed is in the range of 5~12 m/s. It is also obvious from the figure that the MVE method, FCN1, FCN2, FCN3, FCN4, CNN1 and CNN2 model architectures have poor response capability to wind speed in the range of 0–5 m/s. However, CNN3 and GloWS-Net model architectures do not have this problem. This shows that after adding DDM images to the input layer of the CNN3 and GloWS-Net wind models, the architecture incorporating the convolution layer has better performance than those with only the full connection layer because they use the patterns in DDM. In addition, except

for CNN3 and GloWS-Net, other models show overestimation at very low wind speeds (<5 m/s) and underestimation at high wind speeds (15–30 m/s). The correlation between the wind speed retrieved by CNN1 and CNN2 models and the ERA5 data is the worst, which clearly indicates that more auxiliary parameters need to be included in CNN1 and CNN2 to obtain better results.

(3) It is worth mentioning that, compared with the two most advanced deep learning model architectures (i.e., MCNN and CyGNSSnet) currently used for spaceborne GNSS-R wind speed retrieval [40,41], the GloWS-Net model architecture proposed in this paper performs very well in the case of high wind speed, that is, the GloWS-Net model significantly mitigates the underestimation phenomenon at high wind speed. Generally, for marine disasters caused by marine events with large wind speeds such as hurricanes and typhoons, high-precision wind speed estimation results at high wind speeds are very helpful for monitoring those disasters. Although limited to the current level of spaceborne GNSS-R technology, wind speed prediction under strong winds is still facing great challenges [64,65]. However, the excellent performance of the GloWS-Net model at high wind speeds is promising for future marine disaster monitoring.

Figure 6 shows the RMSE and MAE of different models for different wind speed ranges. It can be seen that CNN3 and GloWS-Net outperform other architectures especially for challenging high wind speeds (>20 m/s). In addition, when the wind speed is greater than 8 m/s, it can be seen that including SWH, wave direction and rainfall information enhances the model performance. However, when the wind speed is greater than 20 m/s, the CNN3 model shows equivalent performance as the GloWS-Net model architecture. The reason may be that strong wind speed is accompanied by heavy rainfall or strong wind speed causes dramatic changes in sea conditions. Even if rainfall, swell and wave direction information is introduced into the GloWS-Net model framework, it is difficult to correct these influencing factors completely. Therefore, future research needs to further optimize the GloWS-Net model architecture.

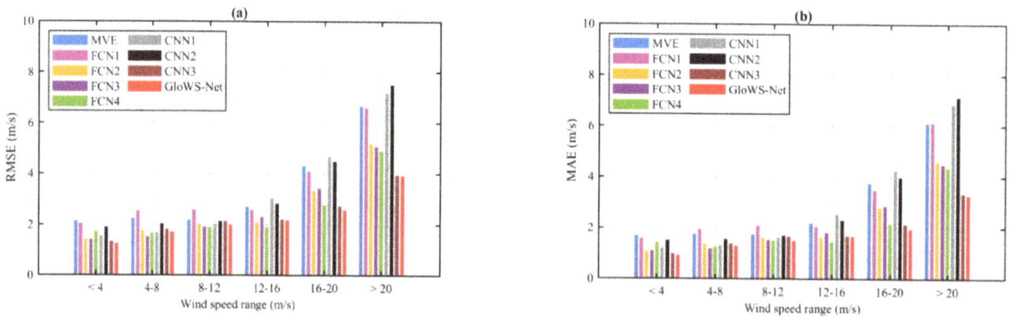


**Figure 6.** When ERA5 wind speed is used as reference for (**a**) RMSE and (**b**) MAE for different wind speed ranges.

In order to compare the global performance of wind speed retrieved by GloWS-Net and the traditional model, we selected the period from July to August 2021 in the test data set for analysis. Figure 7 shows the ERA5 wind speed and the results retrieved using the MVE method, FCN4 and GloWS-Net. Figure 8 shows the deviation distribution histogram between ERA5 wind speed and the results obtained by the MVE method, FCN4 as well as GloWS-Net. In the figure, the average deviation (μ), standard deviation (σ), mean absolute error (MAE) and 80% quantile (Qua) of the deviation are also given. The blue bar chart depicts the error distribution, the red dotted line represents the fitting curve of the probability density function of the error and the green dotted line marks the wind speed deviation of 0 m/s. It can be observed from Figure 7 that the performance of FCN4 and GloWS-Net is better than the MVE method in retrieving global wind speed, and the results from the MVE method are significantly different from the ERA5 data in multiple sea areas around the world (as shown by the magenta rectangle mark in the figure). Comparing the wind speed retrieval results of FCN4 and GloWS-Net with the ERA5 data, it is found that the performance of the GloWS-Net model is better. For example, in the sea area within the longitude of 60° E–120° E and the latitude of 10° S–40° S, the FCN4 model shows an underestimated wind speed, while the retrieved wind speed of the GloWS-Net model is highly consistent with the ERA5 data. From Figure 8, it can also be seen that the performance of the GloWS-Net model is better than that of the FCN4 model. The deviation between the GloWS-Net model wind speed results and the ERA5 data is very concentrated (80% of the wind speed deviation is less than 2.28 m/s) and near the deviation line of 0 m/s, while the wind speed deviation of FCN4 is on the left side of the green line in the histogram. The negative deviation is more obvious, and 80% of the wind speed deviation is less than 2.34 m/s. From the wind speed deviation histograms of the three models, the global wind speed retrieved by the MVE method is the worst, and 80% of the wind speed deviation is 3.20 m/s. The above analysis further confirms that the GloWS-Net model architecture has strong advantages in retrieving global sea surface wind speed.

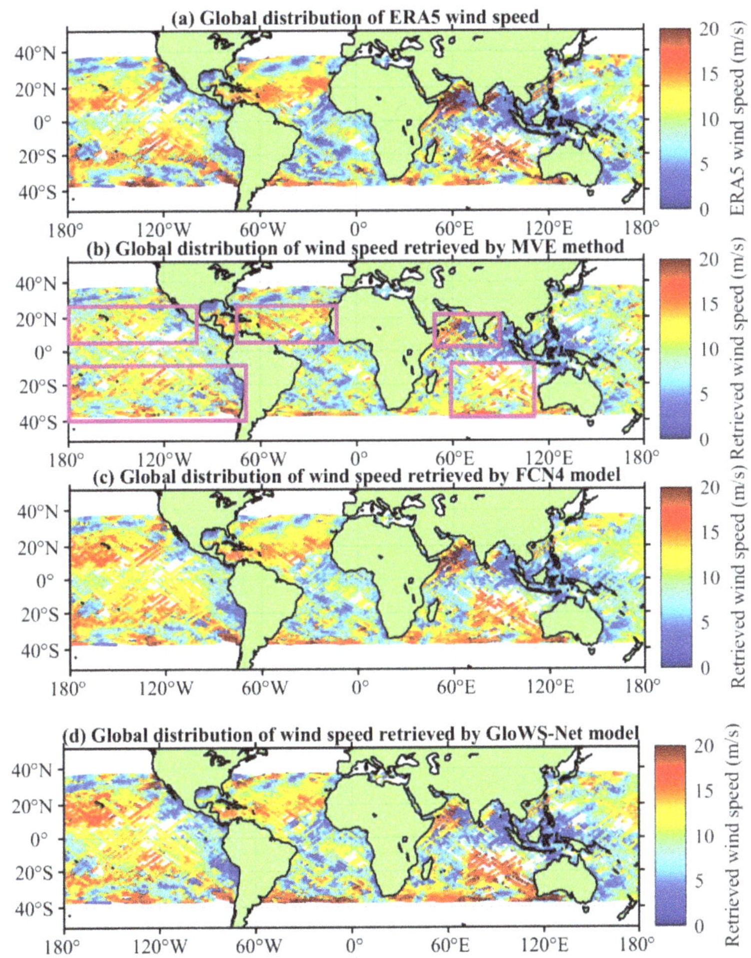


**Figure 7.** Global distribution of ERA5 wind speed data and retrieved wind speed by different models.

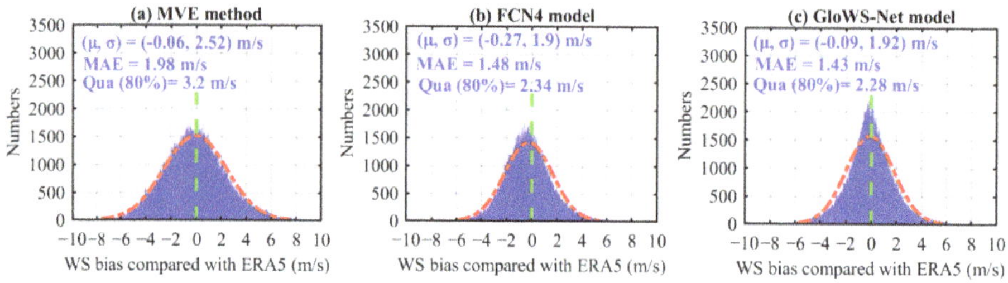


**Figure 8.** Deviation distribution histograms between ERA5 wind speed and retrieved wind speed by different models.

## 4.2. Comparison with CCMP Data

It is necessary to evaluate the wind speed retrieval performance of the proposed deep model through independent wind speed measurement. The CCMP wind speed is closer to conventional in situ measurements from ships than the ERA5 product. In [66], CCMP winds and Tropical Atmosphere Ocean (TAO) mooring observations were compared; a good agreement with a root mean square error (RMSE) of 1 m/s and a correlation coefficient of 0.95 were obtained. The CCMP wind product is a newly released global ocean wind data set and suitable for scientific study at various temporal and spatial resolutions, which is widely used to verify the retrieval of wind speed by spaceborne GNSS-R [55,67,68]. Therefore, the wind speed of CCMP was also collected and compared with the retrieved wind speed. Figure 9 shows the correlation between the ERA5 wind speed data and CCMP wind speed data used in the test model in this study, as well as the probability density function (PDF) distribution curve of ERA5 and CCMP wind speeds. These data cover a wind speed range of 0 to 30 m/s, but only a few samples have wind speeds of 20–30 m/s, which may lead to low accuracy at high wind speeds. We can see that the PDF of the two data sets is almost the same, except that ERA5 wind speed is slightly higher than CCMP at 3–6 m/s and vice versa at 14–19 m/s. This may be due to the differences in the platforms, hardware and algorithms used to generate the two data sets. In addition, the RMSE and CC between the two data sets are 1.46 m and 0.95, respectively, which are highly correlated.

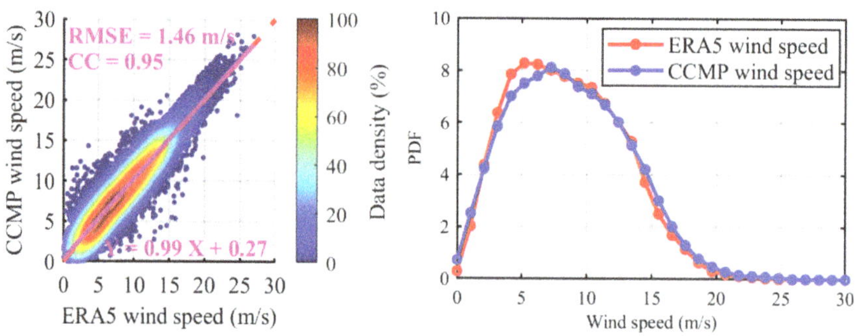


**Figure 9.** Scatter density plots of ERA5 and CCMP wind speed (**left**) and probability density function (PDF) distribution using ERA5 (the red line) and CCMP (the blue line) products as reference in the testing set (**right**).

Table 6 shows the accuracy statistics by comparing the retrieved wind speed data of different models with the CCMP data. The following conclusions can be drawn from the table:

**Table 6.** Accuracy of wind speed retrieved from different models using CCMP data as reference.

| Models | <10 m/s | 10–15 m/s | >15 m/s | 0–30 m/s | | | |
|---|---|---|---|---|---|---|---|
| | **RMSE (m/s)** | **RMSE (m/s)** | **RMSE (m/s)** | **RMSE (m/s)** | **MAE (m/s)** | **CC** | **MAPE (%)** |
| NBRCS | 2.23 | 2.58 | 4.53 | 2.61 | 2.06 | 0.81 | 30.89 |
| LES | 2.45 | 2.63 | 4.78 | 2.87 | 2.15 | 0.68 | 42.36 |
| MVE | 2.47 | 2.52 | 4.47 | 2.71 | 2.12 | 0.79 | 41.84 |
| FCN1 | 2.71 | 2.63 | 4.33 | 2.86 | 2.22 | 0.78 | 42.87 |
| FCN2 | 1.98 | 2.18 | 3.59 | 2.22 | 1.71 | 0.87 | 31.69 |
| FCN3 | 1.80 | 2.27 | 3.68 | 2.16 | 1.66 | 0.87 | 31.45 |
| FCN4 | 2.00 | 2.07 | 3.16 | 2.14 | 1.67 | 0.88 | 36.11 |
| CNN1 | 1.91 | 2.66 | 4.76 | 2.49 | 1.92 | 0.84 | 33.80 |
| CNN2 | 2.26 | 2.52 | 4.62 | 2.60 | 2.02 | 0.80 | 37.60 |
| CNN3 | 2.05 | 2.33 | 2.97 | 2.23 | 1.70 | 0.87 | 29.79 |
| GloWS-Net | 1.97 | 2.34 | 2.80 | 2.16 | 1.65 | 0.88 | 29.70 |

(1) When the wind speed is within the range of 0–30 m/s, the RMSE of the FCN4 model architecture is the lowest (2.14 m/s), followed by the proposed GloWS-Net model architecture (RMSE = 2.16 m/s) and FCN3 model architecture. The results of comparison between retrieved wind speed and CCMP are similar to those of ERA5, the difference of RMSE between the GloWS-Net model architecture and FCN model is very small, but MAPE has been significantly improved by 17.75%. Moreover, in terms of the four indicators (MRSE, MAE, CC and MAPE), the performance of the proposed GloWS-Net model architecture is much better than the model combining the results from NBRCS and LES based on the MVE method, with an increase of 20.27%, 22.21%, 11.20% and 29.02%, respectively. Compared with the NBRCS method, it is increased by 17.20%, 20.11%, 8.89% and 3.86%, respectively. Compared with the LES method, it is increased by 24.71%, 23.45%, 29.70% and 29.90%, respectively.

(2) In the case of strong wind (>15 m/s), except for the GloWS-Net model architecture, the RMSE of wind speed retrieved from other models is large. The RMSE of the GloWS-Net model is 2.8 m/s, and the retrieval accuracy is 37.42% and 11.58% higher than that of the MVE and FCN4 models, respectively.

(3) Under medium (10–15 m/s) and low (<10 m/s) wind speed, the results of the comparison between retrieved wind speed and CCMP are similar to those of ERA5. Among them, the GloWS-Net and FCN4 model retrieved wind speed has good correlation with the CCMP data.

Figure 10 shows the wind speed scatter density plots of different models and CCMP. It can be seen that the comparison results between the wind speed retrieved by different models and CCMP are consistent with those with the ERA5 data. This further shows that the GloWS-Net model proposed in this paper is reliable and has high generalization ability, which means it is good for practical application.

It is also necessary to evaluate the retrieval performance of different retrieval models in different wind speed ranges when the CCMP wind speed data are used as reference. Figure 11 shows the RMSE and MAE for different wind speed ranges. It can be seen that CNN3 and GloWS-Net outperform other architectures especially for challenging high wind speeds (> 20 m/s). Furthermore, when the wind speed is greater than 8 m/s, it can be seen that including SWH, wave direction and rainfall information also improves the GloWS-Net model performance. When the wind speed is greater than 20 m/s, the GloWS-Net model shows the best retrieval performance.

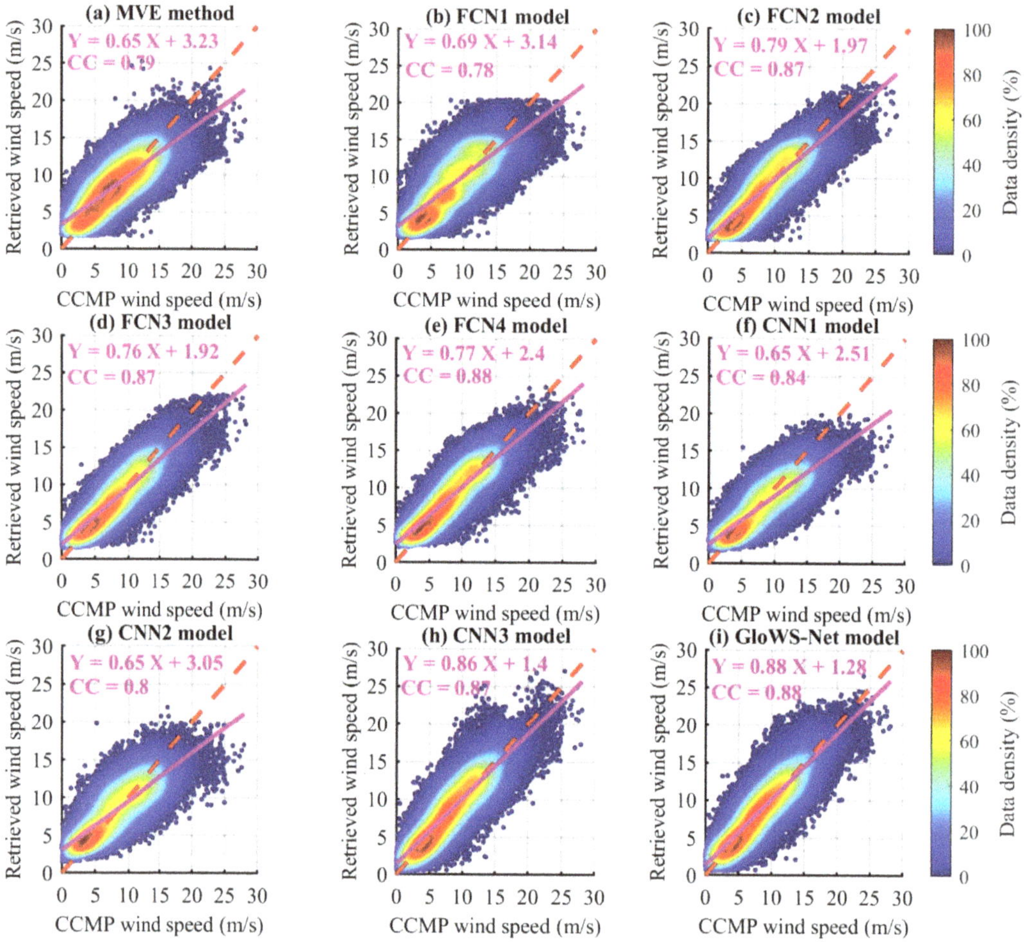


**Figure 10.** Scatter density plots of retrieved wind speed by different models and CCMP wind speed.

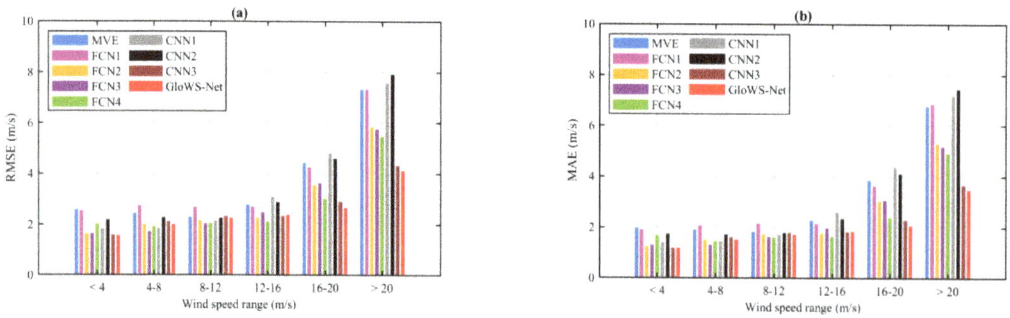


**Figure 11.** (**a**) RMSE and (**b**) MAE for different wind speed ranges when CCMP wind speed is used as reference.

Figure 12 shows the global deviation distribution histogram between CCMP wind speed and the results obtained by the MVE method and FCN4 as well as GloWS-Net. From Figure 12, it can also be seen that the performance of the GloWS-Net model is better than that of the FCN4 model. The deviation between the GloWS-Net model wind speed results and the ERA5 data is very concentrated (80% of the wind speed deviation is less than 2.60 m/s) and near the deviation line of 0 m/s, while 80% of the wind speed deviation of FCN4 model is less than 2.62 m/s, the global wind speed retrieved by the MVE method is the worst, and 80% of the wind speed deviation is 3.39 m/s. The above analysis further proves that the GloWS-Net model architecture has the best generalization ability in retrieving global sea surface wind speed.

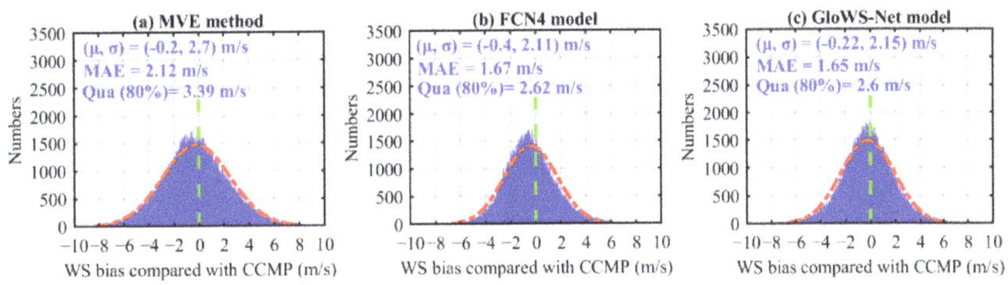


**Figure 12.** Deviation distribution histograms between CCMP wind speed and retrieved wind speed by different models.

*4.3. Discussion*

Previous research has successfully utilized a fully connected network (FCN) model for wind speed retrieval [37–43]. Inspired by image processing technology, Asgarimehr and Guo et al. studied spaceborne GNSS-R wind speed retrieval based on deep learning methods (CyGNSSnet and MCNN) [40,41]. Although both models employ the features extracted from BRCS DDM, MCNN also uses effective scattering area as the input image. However, their research shows that the GNSS-R wind speed retrieval model can benefit from the convolution layer. Because the image information in the DDM is integrated, the architecture that combines the convolution layer and the full connection layer has better performance than the architecture that only has the full connection layer. However, the retrieval results of both their models and the GloWS-Net model architecture proposed in this paper show that, mainly in the case of high wind speed, the model architecture integrating layer that takes DDM as input with the fully connection layer that takes auxiliary parameters as input can obtain more obvious accuracy improvement than the FCN model, but their RMSE improvement in the whole wind speed range of 0–30 m/s is not significant. This may require further optimization of the convolutional layer architecture or training of special models for different wind speed ranges [50]. However, the GloWS-Net model architecture proposed in this paper is superior to the FCN model architecture in MAE and MAPE, especially in MAPE. From this point of view, GloWS-Net achieves better performance improvement than CyGNSSnet and MCNN for the wind speed range of 0–30 m/s.

Although CNN extracts features directly from the input, research shows that the prediction accuracy of CNN can be improved by using traditional features. For example, the combination of CNN and handmade features enhances image classification [69,70]. Similarly, theoretical remote sensing knowledge is increasingly combined with deep learning to further improve its performance [71]. The analysis in this paper also shows that traditional CNN models (such as CNN1 and CNN2) without full connection layer to combine auxiliary parameters and DDM result in an RMSE of 2.46 m/s. The performance of fully connected architectures (such as FCN1) with NBRCS and LES as input characteristics is inferior to CNN1 and CNN2, resulting in an RMSE of 2.66 m/s. However, under high wind speed (>15 m/s), the performance of FCN1 is better than that of CNN1 and CNN2. Therefore, the use of convolution layers to directly extract features from DDM, combined with more auxiliary feature parameters including NBRCS and LES, can provide the best performance (such as CNN3 and GloWS-Net models), and RMSE is better than 1.92 m/s. By comparing the performance of FCN4 and GloWS-Net, it can be concluded that adding more auxiliary parameters does not necessarily reduce the RMSE of the test data. Although previous research and the results in this paper show that these auxiliary parameters can improve the retrieval performance of the model, and based on this fact, they have been selected as the input of the full connection layer. Therefore, it is necessary to further optimize the previously proposed model and GloWS-Net model architecture.

MVE is a classical method for retrieving sea surface wind speed. According to the statistical results in Tables 5 and 6, the GloWS-Net model framework is superior to the MVE method in all wind speed ranges. Especially at high wind speeds (>15 m/s), RMSE is decreased by 37.45%, which shows significant improvement compared with previous studies. Figures 8 and 12 also confirm that the global performance of the GloWS-Net model framework is better than MVE and FCN4, and the deviation has improved globally. Compared with the recent model based on the full connection layer for processing CYGNSS data [36,38], the GloWS-Net model architecture also shows encouraging performance. However, the differences in data length, version, quality control, filtering and validation methods should be considered to ensure the estimation of new data sets will have the same accuracy.

Both MVE and GloWS-Net underestimated high wind speed, which may be caused by the sensitivity saturation of DDM observed under strong wind conditions [32]. This is a common problem of radar scatterometer whose performance will be reduced under high wind speed [72]. However, compared with previous CyGNSSnet and MCNN models, the GloWS-Net model has significantly improved the wind speed retrieval performance in the case of large wind speed, showing good consistence with the ERA5 data. However, it is still challenging to use the GloWS-Net model to retrieve high wind speed. Due to the relatively small number of high wind speed samples involved in the training model, the performance of the deep learning algorithm is limited when the wind speed is greater than 15 m/s.

Figure 7 shows the area with overestimated wind speeds, especially in the Asia Pacific region with longitude between 50°W–0°. This is consistent with the situation reported by Asgarimehr et al. [40]. This area is strongly affected by the L-band signal of the Quasi Zenith Satellite System (QZSS), which may be a potential source of Radio Frequency Interference (RFI). RFI caused by other L-band signals, especially satellite enhancement system (SBAS) signals, has been considered as the root cause of the reduction of signal to noise ratio and the wind speed overestimation of GNSS-R [73]. Asgarimehr et al. also reported similar overestimation in the equatorial region in TDS-1 satellite measurements [74]. Please note that the similarity of wind speed retrieved by MVE, CyGNSSnet and GloWS-Net confirms that overestimation is related to the data quality but not due to the retrieval methods.

## 5. Conclusions

This paper studies the method of retrieving sea surface wind speed using spaceborne GNSS-R data from CYGNSS mission observations, and proposes a GloWS-Net deep learning network model architecture for spaceborne GNSS-R global sea surface wind speed retrieval. Extensive test results show that the GloWS-Net model architecture can obtain the same retrieval accuracy as the FCN model with respect to RMSE, but MAPE has been significantly improved. When compared with ERA5 wind speed and CCMP wind speed, the MAPE improved by 16.56% and 17.75%, respectively. Moreover, when using ERA5 winds as ground truth, compared with the MVE method, RMSE, MAE, CC and MAPE are improved by 23.98%, 27.95%, 11.02% and 32.52%, respectively, when using CCMP winds as ground truth, RMSE, MAE, CC and MAPE are improved by 20.27%, 22.21%, 11.20% and 29.02%, respectively. Compared with previous FCN, CNN, CyGNSSnet and MCNN models, the proposed wind speed retrieval method based on the GloWS-Net model has achieved promising overall performance. The research results also show that the GloWS-Net model architecture has strong advantages in retrieving global sea surface wind speed. In particular, the GloWS-Net model significantly reduces the underestimate phenomenon at high wind speeds. Compared with CyGNSSnet and MCNN, it also performs very well. Being limited by current spaceborne GNSS-R technology level, it still faces a huge challenge to precisely predict wind speed under strong winds. However, the excellent performance of the GloWS-Net model at high wind speeds is promising for future marine disaster monitoring. As we know, buoy data have been widely used as ground truth for validation. Due to limited access and spatial coverage of the buoy (NDBC, TAO, etc.) data, the ERA5 and CCMP data have been employed as the comparison data in this study. In the future, further validation using buoy data can be conducted once such high quality data are available.

**Author Contributions:** All authors have made significant contributions to this manuscript. J.B. partly designed the improved method, analyzed the data, wrote the initial version of paper and validated the improved method; K.Y., X.Z. and W.H. conceived the improved method, wrote the revised version of the paper and provided supervision; J.B., K.Y., X.Z., J.N., Y.L. and W.H. checked and revised this paper. All authors have read and agreed to the published version of the manuscript.

**Funding:** This work was supported in part by the National Natural Science Foundation of China under Grants 42161067 and 42174022, in part by the major scientific and technological projects of Yunnan Province: Research on Key Technologies of ecological environment monitoring and intelligent management of natural resources in Yunnan under Grant 202202AD080010, in part by the Postgraduate Research and Practice Innovation Program of Jiangsu Province under Grant KYCX20_2003, in part by the Future Scientists Program of China University of Mining and Technology under Grant 2020WLKXJ049, in part by the China Scholarship Council (CSC) through a State Scholarship Fund (No.202106420009), and in part by the Programme of Introducing Talents of Discipline to Universities, Plan 111, Grant No. B20046.

**Data Availability Statement:** Not applicable.

**Acknowledgments:** We would like to thank NASA for providing the CYGNSS data, the European Center for Medium-Range Weather Forecasts (ECMWF) for providing wind speed, wave direction and swell height data, the NASA for providing GPM-IMERG rainfall data, and the developers of Cross-Calibrated Multi-Platform (CCMP) wind product for providing the data freely available to the public. The authors also thank the anonymous reviewers for their in-depth reviews and helpful suggestions that have largely contributed to improving this paper.

**Conflicts of Interest:** The authors declare no conflict of interest.

## References

1. Li, C.; Huang, W.; Gleason, S. Dual Antenna Space-Based GNSS-R Ocean Surface Mapping: Oil Slick and Tropical Cyclone Sensing. *IEEE J. Sel. Top. Appl. Earth Observ. Remote Sens.* **2015**, *8*, 425–435. [CrossRef]
2. Mayers, D.; Ruf, C. Tropical cyclone center fix using CYGNSS winds. *J. Appl. Meteorol. Climatol.* **2019**, *58*, 1993–2003. [CrossRef]
3. Wang, S.; Shi, S.; Ni, B. Joint Use of Spaceborne Microwave Sensor Data and CYGNSS Data to Observe Tropical Cyclones. *Remote Sens.* **2020**, *12*, 3124. [CrossRef]
4. Foti, G.; Gommenginger, C.; Srokosz, M. First Spaceborne GNSS-Reflectometry Observations of Hurricanes from the UK TechDemoSat-1 Mission. *Geophys. Res. Lett.* **2017**, *44*, 12358–12366. [CrossRef]
5. Ruf, C.; Atlas, R.; Chang, P.; Clarizia, M.P.; Garrison, J.; Gleason, S.; Katzberg, S.; Jelenak, Z.; Johnson, J.; Majumdar, S.; et al. New Ocean Winds Satellite Mission to Probe Hurricanes and Tropical Convection. *Bull. Am. Meteorol. Soc.* **2016**, *97*, 385–395. [CrossRef]
6. Jing, C.; Niu, X.; Duan, C.; Lu, F.; Di, G.; Yang, X. Sea Surface Wind Speed Retrieval from the First Chinese GNSS-R Mission: Technique and Preliminary Results. *Remote Sens.* **2019**, *11*, 3013. [CrossRef]
7. Yang, G.; Bai, W.; Wang, J.; Hu, X.; Zhang, P.; Sun, Y.; Xu, N.; Zhai, X.; Xiao, X.; Xia, J.; et al. FY3E GNOS II GNSS Reflectometry: Mission Review and First Results. *Remote Sens.* **2022**, *14*, 988. [CrossRef]
8. Clarizia, M.P.; Ruf, C.; Cipollini, P.; Zuffada, C. First spaceborne observation of sea surface height using GPS-Reflectometry. *Geophys. Res. Lett.* **2016**, *43*, 767–774. [CrossRef]
9. Li, W.; Rius, A.; Fabra, F.; Cardellach, E.; Ribo, S.; Martin-Neira, M. Revisiting the GNSS-R Waveform Statistics and Its Impact on Altimetric Retrievals. *IEEE Trans. Geosci. Remote Sens.* **2018**, *56*, 2854–2871. [CrossRef]
10. Alonso-Arroyo, A.; Zavorotny, V.U.; Camps, A. Sea Ice Detection Using U.K. TDS-1 GNSS-R Data. *IEEE Trans. Geosci. Remote Sens.* **2017**, *55*, 4989–5001. [CrossRef]
11. Yan, Q.; Huang, W. Spaceborne GNSS-R Sea Ice Detection Using Delay-Doppler Maps: First Results from the U.K. TechDemoSat-1 Mission. *IEEE J. Sel. Top. Appl. Earth Observ. Remote Sens.* **2016**, *9*, 4795–4801. [CrossRef]
12. Zhu, Y.; Tao, T.; Zou, J.; Yu, K.; Wickert, J.; Semmling, M. Spaceborne GNSS Reflectometry for Retrieving Sea Ice Concentration Using TDS-1 Data. *IEEE Geosci. Remote Sens. Lett.* **2020**, *18*, 612–616. [CrossRef]

13. Yan, Q.; Huang, W. Sea Ice Sensing From GNSS-R Data Using Convolutional Neural Networks. *IEEE Geosci. Remote Sens. Lett.* **2018**, *15*, 1510–1514. [CrossRef]
14. Yan, Q.; Huang, W. Sea Ice Thickness Measurement Using Spaceborne GNSS-R: First Results with TechDemoSat-1 Data. *IEEE J. Sel. Top. Appl. Earth Observ. Remote Sens.* **2020**, *13*, 577–587. [CrossRef]
15. Yan, Q.; Huang, W. Tsunami Detection and Parameter Estimation From GNSS-R Delay-Doppler Map. *IEEE J. Sel. Top. Appl. Earth Observ. Remote Sens.* **2016**, *9*, 4650–4659. [CrossRef]
16. Warnock, A.M.; Ruf, C.S.; Morris, M. Storm surge prediction with cygnss winds. In Proceedings of the 2017 IEEE International Geoscience and Remote Sensing Symposium (IGARSS), Fort Worth, TX, USA, 23–28 July 2017; pp. 2975–2978.
17. Li, X.; Yang, D.; Han, G.; Yang, L.; Wang, J.; Yang, J.; Chen, D.; Zheng, G. Exploiting the Potential of Coastal GNSS-R for Improving Storm Surge Modeling. *IEEE Geosci. Remote Sens. Lett.* **2021**, *18*, 1134–1138. [CrossRef]
18. Peng, Q.; Jin, S. Significant Wave Height Estimation from Space-Borne Cyclone-GNSS Reflectometry. *Remote Sens.* **2019**, *11*, 584. [CrossRef]
19. Yang, S.; Jin, S.; Jia, Y.; Ye, M. Significant Wave Height Estimation from Joint CYGNSS DDMA and LES Observations. *Sensors* **2021**, *21*, 6123. [CrossRef]
20. Bu, J.; Yu, K. Significant Wave Height Retrieval Method Based on Spaceborne GNSS Reflectometry. *IEEE Geosci. Remote Sens. Lett.* **2022**, *19*, 1503705. [CrossRef]
21. Bu, J.; Yu, K. A New Integrated Method of CYGNSS DDMA and LES Measurements for Significant Wave Height Estimation. *IEEE Geosci. Remote Sens. Lett.* **2022**, *19*, 1505605. [CrossRef]
22. Wang, F.; Yang, D.; Yang, L. Retrieval and Assessment of Significant Wave Height from CYGNSS Mission Using Neural Network. *Remote Sens.* **2022**, *14*, 3666. [CrossRef]
23. Yu, K.; Han, S.; Bu, J.; An, Y.; Zhou, Z.; Wang, C.; Tabibi, S.; Cheong, J.W. Spaceborne GNSS Reflectometry. *Remote Sens.* **2022**, *14*, 1605. [CrossRef]
24. Bu, J.; Yu, K.; Park, H.; Huang, W.; Han, S.; Yan, Q.; Qian, N.; Lin, Y. Estimation of Swell Height Using Spaceborne GNSS-R Data from Eight CYGNSS Satellites. *Remote Sens.* **2022**, *14*, 4634. [CrossRef]
25. Asgarimehr, M.; Zavorotny, V.; Wickert, J.; Reich, S. Can GNSS Reflectometry Detect Precipitation Over Oceans? *Geophys. Res. Lett.* **2018**, *45*, 12585–12592. [CrossRef]
26. Bu, J.; Yu, K.; Ni, J.; Yan, Q.; Han, S.; Wang, J.; Wang, C. Machine learning-based methods for sea surface rainfall detection from CYGNSS delay-doppler maps. *GPS Solut.* **2022**, *26*, 132. [CrossRef]
27. Bu, J.; Yu, K. Sea Surface Rainfall Detection and Intensity Retrieval Based on GNSS-Reflectometry Data from the CYGNSS Mission. *IEEE Trans. Geosci. Remote Sens.* **2022**, *60*, 5802015. [CrossRef]
28. Bu, J.; Yu, K.; Han, S.; Qian, N.; Lin, Y.; Wang, J. Retrieval of Sea Surface Rainfall Intensity Using Spaceborne GNSS-R Data. *IEEE Trans. Geosci. Remote Sens.* **2022**, *60*, 5803116. [CrossRef]
29. Yan, Q.; Huang, W.; Jin, S.; Jia, Y. Pan-tropical soil moisture mapping based on a three-layer model from CYGNSS GNSS-R data. *Remote Sens. Environ.* **2020**, *247*, 111944. [CrossRef]
30. Carreno-Luengo, H.; Luzi, G.; Crosetto, M. Above-Ground Biomass Retrieval over Tropical Forests: A Novel GNSS-R Approach with CyGNSS. *Remote Sens.* **2020**, *12*, 29. [CrossRef]
31. Clarizia, M.P.; Ruf, C.S.; Jales, P.; Gommenginger, C. Spaceborne GNSS-R Minimum Variance Wind Speed Estimator. *IEEE Trans. Geosci. Remote Sens.* **2014**, *52*, 6829–6843. [CrossRef]
32. Ruf, C.S.; Gleason, S.; McKague, D.S. Assessment of CYGNSS Wind Speed Retrieval Uncertainty. *IEEE J. Sel. Top. Appl. Earth Observ. Remote Sens.* **2019**, *12*, 87–97. [CrossRef]
33. Clarizia, M.P.; Ruf, C.S. Wind Speed Retrieval Algorithm for the Cyclone Global Navigation Satellite System (CYGNSS) Mission. *IEEE Trans. Geosci. Remote Sens.* **2016**, *54*, 4419–4432. [CrossRef]
34. Rodriguez-Alvarez, N.; Garrison, J.L. Generalized Linear Observables for Ocean Wind Retrieval from Calibrated GNSS-R Delay–Doppler Maps. *IEEE Trans. Geosci. Remote Sens.* **2016**, *54*, 1142–1155. [CrossRef]
35. Hammond, M.L.; Foti, G.; Rawlinson, J.; Gommenginger, C.; Srokosz, M.; King, L.; Unwin, M.; Roselló, J. First Assessment of Geophysical Sensitivities from Spaceborne Galileo and BeiDou GNSS-Reflectometry Data Collected by the UK TechDemoSat-1 Mission. *Remote Sens.* **2020**, *12*, 2927. [CrossRef]
36. Liu, Y.; Collett, I.; Morton, Y.J. Application of Neural Network to GNSS-R Wind Speed Retrieval. *IEEE Trans. Geosci. Remote Sens.* **2019**, *57*, 9756–9766. [CrossRef]
37. Asgarimehr, M.; Zhelavskaya, I.; Foti, G.; Reich, S.; Wickert, J. A GNSS-R Geophysical Model Function: Machine Learning for Wind Speed Retrievals. *IEEE Geosci. Remote Sens. Lett.* **2020**, *17*, 1333–1337. [CrossRef]
38. Reynolds, J.; Clarizia, M.P.; Santi, E. Wind Speed Estimation from CYGNSS Using Artificial Neural Networks. *IEEE J. Sel. Top. Appl. Earth Observ. Remote Sens.* **2020**, *13*, 708–716. [CrossRef]
39. Chu, X.; He, J.; Song, H.; Qi, Y.; Sun, Y.; Bai, W.; Li, W.; Wu, Q. Multimodal Deep Learning for Heterogeneous GNSS-R Data Fusion and Ocean Wind Speed Retrieval. *IEEE J. Sel. Top. Appl. Earth Observ. Remote Sens.* **2020**, *13*, 5971–5981. [CrossRef]
40. Asgarimehr, M.; Arnold, C.; Weigel, T.; Ruf, C.; Wickert, J. GNSS reflectometry global ocean wind speed using deep learning: Development and assessment of CyGNSSnet. *Remote Sens. Environ.* **2022**, *269*, 112801. [CrossRef]
41. Guo, W.; Du, H.; Guo, C.; Southwell, B.J.; Cheong, J.W.; Dempster, A.G. Information fusion for GNSS-R wind speed retrieval using statistically modified convolutional neural network. *Remote Sens. Environ.* **2022**, *272*, 112934. [CrossRef]

42. Li, X.; Yang, D.; Yang, J.; Zheng, G.; Han, G.; Nan, Y.; Li, W. Analysis of coastal wind speed retrieval from CYGNSS mission using artificial neural network. *Remote Sens. Environ.* **2021**, *260*, 112454. [CrossRef]
43. Liu, X.; Bai, W.; Xia, J.; Huang, F.; Yin, C.; Sun, Y.; Du, Q.; Meng, X.; Liu, C.; Hu, P.; et al. FA-RDN: A Hybrid Neural Network on GNSS-R Sea Surface Wind Speed Retrieval. *Remote Sens.* **2021**, *13*, 4820. [CrossRef]
44. Gleason, S. Space-Based GNSS Scatterometry: Ocean Wind Sensing Using an Empirically Calibrated Model. *IEEE Trans. Geosci. Remote Sens.* **2013**, *51*, 4853–4863. [CrossRef]
45. Gleason, S.; Ruf, C.S.; Clarizia, M.P.; Brien, A.J.O. Calibration and Unwrapping of the Normalized Scattering Cross Section for the Cyclone Global Navigation Satellite System. *IEEE Trans. Geosci. Remote Sens.* **2016**, *54*, 2495–2509. [CrossRef]
46. Jing, C.; Yang, X.; Ma, W.; Yu, Y.; Dong, D.; Li, Z.; Xu, C. Retrieval of sea surface winds under hurricane conditions from GNSS-R observations. *Acta Oceanol. Sin.* **2016**, *35*, 91–97. [CrossRef]
47. Bu, J.; Yu, K.; Zhu, Y.; Qian, N.; Chang, J. Developing and Testing Models for Sea Surface Wind Speed Estimation with GNSS-R Delay Doppler Maps and Delay Waveforms. *Remote Sens.* **2020**, *12*, 3760. [CrossRef]
48. Clarizia, M.P.; Ruf, C.S. Statistical Derivation of Wind Speeds from CYGNSS Data. *IEEE Trans. Geosci. Remote Sens.* **2020**, *58*, 3955–3964. [CrossRef]
49. Clarizia, M.P.; Ruf, C.S. Bayesian Wind Speed Estimation Conditioned on Significant Wave Height for GNSS-R Ocean Observations. *J. Atmos. Ocean. Technol.* **2017**, *34*, 1193–1202. [CrossRef]
50. Wang, C.; Yu, K.; Qu, F.; Bu, J.; Han, S.; Zhang, K. Spaceborne GNSS-R Wind Speed Retrieval Using Machine Learning Methods. *Remote Sens.* **2022**, *14*, 3507. [CrossRef]
51. Li, C.; Huang, W. An Algorithm for Sea-Surface Wind Field Retrieval From GNSS-R Delay-Doppler Map. *IEEE Geosci. Remote Sens. Lett.* **2014**, *11*, 2110–2114. [CrossRef]
52. Foti, G.; Gommenginger, C.; Jales, P.; Unwin, M.; Shaw, A.; Robertson, C.; Rosello, J. Spaceborne GNSS reflectometry for ocean winds: First results from the UK TechDemoSat-1 mission. *Geophys. Res. Lett.* **2015**, *42*, 5435–5441. [CrossRef]
53. Bu, J.; Yu, K.; Han, S.; Wang, C. Multi-Observable Wind Speed Retrieval Based on Spaceborne GNSS-R Delay Doppler Maps. In Proceedings of the IGARSS 2021-2021 IEEE International Geoscience and Remote Sensing Symposium, Brussels, Belgium, 11–16 July 2021; pp. 7580–7583.
54. Ruf, C.S.; Balasubramaniam, R. Development of the CYGNSS Geophysical Model Function for Wind Speed. *IEEE J. Sel. Top. Appl. Earth Observ. Remote Sens.* **2019**, *12*, 66–77. [CrossRef]
55. Guo, W.; Du, H.; Cheong, J.W.; Southwell, B.J.; Dempster, A.G. GNSS-R Wind Speed Retrieval of Sea Surface Based on Particle Swarm Optimization Algorithm. *IEEE Trans. Geosci. Remote Sens.* **2022**, *60*, 4202414. [CrossRef]
56. Atlas, R.; Hoffman, R.N.; Ardizzone, J.; Leidner, S.M.; Jusem, J.C.; Smith, D.K.; Gombos, D. A cross-calibrated, multiplatform ocean surface wind velocity product for meteorological and oceanographic applications. *Bull. Am. Meteorol. Soc.* **2011**, *92*, 157–174. [CrossRef]
57. Mears, C.A.; Scott, J.; Wentz, F.J.; Ricciardulli, L.; Leidner, S.M.; Hoffman, R.; Atlas, R. A near-real-time version of the cross-calibrated multiplatform (CCMP) ocean surface wind velocity data set. *J. Geophys. Res. Ocean.* **2019**, *124*, 6997–7010. [CrossRef]
58. Marchan-Hernandez, J.F.; Valencia, E.; Rodriguez-Alvarez, N.; Ramos-Perez, I.; Bosch-Lluis, X.; Camps, A.; Eugenio, F.; Marcello, J. Sea-State Determination Using GNSS-R Data. *IEEE Geosci. Remote Sens. Lett.* **2010**, *7*, 621–625. [CrossRef]
59. Zhu, Y.; Yu, K.; Zou, J.; Wickert, J. Sea Ice Detection Based on Differential Delay-Doppler Maps from UK TechDemoSat-1. *Sensors* **2017**, *17*, 1614. [CrossRef]
60. Said, F.; Jelenak, Z.; Park, J.; Soisuvarn, S.; Chang, P.S. A 'track-wise' wind retrieval algorithm for the CYGNSS mission. In Proceedings of the IGARSS 2019-2019 IEEE International Geoscience and Remote Sensing Symposium, Yokohama, Japan, 28 July–2 August 2019; pp. 8711–8714.
61. Said, F.; Jelenak, Z.; Park, J.; Chang, P.S. The NOAA Track-Wise Wind Retrieval Algorithm and Product Assessment for CyGNSS. *IEEE Trans. Geosci. Remote Sens.* **2022**, *60*, 4202524. [CrossRef]
62. Balasubramaniam, R.; Ruf, C. Characterization of rain impact on L-Band GNSS-R ocean surface measurements. *Remote Sens. Environ.* **2020**, *239*, 111607. [CrossRef]
63. Ertuğrul, Ö.F. A novel type of activation function in artificial neural networks: Trained activation function. *Neural Netw.* **2018**, *99*, 148–157. [CrossRef]
64. Saïd, F.; Katzberg, S.J.; Soisuvarn, S. Retrieving hurricane maximum winds using simulated CYGNSS power-versus-delay waveforms. *IEEE J. Sel. Top. Appl. Earth Observ. Remote Sens.* **2017**, *10*, 3799–3809. [CrossRef]
65. Ruf, C.; Unwin, M.; Dickinson, J.; Rose, R.; Rose, D.; Vincent, M.; Lyons, A. CYGNSS: Enabling the Future of Hurricane Prediction [Remote Sensing Satellites]. *IEEE Geosci. Remote Sens. Mag.* **2013**, *1*, 52–67. [CrossRef]
66. Wang, X.; Shum, C.; Johnson, J. Analysis of Surface Wind Diurnal Cycles in Tropical Regions using Mooring Observations and the CCMP Product. In Proceedings of the Conference on Hurricanes and Tropical Meteorology American Meteorological Society, Washington, DC, USA, 1–4 April 2014; p. 7A.7.
67. Li, X.; Yang, D.; Yang, J.; Han, G.; Zheng, G.; Li, W. Validation of NOAA CyGNSS Wind Speed Product with the CCMP Data. *Remote Sens.* **2021**, *13*, 1832. [CrossRef]
68. Dong, Z.; Jin, S. Evaluation of Spaceborne GNSS-R Retrieved Ocean Surface Wind Speed with Multiple Datasets. *Remote Sens.* **2019**, *11*, 2747. [CrossRef]

69. Tianyu, Z.; Zhenjiang, M.; Jianhu, Z. Combining cnn with hand-crafted features for image classification. In Proceedings of the 2018 14th IEEE International Conference on Signal Processing (ICSP), Beijing, China, 12–16 August 2018; pp. 554–557.
70. Zhang, S.; Li, C.; Qiu, S.; Gao, C.; Zhang, F.; Du, Z.; Liu, R. EMMCNN: An ETPS-Based Multi-Scale and Multi-Feature Method Using CNN for High Spatial Resolution Image Land-Cover Classification. *Remote Sens.* **2020**, *12*, 66. [CrossRef]
71. Nabi, M.; Senyurek, V.; Gurbuz, A.C.; Kurum, M. Deep Learning-Based Soil Moisture Retrieval in CONUS Using CYGNSS Delay–Doppler Maps. *IEEE J. Sel. Top. Appl. Earth Observ. Remote Sens.* **2022**, *15*, 6867–6881. [CrossRef]
72. Zeng, L.; Brown, R.A. Scatterometer observations at high wind speeds. *J. Appl. Meteorol.* **1998**, *37*, 1412–1420. [CrossRef]
73. Querol, J.; Alonso-Arroyo, A.; Onrubia, R.; Pascual, D.; Park, H.; Camps, A. SNR degradation in GNSS-R measurements under the effects of radio-frequency interference. *IEEE J. Sel. Top. Appl. Earth Observ. Remote Sens.* **2016**, *9*, 4865–4878. [CrossRef]
74. Asgarimehr, M.; Wickert, J.; Reich, S. TDS-1 GNSS reflectometry: Development and validation of forward scattering winds. *IEEE J. Sel. Top. Appl. Earth Observ. Remote Sens.* **2018**, *11*, 4534–4541. [CrossRef]

*remote sensing*

MDPI

*Communication*

# A Coastal Experiment for GNSS-R Code-Level Altimetry Using BDS-3 New Civil Signals

Fan Gao [1,2], Tianhe Xu [2,*], Xinyue Meng [1], Nazi Wang [2], Yunqiao He [1] and Baojiao Ning [1]

[1] School of Space Science and Physics, Shandong University, Weihai 264209, China; gaofan@sdu.edu.cn (F.G.); 201916534@mail.sdu.edu.cn (X.M.); heyunqiao@mail.sdu.edu.cn (Y.H.); 202017724@mail.sdu.edu.cn (B.N.)

[2] Institute of Space Science, Shandong University, Weihai 264209, China; wnz@sdu.edu.cn

* Correspondence: thxu@sdu.edu.cn

**Abstract:** High temporal and spatial resolutions are the key advantages of the global navigation satellites system-reflectometry (GNSS-R) technique, while low precision and instabilities constrain its development. Compared with conventional Ku/C band nadir-looking radar altimetry, the precision of GNSS-R code-level altimetry is restricted by the smaller bandwidth and the lower transmitted power of the signals. Fortunately, modernized GNSS broadcast new open-available ranging codes with wider bandwidth. The Chinese BDS-3 system was built on 31 July 2020; its inclined geostationary orbit and medium circular orbit satellites provide B1C and B2a public navigation service signals in the two frequency bands of B1 and B2. In order to investigate their performance on GNSS-R code-level altimetry, a coastal experiment was conducted on 5 November 2020 at a trestle of Weihai in the Shandong province of China. The raw intermediate frequency data with a 62 MHz sampling rate were collected and post-processed to solve the sea surface height every second continuously for over eight hours. The precisions were evaluated using the measurements from a 26 GHz radar altimeter mounted on the same trestle near our GNSS-R setup. The results show that a centimeter-level accuracy of GNSS-R altimetry—based on B1C code after the application of the moving average—can be achieved, while for B2a code, the accuracy is about 10 to 20 cm.

**Keywords:** ocean altimetry; global Navigation satellite systems reflectometer (GNSS-R); BDS; B1C; B2a

**Citation:** Gao, F.; Xu, T.; Meng, X.; Wang, N.; He, Y.; Ning, B. A Coastal Experiment for GNSS-R Code-Level Altimetry Using BDS-3 New Civil Signals. *Remote Sens.* **2021**, *13*, 1378. https://doi.org/10.3390/rs13071378

Academic Editors: Hugo Carreno-Luengo, Dallas Masters and Chun-Liang Lin

Received: 15 February 2021
Accepted: 31 March 2021
Published: 3 April 2021



## 1. Introduction

Satellite radar altimetry and tide gauges have been used to monitor large-scale global sea surface heights (SSH) in the past several decades; these have contributed much to Earth sciences [1,2]. However, their spatial and temporal resolutions cannot meet the requirements for probing mesoscale features in the ocean height. To solve this problem, global navigation satellites system-reflectometry (GNSS-R) was proposed as a multi-static radar means with the prospect of providing additional high-density SSH measurements [3]. Essentially, the performance of this technique relies on the accuracy of the relative path delay between the direct and reflected signals. As GNSS signals are not dedicated for altimetry, the precision of GNSS-R code-level altimetry is restricted by their smaller bandwidth and lower transmitted power [4].

Many experiments have been performed on different platforms to test the precision and accuracy of GNSS-R code-level altimetry. The results of a bridge-based experiment showed that global positioning system (GPS) C/A code and P-code provided the water surface reflector height with accuracies of 3 and 0.3 m, respectively [5]. The experiment was enhanced; its results indicated a significant improvement in GNSS-R altimetric performance with 7.5 cm uncertainty [6]. The feasibility of code-level altimetry based on BDS B1I signals using coastal GNSS-R setups was also verified [7]. The first airborne GNSS-R ocean altimetry experiment was performed in 2002, with results showing that the root-mean-square residual height was at the meter level for GPS C/A code and decimeter level for

P-code [8,9]. Another airborne experiment was conducted to investigate the performance of code-delay altimetry using clean-replica and interferometric approaches based on GPS L1 signals [10]. GNSS-R airborne GPS L5 signals were also used for altimetry analysis; precision within meter and sub-meter levels was achieved [11]. Apart from the cases for the lower-altitude of the receivers, an accuracy of two to three meters can be achieved for space-borne GNSS-R code-level ocean altimetry, based on GPS C/A and BDS code, using the data from TDS-1 and CYGNSS missions [12–14]. In addition, by analyzing the signals reflected from the lake surface, the accuracy can be reached at the sub-meter level on board [15].

In 2020, China finished constructing its BDS-3, which can transmit B1C and B2a civil code signals with wider bandwidths at center frequencies of 1575.42 and 1176.45 MHz, respectively [16]. Hence, it can be expected that the precision of GNSS-R code-level altimetry can be improved by using the new BDS civil codes. In order to demonstrate the potential of BDS B1C and B2a signals for GNSS-R altimetry, we performed a static coastal experiment on a trestle bridge. The raw intermediate frequency (IF) data produced by GNSS-R setups and other precise auxiliary measurements obtained by geodetic GNSS setups, radar altimeter, and electronic total station were collected. These data were post-processed to solve the SSH every second continuously for over eight hours.

The remainder of this paper is organized as follows. In Section 2, we briefly review the characteristics of the two new BDS-3 civil signals and the basic principle of our work. In Section 3, details of our coastal GNSS-R altimetry experiment and the setups that were used are described. In Section 4, we analyze the solutions and evaluate their accuracy by comparing them with the measurements of the radar altimeter. Finally, the main results are summarized and the problems that remain unsolved in this work are discussed.

## 2. Materials and Methods

BDS-3 B1C and B2a signals for fundamental positioning, navigation, and timing service are broadcast from 24 medium circular orbit (MEO) satellites and three inclined geostationary orbit (IGSO) satellites. Their pseudo random noise code (PRN) numbers range from 19 to 46, in which 38, 39, and 40 are the PRNs of IGSO. The two kinds of new civil ranging codes have the same length of 10230, while the chip rates of B1C and B2a are 1.023 Mbps and 10.23 Mbps, respectively [17,18]. These are separately modulated on carrier signals with center frequencies of 1575.42 MHz and 1176.45 MHz, and different ranging codes are modulated on the data component and pilot component. The power ratio of the data component to the pilot component is 1:1 for B2a while the ratio is 1:3 for B1C. In addition, both the components of B2a adopt BPSK(10) modulation while those for B1C adopt more advanced BOC and QMBOC modulation.

The expression of the modulated B2a signal can be described as:

$$S_{B2a}^{(i)}(t) = \sqrt{2P_{B2a}}\left[D_{B2a\_d}^{(i)}(t)C_{B2a\_d}^{(i)}(t)\cos(2\pi ft) - C_{B2a\_p}^{(i)}(t)\sin(2\pi ft)\right] \tag{1}$$

where $S_{B2a}^{(i)}$ stands for the signals from satellite $i$, $D_{B2a\_d}^{(i)}$ is the modulation data, $C_{B2a\_d}^{(i)}$ stands for the ranging code on the data component, $C_{B2a\_p}^{(i)}$ is the ranging code on the pilot component, $P_{B2a}$ is the B2a signal power, and $f$ is the carrier frequency. In Equation (1), the pilot channel is data-less, and so can be used for estimating the ranging information better without the problem of sign transitions [17].

The expression of the modulated B1C signal can be described as:

$$S_{B1C}^{(i)}(t) = \sqrt{2P_{B1C}}\left[\frac{1}{2}D_{B1C}^{(i)}(t)C_{B1C\_d}^{(i)}(t)\cos(2\pi ft) - \frac{\sqrt{3}}{2}C_{B1C\_p}^{(i)}(t)\sin(2\pi ft)\right] \tag{2}$$

where $S_{B1C}^{(i)}$ stands for the signals from satellite $i$; $D_{B1C}^{(i)}$ is the modulation data; $C_{B1C\_d}^{(i)}$ stands for the ranging code on the data component; $P_{B1C}$ is the B1C signal power; and

$C_{B1C\_p}^{(i)}$ is the ranging code on the pilot component, which employs QMBOC(6,1,4/33) modulation [18]. In Equation (2), we can see that not only is the pilot channel free of data information, but that the modulation is also more complicated. These improvements will enhance the ranging ability of the B1C code.

Raw IF data of direct and reflected signals were processed to obtain the code-level path delay measurements using a software-defined receiver (SDR) modified from an open source code using MATLAB [19]. Figure 1 shows a brief flow chart of this GNSS-R SDR. Firstly, both the direct and reflected signals are cross-correlated separately with the pilot and data local replicas. The coherent span for both B1C and B2a codes is 10 milliseconds; each signal produces two waveforms because of its pilot and data components. In order to increase the signal-to-noise ratio, the two values added incoherently by power ratios are the direct signal waveform and reflected signal waveform. Then, the code-level path delays are computed from the positions of waveform peaks by applying cubic spline interpolations. Furthermore, in order to increase the stability and precision, we computed 21 path delays in one second at intervals of 50 milliseconds and selected their median value for further processing. Finally, the reflector heights from the sea surface are calculated in accordance with the geometry of ground-based GNSS-R altimetry.

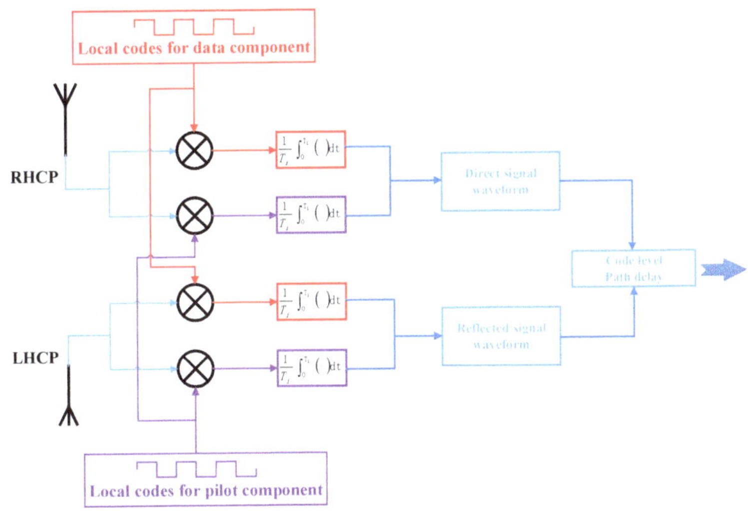


**Figure 1.** Basic concept of deriving code-level path delay from waveforms of direct and reflected signals using the data component (red) and the pilot component (purple).

## 3. Experiments

Our GNSS-R IF data collection system is mainly composed of two dual-circularly polarized crossed dipole antennas and one raw data recorder with four radio frequency signal input ports. Two of them are for the direct and reflected signals from the B1 band while the other two are for the B2 band. The bandwidths of the recorder are 20.46 MHz; its central frequencies are at 1529 MHz and 1130 MHz for the two kinds of signals, respectively. The IF data are quantified with 2 bits and recorded at a sampling rate of 62 MHz continuously in most cases. The data are then transferred to a laptop through a USB 3.0 cable.

Apart from GNSS-R setups, a geodetic GNSS receiver, chock-ring antenna, and an electronic total station were used to obtain the geodetic height of the GNSS-R antennas. An independent 26 GHz radar altimeter was installed on the trestle, which can provide vertical distance from the sea surface to its phase center with 3 mm accuracy every second. We measured the precise height differences among the phase centers of the chock-ring GNSS antenna, GNSS-R antenna, and radar altimeter using the electronic total station. Figure 2 shows the relevant photos.

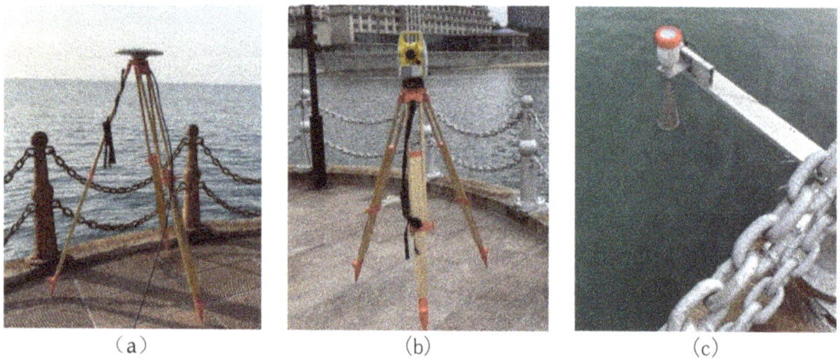

**Figure 2.** Photos of the geodetic GNSS chock-ring antenna (**a**), electronic total station (**b**), and the monostatic radar altimeter (**c**).

We performed the experiment on a shore trestle bridge located at Weihai in Shandong province with latitude and longitude coordinates (37°32′2.62″ N, 122°2′44.11″ E) on November 5, 2020. An upward GNSS-R antenna was used for receiving direct signals with right-handed circular polarization while the down-looking one was for reflected signals with left-handed circular polarization. As the operating band of the antennas ranged from 1.16 to 1.62 GHz, it can cover B1C and B2a signals. Before entering the recorder, both of the direct and reflected signals are spilt into two channels by two two-way power dividers. One is for B1C and another is for B2a. The configuration of antennas and the satellite image of the trestle bridge are shown in Figure 3 and the tilt angles of the antennas are both 30° to horizontal. Their phase centers are in a single plumb line. The height of the antennas above the sea surface ranged from about three to five meters during our experiment. In addition, as the beam width angle of the antennas is 60°, the available satellite elevation ranges from 30° to 90° for this experiment. The two antennas faced south to receive more reflected signals, and so the available satellite azimuth ranged from 150° to 210°.

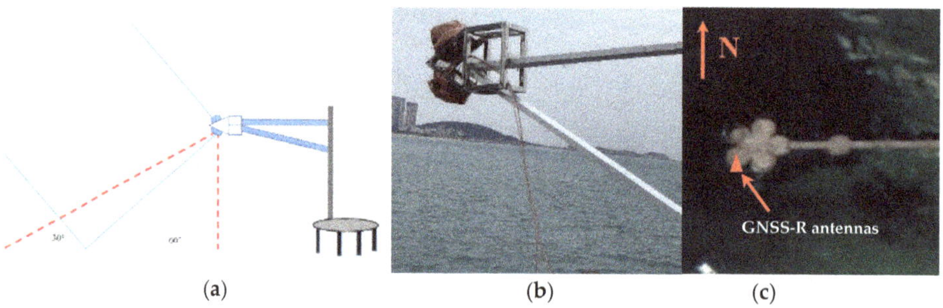


**Figure 3.** The configuration of antennas for our GNSS-R altimetry experiment (**a**), the photograph of the antennas' arrangement (**b**) and top view of the trestle bridge (**c**) during the experiment.

## 4. Results

The signals from five different satellites, including four MEOs and an IGSO, were used for computing reflector heights at different periods. The SSHs were separately derived, based on B2a and B1C signals from 13:30 to 22:00 on 5 November (local time). In general, continuous GNSS-R altimetry solutions were achieved for more than eight hours. As the reflector heights could be derived by using the code-level path delay measurements of one satellite, we selected signals from only one satellite over a certain period of time. The sea surface was smooth in general during the experiment, so the path delay measurements were derived from the peak point positions of the direct and reflected waveforms. As the coherence time for both B1C and B2a codes was 10 milliseconds, each time slot of 10 milliseconds produced an estimate of the path delay. Considering that our SDR runs very slow, we computed a delay measurement every 50 milliseconds to save time. So, there are 21 delays in one second, and their median value was chosen for altimetry retrievals. Figures 4 and 5 show SSHs derived from B1C and B2a code-level delay measurements, respectively. We find that the solutions of both B1C and B2a can reflect the trend of the sea surface change, compared with the measurements of the radar altimeter. However, the noise level of B2a is larger than that of B1C. It should be noted that the gaps around 14:10 in Figures 4 and 5 were caused by an accidental interruption in the power supply at the beginning of our experiment.

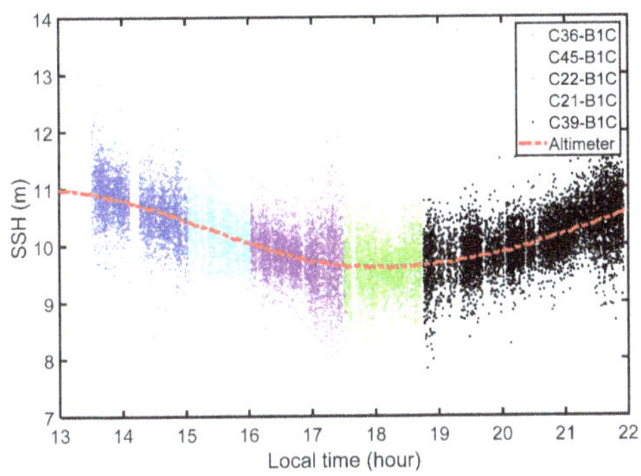


**Figure 4.** SSHs derived from B1C code-level delay and radar altimeter measurements for more than eight hours.

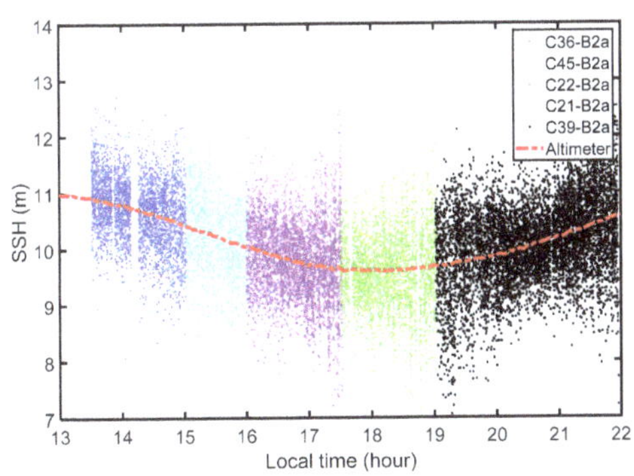


**Figure 5.** SSHs derived from B2a code-level delay and radar altimeter measurements for more than eight hours.

In order to evaluate the precision of GNSS-R altimetry based on the two kinds of new BDS civil codes, we differentiated between the solutions and radar altimeter measurements. Figures 6 and 7, respectively, show their height difference sequence with the satellite elevation angles for B1C and B2a signals. The root mean square (RMS) values of the two sequences are 0.394 m and 0.668 m for B1C and B2a, which are better than the solutions derived from GPS C/A and BDS B1I code [20]. It is worth noting that the divergence is a minimum of between 60° and 70° in both the cases. This is because the signals with elevation angles around 60° have small incidence angles for both upward- and downward-looking antennas, thanks to their 30° tilted angles. The gain of antennas is maximal in these directions, indicating that higher gain of antenna will help improve the precision of the solutions. On the other hand, both the direct and reflected signals with higher elevations have higher power, so that the divergence is a minimum of around 65° instead of 60°.

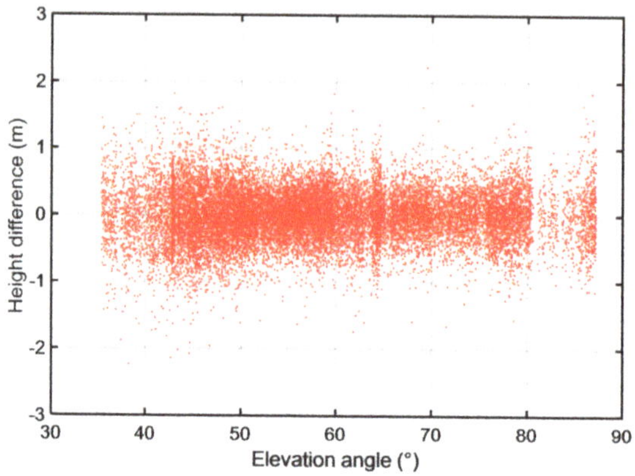


**Figure 6.** Differences between measured SSHs using monostatic radar and B1C signals at different elevations.

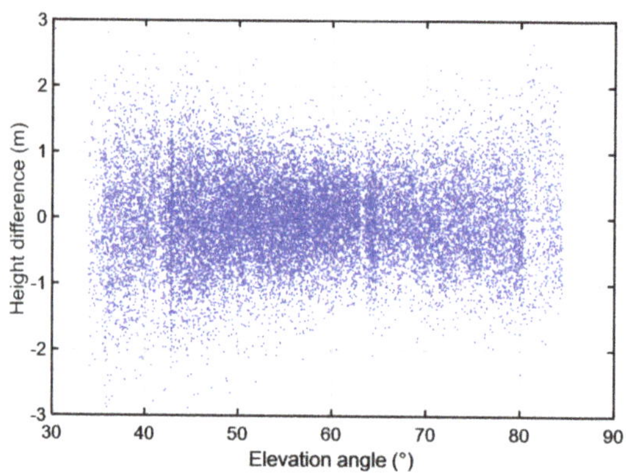


**Figure 7.** Differences between measured SSHs using monostatic radar and B2a signals at different elevations.

The above analysis shows that the precision of GNSS-R code-delay altimetry achieved from B1C is better than that from B2a. In this study, the coherent time is 10 milliseconds for both signals. The code rate of B2a signals is 10 times that of B1C, but the complicated code construction of B1C produces its wider bandwidth compared to B2a. In addition, the results of a positioning experiment using BDS-3 signals showed that B2a signals have relatively poor quality, although they have stronger power than the other open available ranging code [21]. Affected by the above factors, GNSS-R code-delay altimetry based on B2a signals from our experiment has worse precision than that based on B1C signals.

Since our altimetry solutions are derived from the differential measurement of the direct and reflected code ranges, we investigated the cross-correlated waveforms of the two new BDS-3 civil signals for further exploration. Figures 8 and 9, respectively, show the waveforms of B1C and B2a codes for direct and reflected signals. During the experiment, the sea surface had no appreciable roughness and the reflector heights ranged from three to five meters, so that the path delays can be calculated from the peak positions of the waveforms [6]. From Figures 8 and 9, the direct and reflected B2a waveforms are about half of the B1C ones. This may be caused by its poor signal quality and narrower bandwidth.

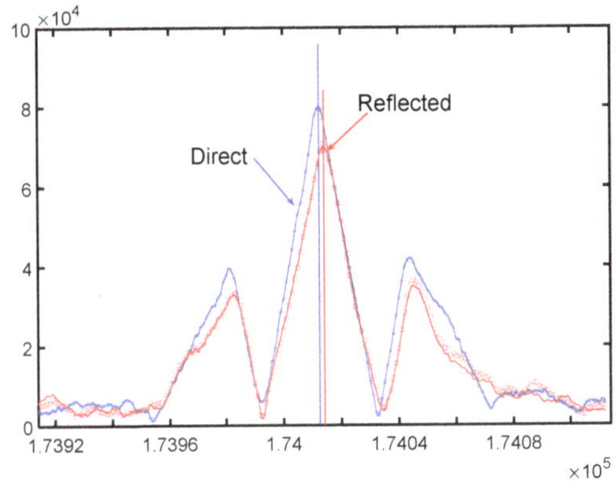


**Figure 8.** Cross-correlated waveforms of B1C code for direct (blue) and reflected (red) signals.

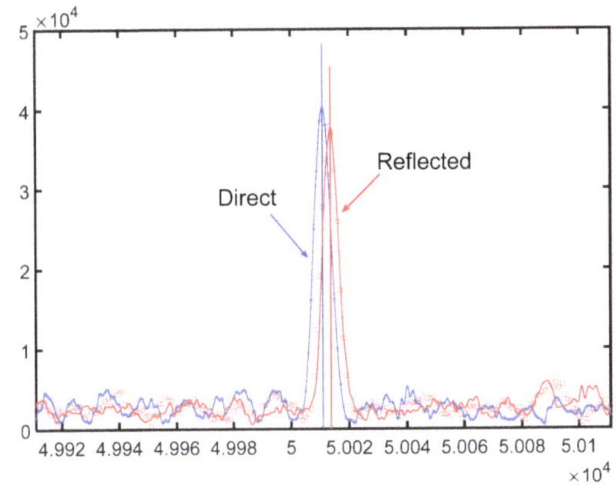


**Figure 9.** Cross-correlated waveforms of B2a code for direct (blue) and reflected (red) signals.

Centimeter-level SSH measurements are widely required for many geoscience applications. Obviously, the original solutions derived from B1C and B2a cannot directly satisfy this requirement. However, as BDS-3 has completed its full operations, an adequate number of satellites could be observed for GNSS-R altimetry during our experiment. Their SSH measurements could be obtained continuously and so the change of actual SSH was a steady dynamic process. In this paper, moving averages with windows of one minute and five minutes were applied to smoothing solutions derived from B1C and B2a signals. In Figures 10 and 11, the red points stand for the SSH obtained from radar altimeter; the blue ones stand for those after applying a one-minute moving average; green ones stand for those after applying a five-minute moving average. The results indicated that the precision improved a lot in both cases.

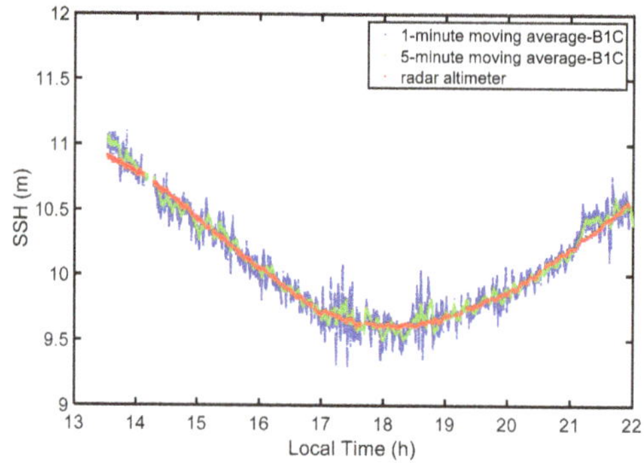


**Figure 10.** SSHs derived from B1C code-level delay measurements with moving average and radar altimeter for more than eight hours.

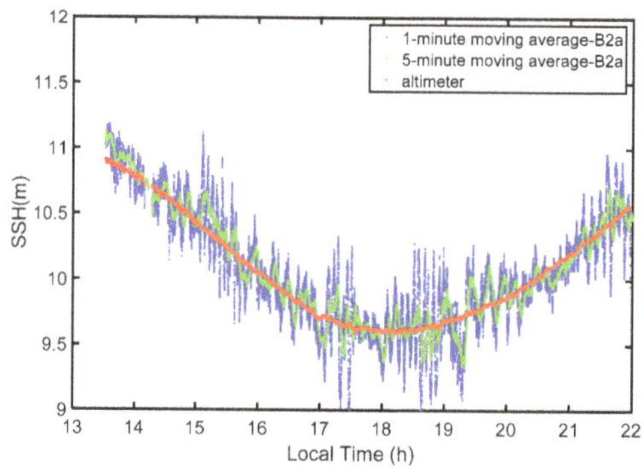


**Figure 11.** SSHs derived from B2a code-level delay measurements with moving average and radar altimeter for more than eight hours.

In order to evaluate the performance of the filters, we differentiated between the solutions and measurements of the radar altimeter. Figures 12 and 13 show the residuals. The RMSs of B1C-based SSH are 0.090 m and 0.053 m for one-minute and five-minute moving averages while those for B2a case are 0.199 m and 0.111 m. The final results show that centimeter-level SSH can be achieved using the B1C signal, while the precision for the B2a case can only reach the decimeter level.

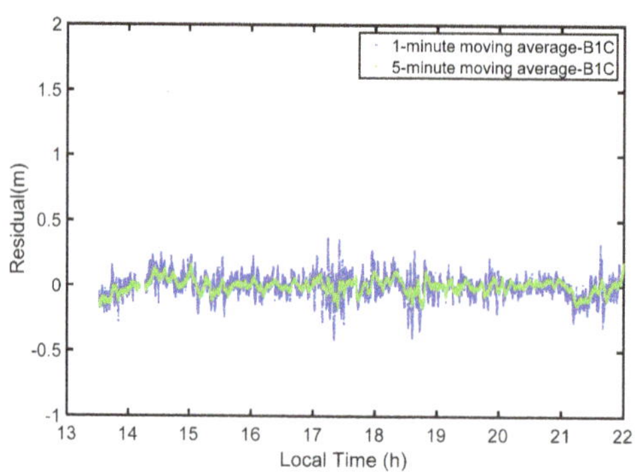


**Figure 12.** Differences between measured SSHs using monostatic radar and smoothed solutions from B1C signals.

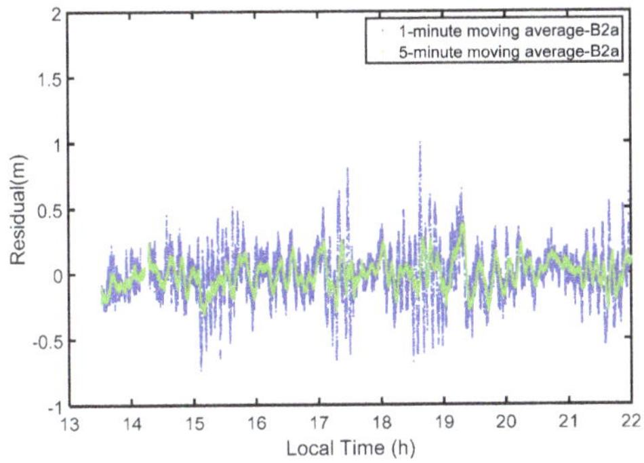


**Figure 13.** Differences between measured SSHs using monostatic radar and smoothed solutions from B2a signals.

## 5. Discussion

GNSS-R is a promising and low-cost technique for ocean altimetry on different platforms. New GNSS signals, designed with better performance, bring opportunities for improving the accuracy and precision. We tested the performance of GNSS-R code-level altimetry based on new BDS-3 civil codes by conducting a coastal experiment for the first time. The solutions derived from B1C and B2a signals are achieved at one second intervals for a period of about eight hours. The final results show that the centimeter-level precision of GNSS-R altimetry based on B1C codes can be achieved; it is similar to that of the tide gauge.

Our results demonstrated that the precision of solutions from the two new civilcodesare higher than those from conventional GPS C/A and BDS B1I. Furthermore, the solutions from B1C are better than those from B2a. The poor performance of the single-frequency band B2a was attributed to its poor signal quality and narrower bandwidth. In addition, we also found that the precision of the solutions can be affected by signal power.

One of a main feature of BDS-3 is its hybrid constellation, in which the GEO satellites can provide stable geometries for GNSS-R observations. However, in this paper, no solution was retrieved from the signals of the BDS-3 GEO satellite. They provide fundamental PNT service on the legacy B1I and B3I signals, while new B1C and B2a signals are used for providing SABS service. Unfortunately, their SABS services are still in testing. As the GNSS-R code-level altimetry performance of BDS-3 B1C and B2a signals is studied, we could not test BDS-3 GEO signals during this experiment.

In this work, the performance of GNSS-R altimetry based on the B1C and B2a signals was only tested on a very low platform when the sea surface was in a good condition. We plan to conduct experiments on higher platforms, such as using an unmanned aerial vehicle and/or plane to investigate their characteristics. We should find a proper and safe place to conduct experiments for different sea states.

The monostatic radar altimeter can accurately measure the vertical distance from its phase center to the sea surface, which enabled us to obtain the precise reflector height. In this work, we solved a bias for each satellite using the precise reflector height values. This is because the biases are caused not only by the sea surface roughness and the electromagnetic characters, but are also affected by the signal bandwidth and instrumental reasons [22–24]. So, further explorations for the biases require much more data over a long time. However, as our SDR runs very slowly and the IF data are too large to be stored, we cannot process and analyse long-time data in this work.

**Author Contributions:** Conceptualization, F.G., T.X. and N.W.; methodology, F.G., X.M.; software, F.G., X.M. and Y.H.; validation, N.W., X.M. and Y.H. and B.N.; formal analysis, F.G.; investigation, F.G., X.M. and Y.H. and B.N.; resources, F.G., X.M. and Y.H. and B.N.; data curation, X.M. and Y.H. and B.N.; writing—original draft preparation, F.G.; writing—review and editing, F.G., T.X. and N.W.;

visualization, F.G.; supervision, T.X.; project administration, T.X.; funding acquisition, F.G., T.X. and N.W. All authors have read and agreed to the published version of the manuscript.

**Funding:** This research was jointly funded by the National Key Research and Development Program of China (2016YFB0501701) and the Program of the National Natural Science Foundation of China (41604003, 41704017, 41704018).

**Data Availability Statement:** The datasets analyzed in this study are managed by Institute of Space Science, Shandong University and can be made available by the corresponding author on request.

**Acknowledgments:** The authors thank the staff of Weihai Golden Bay Hotel who kindly provided help during the experiment.

**Conflicts of Interest:** The authors declare no conflict of interest.

## References

1. Ablain, M.; Cazenave, A.; Larnicol, G.; Balmaseda, M.; Cipollini, P.; Faugère, Y.; Fernandes, M.J.; Henry, O.; Johannessen, J.A.; Knudsen, P. Improved sea level record over the satellite altimetry era (1993–2010) from the Climate Change Initiative project. *Ocean Sci. Discuss.* **2015**, *11*, 2029–2071. [CrossRef]
2. Xu, Q.; Tu, K.; Cheng, Y.; Wang, W.; Jia, Y.; Ye, X. Satellite Altimetry and Tide Gauge Observed Teleconnections between Long-Term Sea Level Variability in the U.S. East Coast and the North Atlantic Ocean. *Remote Sens.* **2019**, *11*, 2816. [CrossRef]
3. Martín-Neira, M. A Passive reflectometry and interferometry system (PARIS): Application to ocean altimetry. *ESA J.* **1993**, *17*, 331–335.
4. Pascual, D.; Camps, A.; Martin, F.; Park, H.; Arroyo, A.A.; Onrubia, R. Precision bounds in GNSS-R ocean altimetry. *IEEE J. Sel. Top. Appl. Earth Observ. Remote Sens.* **2014**, *7*, 1416–1423. [CrossRef]
5. Martin-Neira, M.; Caparrini, M.; Font-Rossello, J.; Lannelongue, S.; Vallmitjana, C.S. The PARIS concept: An experimental demonstration of sea surface altimetry using GPS reflected signals. *IEEE Trans. Geosci. Remote Sens.* **2001**, *39*, 142–150. [CrossRef]
6. Rius, A.; Nogués-Correig, O.; Ribó, S.; Cardellach, E.; Oliveras, S.; Valencia, E.; Park, H.; Tarongí, J.M.; Camps, A.; van der Marel, H.; et al. Altimetry with GNSS-R interferometry: First proof of concept experiment. *GPS Solut.* **2012**, *16*, 231–241. [CrossRef]
7. Zhang, Y.; Tian, L.; Meng, W.; Gu, Q.; Han, Y.; Hong, Z. Feasibility of code-level altimetry using coastal BeiDou reflection (BeiDou-R) setups. *IEEE J. Sel. Top. Appl. Earth Observ. Remote Sens.* **2015**, *8*, 4130–4140. [CrossRef]
8. Lowe, S.T.; Zuffada, C.; Labrecque, J.L.; Lough, M.; Young, L.E. An ocean-altimetry measurement using reflected GPS signals observed from a low-altitude aircraft. In Proceedings of the IEEE 2000 International Geoscience and Remote Sensing Symposium, Honolulu, HI, USA, 24–28 July 2000.
9. Lowe, S.T.; Zuffada, C.; Chao, Y.; Kroger, P.; Young, L.E.; Labrecque, J.L. 5-cm-precision aircraft ocean altimetry using GPS reflections. *Geophys. Res. Lett.* **2002**, *29*, 1375. [CrossRef]
10. Cardellach, E.; Rius, A.; Martin-Neira, M.; Fabra, F.; Nogues-Correig, O.; Ribo, S.; Kainulainen, J.; Camps, A.; D"Addio, S. Consolidating the precision of interferometric GNSS-R ocean altimetry using airborne experimental data. IEEE Trans. *Geosci. Remote Sens.* **2014**, *52*, 4992–5004. [CrossRef]
11. Li, W.; Rius, A.; Fabra, F.; Cardellach, E.; Martin-Neira, M. Revisiting the GNSS-R Waveform Statistics and Its Impact on Altimetric Retrievals. *IEEE Trans. Geosci. Remote Sens.* **2018**, *56*, 2854–2871. [CrossRef]
12. Clarizia, M.P.; Ruf, C.; Cipollini, P.; Zuffada, C. First spaceborne observation of sea surface height using GPS-reflectometry. *Geophys. Res. Lett.* **2016**, *43*, 767–774. [CrossRef]
13. Li, W.; Cardellach, E.; Fabra, F.; Ribó, S.; Rius, A. Assessment of Spaceborne GNSS-R Ocean Altimetry Performance Using CYGNSS Mission Raw Data. *IEEE Trans. Geosci. Remote Sens.* **2020**, *58*, 238–250. [CrossRef]
14. Li, W.; Cardellach, E.; Serni, R.; Rius, A.; Zhou, B. First spaceborne demonstration of BeiDou-3 signals for GNSS reflectometry from CYGNSS constellation. *Chin. J. Aeronaut.* **2021**, (in press). [CrossRef]
15. Li, W.; Cardellach, E.; Fabra, F.; Ribó, S.; Rius, A. Lake Level and Surface Topography Measured with Spaceborne GNSS-Reflectometry from CYGNSS Mission: Example for the Lake Qinghai. *Geophys. Res. Lett.* **2018**, *45*, 13,332–13,341. [CrossRef]
16. Yang, Y.; Mao, Y.; Sun, B. Basic performance and future developments of BeiDou global navigation satellite system. *Satell. Navig.* **2020**, *1*, 1–8. [CrossRef]
17. BeiDou Navigation Satellite System Signal in Space Interface Control Document Open Service Signal B2a (Version 1.0). Available online: http://www.beidou.gov.cn/xt/gfxz/201712/P020171226742357364174.pdf (accessed on 27 December 2017).
18. BeiDou Navigation Satellite System Signal in Space Interface Control Document Open Service Signal B1C (Version 1.0). Available online: https://www.beidou.gov.cn/xt/gfxz/201712/P020171226741342013031.pdf (accessed on 24 October 2019).
19. Li, Y.; Shivaramaiah, N.C.; Akos, D.M. Design and implementation of an open-source BDS-3 B1C/B2a SDR receiver. *GPS Solut.* **2019**, *23*, 60. [CrossRef]
20. Gao, F.; Xu, T.; Wang, N.; He, Y.; Luo, X. A shipborne experiment using a dual-antenna reflectometry system for GPS/BDS code delay measurements. *J. Geod.* **2020**, *94*, 88. [CrossRef]
21. Yuan, Y.; Mi, X.; Zhang, B. Initial assessment of single and dual-frequency BDS-3 RTK positioning. *Satell. Navig.* **2020**, *1*, 1–7. [CrossRef]

22. Rius, A.; Cardellach, E.; Martin-Neira, M. Altimetric Analysis of the Sea-Surface GPS-Reflected Signals. *IEEE Trans. Geosci. Remote Sens.* **2010**, *48*, 2119–2127. [CrossRef]
23. Zavorotny, V.U.; Gleason, S.; Cardellach, E.; Camps, A. Tutorial on Remote Sensing Using GNSS Bistatic Radar of Opportunity. *IEEE Trans. Geosci. Remote Sens. Mag.* **2014**, *2*, 8–45. [CrossRef]
24. Ghavidel, A.; Schiavulli, D.; Camps, A. Numerical Computation of the Electromagnetic Bias in GNSS-R Altimetry. *IEEE Trans. Geosci. Remote Sens.* **2016**, *54*, 489–498. [CrossRef]

remote sensing

MDPI

*Technical Note*

# A Novel Dual-Branch Neural Network Model for Flood Monitoring in South Asia Based on CYGNSS Data

**Dongmei Song [1,2], Qiqi Zhang [1,*], Bin Wang [1], Cong Yin [3] and Junming Xia [3]**

1 College of Oceanography and Space Informatics, China University of Petroleum (East China), Qingdao 266580, China
2 Laboratory for Marine Mineral Resources, Qingdao National Laboratory for Marine Science and Technology, Qingdao 266071, China
3 National Space Science Center, Chinese Academy of Sciences, Beijing 100190, China
* Correspondence: s20160052@s.upc.edu.cn

**Abstract:** Microwave remote sensing is widely applied in flood monitoring due to its independence from severe weather conditions, which usually restrict the usage of optical sensors. However, it is challenging to track the variation process of flood events in a timely manner by traditional active and passive microwave techniques, since they cannot simultaneously provide measurements with high spatial and temporal resolution. The emerging Global Navigation Satellite System Reflectometry (GNSS-R) technique with high spatio-temporal resolution offers a new solution to the dynamic monitoring of flood inundation. Considering the high sensitivity of GNSS-R signals to flooding, this paper proposes a dual-branch neural network (DBNN) with a convolution neural network (CNN) and a back propagation (BP) neural network for flood monitoring. The CNN module is used to automatically extract the abstract features from delay-Doppler maps (DDMs), while the BP module is fed with GNSS-R typical features, such as surface reflectivity and power ratio, as well as vegetation information from Soil Moisture Active Passive satellite (SMAP) data. In the experiments, the superiority of the DBNN method is firstly demonstrated by comparing it with the surface reflectivity and power ratio methods. Then, the spatio-temporal variation process of the 2020 South Asian flood events is analyzed by the proposed method based on Cyclone Global Navigation Satellite System (CYGNSS) data. The understanding of flood change processes could help enhance the capacity for resisting flood disasters.

**Keywords:** GNSS-R; CYGNSS; SMAP; flood monitoring

**Citation:** Song, D.; Zhang, Q.; Wang, B.; Yin, C.; Xia, J. A Novel Dual-Branch Neural Network Model for Flood Monitoring in South Asia Based on CYGNSS Data. *Remote Sens.* **2022**, *14*, 5129. https://doi.org/10.3390/rs14205129

Academic Editors: Dallas Masters, Chun-Liang Lin and Hugo Carreno-Luengo

Received: 13 August 2022
Accepted: 10 October 2022
Published: 14 October 2022



## 1. Introduction

The arrival of South Asia's annual southwest monsoons usually brings continuous heavy rainfall, leading to significant flood events and other natural disasters in parts of India, Nepal, and Bangladesh. The particularly noticeable flood inundation event in 2020 was South Asia's most significant flood disaster over the past decade, causing huge property losses to local people [1–3]. Due to the highly dynamic nature of floods, rapid and effective flood monitoring is important for early disaster prevention, midterm relief, and post-disaster reconstruction [4].

The existing remote sensing means for flood monitoring mainly include optical and microwave remote sensing. However, optical remote sensing means cannot be used on rainy and cloudy days due to the sensors' inherent characteristics, although they are capable of obtaining Earth surface observations with a satisfactory spatial resolution [4–7]. Conversely, passive microwave sensors, such as radiometers, have the ability to penetrate clouds and heavy fog owing to their long wavelength, which is requisite for flood monitoring as floods often occur during the rainy season. Nevertheless, the low spatial resolution of even dozens of kilometers limits the successful applications of passive microwave sensors for flood monitoring. On the other hand, active microwave remote sensing normally has a higher spatial resolution than passive microwave remote sensing. However, it also has a relatively lower temporal resolution and is still unable to conduct dynamic flood monitoring in a timely manner [8]. The recently developed Global Navigation Satellite System Reflectometry (GNSS-R) is a novel remote sensing technology for physical parameter inversion by means of GNSS signals reflected from the Earth's surface [9–11]. The Cyclone Global Navigation Satellite System (CYGNSS), launched by NASA, provides openly accessed GNSS-R data, which has been successfully employed in the inversion of sea surface wind speed [12–14], soil moisture estimation [15–17], flood dynamics monitoring [7,18,19], and other features. The CYGNSS constellation constitutes eight small satellites, on which the receivers are mounted to capture the direct and reflected signals from the navigation satellites. The average revisit period of CYGNSS is only 7 hours, and the spatial resolution is about 3.5 km $\times$ 0.5 km on the land surface. Compared with other microwave remote sensing technologies, CYGNSS simultaneously provides observations with higher spatio-temporal resolution, which can be more suitable for dynamic flood monitoring [18,20,21].

Research on GNSS-R flood monitoring first began in 2018. Chew produced a flood inundation map using surface reflectivity (SR) on specular points [18]. Based on CYGNSS data, Wei Wan and Wentao Yang also conducted flood monitoring by surface reflectivity in 2019 and 2021, respectively [7,19]. Furthermore, Unnithan produced large-scale, high-resolution flood inundation maps in 2020 by combining the feature of signal-to-noise ratio in delay-Doppler maps (DDMs) with the topographic information [22]. Through further research, Chew proposed a theoretical model for flood monitoring based on changes in surface reflectivity in different land cover types in 2020. The research results showed that surface reflectivity was mainly dependent on surface roughness. When flood events occurred, the surface reflectivity in densely vegetated areas greatly varied, while that of the relatively smooth surfaces changed little, both before and after the flood [20]. Later, Al-Khaldi proposed the power ratio (PR) method in 2021 to detect water bodies using the coherent properties of DDMs from CYGNSS, and found that over 90% of the land surface reflections presented incoherent scattering, while about 80% of the coherent reflections were related to water bodies [23].

Reviewing the current GNSS-R flood monitoring methods, it is found that most of them are only based on a specific GNSS-R physical feature, such as SR, PR, or signal-to-noise ratio. However, one single GNSS-R feature cannot simultaneously represent the dielectric constant and roughness of the reflective surface. For example, SR is calculated from the power of the specular point [7], so it mainly represents the dielectric property of the specular point rather than the roughness property of the reflective surface, which is essential for the flood inversion [24]. Furthermore, the impact of vegetation on GNSS (direct and reflected) signals has not yet been considered in the available literature [24,25]. As the primary observation data of GNSS-R, DDMs contain much useful detailed information, such as SR and PR. Some scholars have utilized DDMs-based features for flood monitoring, such as PR and signal-to-noise ratio. However, the valuable information in DDMs has not yet been fully excavated. In recent years, deep learning (DL) has been widely employed to automatically learn feature representations from data and establish the intrinsic relationship between inputs and outputs [26]. Among a variety of DL algorithms, convolution neural network (CNN) has surpassed most other DL algorithms in two-dimensional image processing due to its local connectivity, weight sharing, and down-sampling strategies, which can reduce the complexity of neural networks and successfully learn feature representations of images [26]. Moreover, the back propagation (BP) neural networks have powerful nonlinear mapping ability, which is especially suitable for solving the complicated internal mapping problem between one-dimensional input vectors and outputs [27]. Therefore, by combining a CNN and a BP neural network in parallel, a dual-branch neural network (DBNN) is constructed for better flood monitoring. In the model, the CNN takes two-dimensional DDMs as input and automatically extracts the deep abstract features in DDMs. The BP neural network is fed with the existing typical GNSS-R features and the vegetation information provided by the Soil Moisture Active Passive satellite (SMAP) [28].

The rest of this paper is organized as follows. Section 2 provides the descriptions of CYGNSS data, SMAP data, and the study area. Section 3 introduces the proposed method in detail. The experimental results and discussion are given in Section 4. Finally, Section 5 concludes the study.

## 2. Data and Study Area

### 2.1. CYGNSS Data

The GNSS-R data in this study were produced from the CYGNSS constellation, a bi-static radar system (shown in Figure 1). This constellation is capable of collecting the near-global (between 38°N and 38°S latitudes) daily reflected L-1 coarse acquisition GPS signals in the form of DDMs [29–32]. CYGNSS data have been widely employed in GNSS-R scientific research and other practical applications [30]. The mean and median revisit periods of the CYGNSS were 7 h and 3 h, respectively. The spatial resolution is related to surface roughness, which is around 3.5 km × 0.5 km on the land surface and about 25 km × 25 km on the rough sea surface [23]. Compared with conventional microwave remote sensing, the CYGNSS has higher spatial and temporal resolution in observing the Earth's surface, which is conducive to in-depth scientific research by GNSS-R techniques [21,24,33].

The data employed in this study is the 3.0 version CYGNSS product at L1 level, which primarily contains DDMs and a range of metadata describing the geometry and instrument parameters in acquisition. CYGNSS data can be acquired in NetCDF format from https://podaac.jpl.nasa.gov (accessed on 10 September 2021). The primary variables utilized are summarized in Table 1; including DDMs, the transmitter/receiver distance to the specular point, transmitter and receiver, antenna gain, etc.

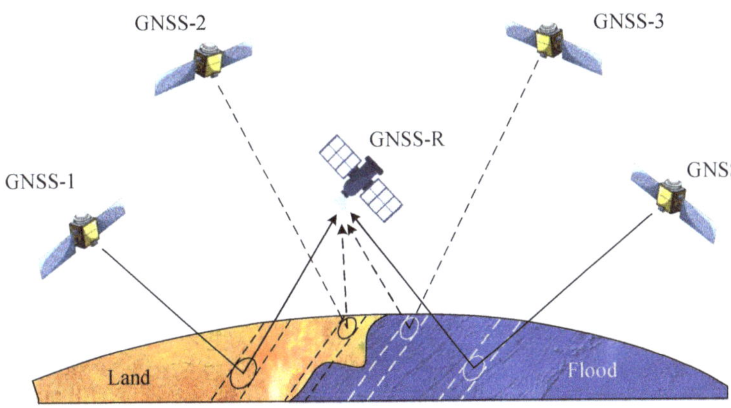


**Figure 1.** Schematic of the GNSS-R technique. GNSS satellites transmit signals to the Earth's surface. The signal is reflected by the Earth's surface and captured by GNSS-R receivers onboard low Earth-orbiting satellites. The specific locations of the specular points depend on the geometric positions of the transmitting and receiving satellites.

**Table 1.** Main parameters of CYGNSS satellite data.

| Parameters | Units | Comment |
|---|---|---|
| raw_counts | watt | DDM bin raw counts |
| gps_tx_power_db_w | dBi | GPS Space Vehicle transmit power |
| rx_to_sp_range | m | Receiver to specular point range |
| tx_to_sp_range | m | Transmitter to specular point range |
| gps_ant_gain_db_i | dBi | GPS Space Vehicle transmit antenna gain |
| sp_rx_gain | dBi | Specular point receiver antenna gain |
| sp_lat | degree | Specular point latitude |
| sp_lon | degree | Specular point longitude |
| sp_inc_angle | degree | Specular point incident Angle |

### 2.2. SMAP Data

The SMAP mission, launched by NASA in January 2015, has been designed to collect L-band signals which are sensitive to land surface soil moisture (~5 cm of depth). The SMAP satellite is equipped with active radar and passive radiometer sensors. However, the radar failed to work in orbit after two months. Hence, the satellite currently relies only on the L-band radiometer to retrieve land surface information, such as soil moisture and vegetation information, by measuring brightness temperature. Since the signals received by SMAP satellite and GNSS signals are both L-band with similar frequencies and sensitivity to the land surface, SMAP data can be employed for comparison analysis with CYGNSS data. The SMAP satellite observes land area between 85°S and 85°N, with a revisit frequency of 2–3 days [28].

The data product utilized in this study is the SMAP Enhanced L3 Radiometer Global Daily 9 km EASE-Grid Soil Moisture, available at https://nsidc.org/data/SPL3SMP_E/versions/4 (accessed on 16 October 2021) [34–36]. The enhanced SMAP product is an interpolated and gridded result of SMAP Radiometer measurements, which are posted to the 9 km Equal-Area Scalable Earth Grid, Version 2.0 (EASE-Grid 2.0). The primary data are summarized in Table 2. It is worth noting that two datasets from SMAP, soil moisture and vegetation opacity, were employed in this study. Soil moisture is used to verify the accuracy of the inversion results, while vegetation opacity information is fed as auxiliary input data to the model for flood inversion.

**Table 2.** Main parameters of SMAP mission data.

| Parameters | Units | Comment |
|---|---|---|
| latitude | degrees_north | Latitude of the center of the Earth based grid cell. |
| longitude | degrees_east | Longitude of the center of the Earth based grid cell. |
| soil_moisture | $cm^3/cm^3$ | Representative soil moisture measurement for the Earth based grid cell. |
| surface_temperature | Kelvins | Temperature at land surface based on GMAO GEOS-5 data |
| vegetation_opacity | unitless | The measured opacity of the vegetation used in retrievals in the grid cell. |
| roughness_coefficient | unitless | A unitless value that is indicative of bare soil roughness used in retrievals within the 9 km grid cell. |

### 2.3. Study Area

South Asia is an ideal area for flood inversion research using GNSS-R technology, due to its frequent large-scale flooding inundation. As shown in Figure 2, the subgraph (a) displays the coverage area of one CYGNSS satellite in a day, and the subgraph (b) represents the intercepted area of South Asia ranging from 4°N–28°N and 66°E–96°E. The northern part of the region is the Himalayan mountains, with an average elevation of more than 6000 meters; the central part is the Great Plain, with dense river networks and numerous irrigation canals; and the southern part is the Deccan Plateau. The total land area of the study area is approximately 3.24 million square kilometers. In addition, the region is situated in a tropical monsoon climate, with the hot season usually from March to May, the rainy season usually from June to October, and the cool season usually from November to the following February. During the rainy season, the southwest monsoons carry large amounts of water vapor as they pass over the warm ocean, leading to extremely abundant rainfall in South Asia. In 2020, the rainy season in South Asia lasted from June to September, causing the region to experience the most severe flooding of the past decade. Therefore, CYGNSS data ranging from May to October 2020 are selected to conduct the flood inversion, as well as the spatio-temporal dynamic analysis by the proposed DBNN model.

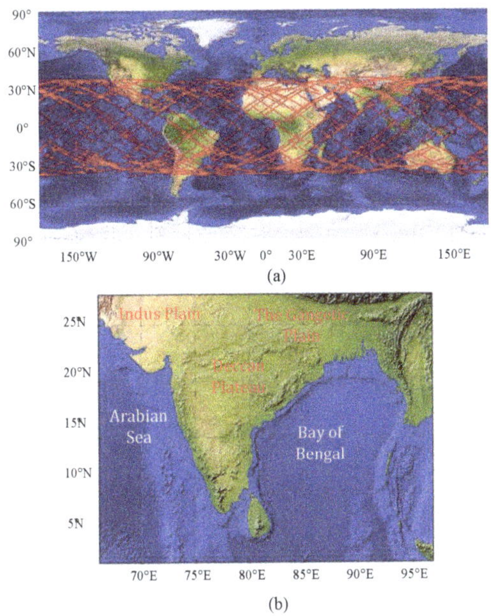


**Figure 2.** (**a**) The coverage area of one CYGNSS satellite in a day. (**b**) The study area in South Asia.

## 3. Method

In this study, a DBNN model for flood monitoring is proposed, which is mainly composed of two parallel subnetworks: the CNN and BP neural network. In this model, two-dimensional DDMs are input into the CNN, which automatically extracts the abstract features from images, while the BP neural network is fed with seven typical GNSS-R features, including surface reflectivity [18], power ratio [23], and the leading edge of slope [37], etc., as well as vegetation information from SMAP data [28]. The output results of the model are the probability values that DDMs belong to the submerged region. The proposed method consists of three steps, as follows: (1) data pre-processing and features extraction; (2) construction and training of DBNN model; and (3) prediction of DBNN model. The process of this method is shown in Figure 3.

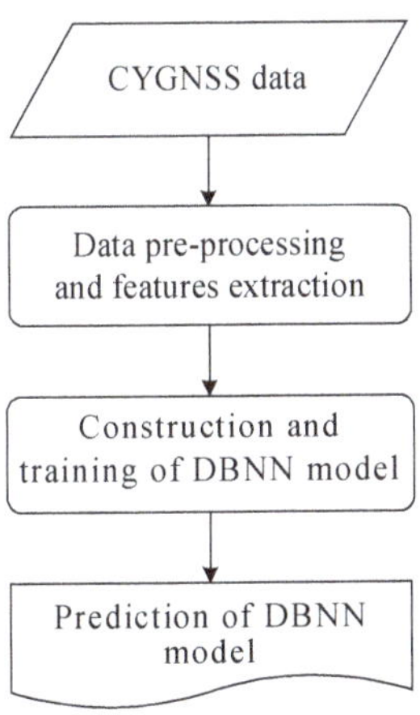


**Figure 3.** Flow chart of the inversion method.

### 3.1. Data Pre-Processing and Features Extraction

In order to obtain good inversion results, only CYGNSS data meeting the following conditions are employed in this study:

(1) If the SNR of DDM is too small, there are fewer signals and more noise in the DDM. Therefore, the DDMs are filtered out with SNR less than 1.5 dB.

(2) The left-handed circular polarization signals received by CYGNSS decrease with the increase of the incident angle. When the incident angle is larger than 65°, the proportion of the left-handed circular polarization signals rapidly decreases; thus, DDMs with the incident angle larger than 65° are filtered out in this study [7,38].

(3) In this study, only DDMs with transmitter and receiver antenna gain greater than 0 are selected according to the references [25,38].

(4) The positions of the peak points move toward the edge of DDMs with altitude, and the DTU10 digital elevation model used by CYGNSS does not sufficiently consider the effect of land topography, which results in significant errors in the estimation of positions of specular points with elevations greater than 600 m [23]. Thus, in this study, only DDMs with peak points occurring in delay bins inside of 7–10 pixels are retained [38].

The power values of some pixels in DDMs, which cannot be mapped to the real Earth's surface, are mainly generated by thermal noise (as shown in the white area in Figure 4). Fortunately, the thermal noise pixels in DDMs can be eliminated by using the method provided by Al-Khaldi [23] after screening CYGNSS data. The DDMs before and after removing the thermal noise pixels are displayed in Figure 5.

The dielectric constant and roughness of the land surface obviously vary when floods occur, so they can be regarded as very useful physical parameters for judging whether floods have occurred or not. Therefore, seven features related to the above two parameters are extracted from DDMs in this study as the inputs of the BP neural network, which include surface reflectivity [18], power ratio [32], leading edge of slope [37], trailing edge of slope [25], peak point power, DDM average [39], and signal-to-noise ratio [12] (as shown in Table 3).

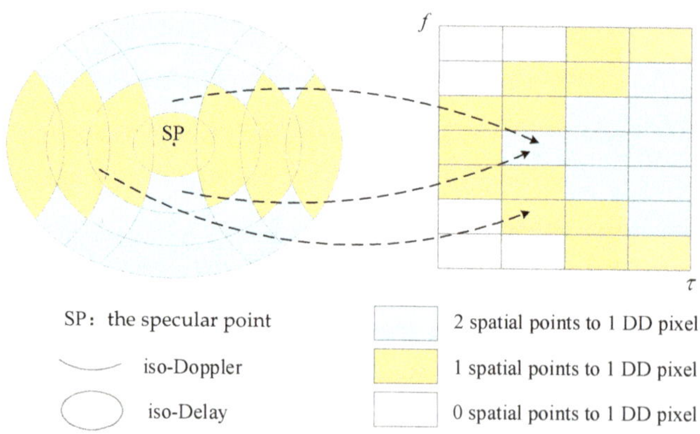


**Figure 4.** Mapping relationship between the spatial coordinate system and delay-Doppler coordinate system. The subfigure on the left depicts the spatial coordinate system. The ellipse and the curve represent a delay isoline and a Doppler shift isoline, respectively. The subfigure on the right represents the delay-Doppler coordinate system, where the blue and yellow delay Doppler pixels in DDMs correspond to one and two spatial points in the left subfigure, respectively, while white pixels represent the thermal noise pixels without corresponding spatial point. SP: the specular point.

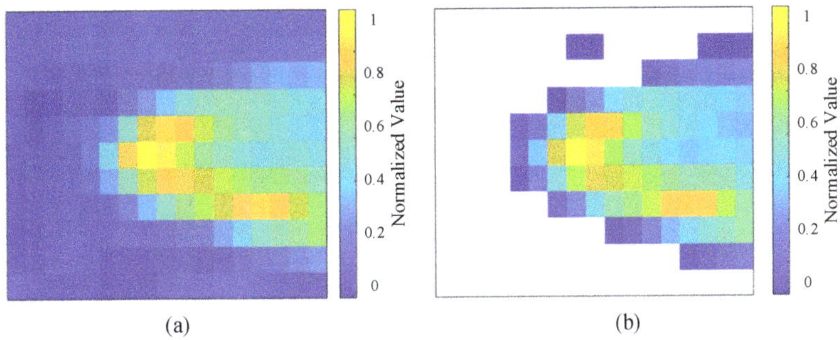


**Figure 5.** (**a**) DDM before removing the thermal noise pixels; (**b**) DDM after removing the thermal noise pixels.

**Table 3.** Features extracted from DDMs.

| Number | Features Symbol | Features Name |
|---|---|---|
| 1 | DDMA | DDM average |
| 2 | TES | Tail edge of slope |
| 3 | LES | Leading edge of slope |
| 4 | SR | Surface reflectivity |
| 5 | PR | Power ratio |
| 6 | SNR | Signal to noise ratio |
| 7 | Peak | DDM peak value |

### *3.2. Construction and Training of DBNN Model*

#### 3.2.1. Construction of DBNN Model

The DBNN model consists of two parallel sub-networks, the CNN module and the BP neural network module, which are followed by a concatenate layer, a full connection layer, and an output layer, as illustrated in Figure 6b. In this model, the CNN module consists of two convolutional layers, made up of 16 and 32 the $3 \times 3$ convolution kernels, respectively, and two pooling layers. Each convolution kernel can be regarded as a feature extractor, which convolves with the input DDMs to generate feature maps. Taking the operation of the $k$th convolution kernel in the first convolution layer as an example, the input DDM is processed by the convolution kernel to generate the feature map $h_k^{(1)}$, which can be expressed as follows:

$$h_k^{(1)} = f\left(\left(W_k^{(1)} * X\right) + b_k^{(1)}\right) \tag{1}$$

where $X$ is the input DDM, $W_k^{(1)}$ and $b_k^{(1)}$ are the weight and bias of the $k$th convolution kernel respectively, $*$ represents the convolution operation, and $f$ denotes the activation function; this study adopts the widely used ReLU function with the following equation:

$$f(z) = \max(0, z) \tag{2}$$

The max pooling layers with size $2 \times 2$ and stride 2 are applied in the CNN module to downsample the feature maps from the convolution layers, so as to reduce the redundant information and retain critical features.

In the DBNN model, the BP module consists of two fully connected layers equipped with 16 and 32 neurons, respectively. Each neuron in the module conducts the weighted summation operation on the inputs, which is subsequently processed by the ReLU activation function to create an output feature. As an example, consider the operation of the $i$th neuron in the first fully connected layer, its output feature $y_i^{(1)}$ can be expressed by the following equation:

$$y_i^{(1)} = f\left(\sum\left(x_j \times w_{ij}^{(1)}\right) + b_i^{(1)}\right) \tag{3}$$

where $x_j$ is the $j$th input feature of the neuron, $w_{ij}^{(1)}$ and $b_i^{(1)}$ are the weight and bias of this neuron, respectively. $f(\cdot)$ denotes the ReLU activation function.

The features output from the CNN module and the BP module are then transferred into the concatenation layer. After being further nonlinearly processed by the full connection layer with 64 neurons, these concatenated features are finally delivered to the output layer. The output layer contains two neurons with the softmax activation function [40–42] and outputs probabilities $p_i$ of the input DDM corresponding to the submerged region and the unsubmerged region, respectively. The probabilities $p_i$ can be expressed as follows:

$$p_i = \frac{\exp(v_i)}{\sum_{j=1}^{k}\exp(v_j)} \; where \; i = 1, 2 \tag{4}$$

where $k$ is the number of neurons in the output layer, set to 2; and $v_1$, $v_2$ are the input values of the softmax function. It should be noted that the softmax function in the neural networks outputs a set of probability values belonging to each classification category, and the summation of all probability values equals 1, where the category corresponding to the largest probability value is the attribution category of the sample. As a binary classification model, the proposed DBNN model only outputs two probability values. Therefore, if there is a probability value greater than 0.5, the corresponding category will be identified as the classification category. Since there are only two probability values output from the DBNN model and their summation is 1, the probability value belonging to the flooded region can be selected as the prediction result of the model. Therefore, samples with probability values greater than 0.5 are regarded as submerged.

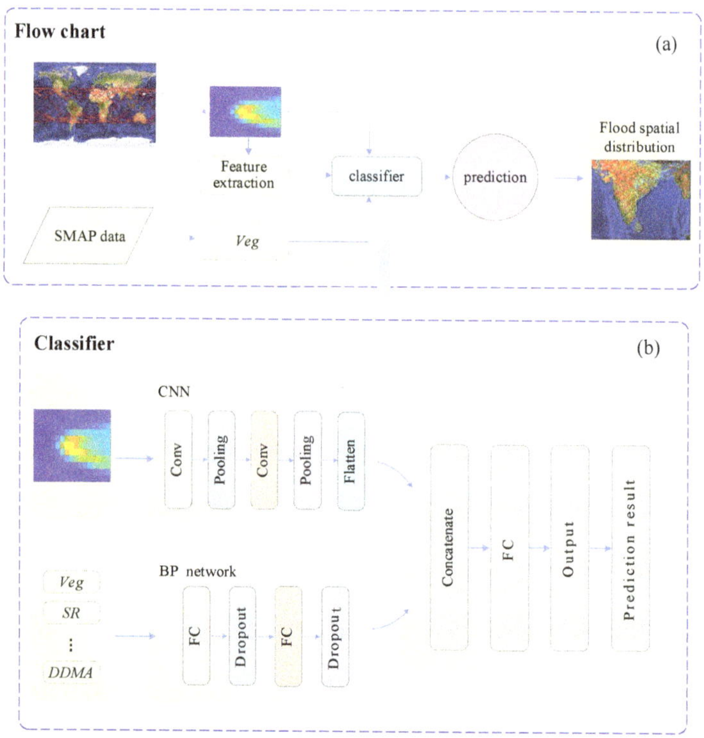


**Figure 6.** (**a**) Flooding monitoring process of DBNN method; (**b**) DBNN model structure. FC: full connection layer.

### 3.2.2. Training of DBNN Model

The dataset in this study is obtained by spatio-temporal matching of CYGNSS data and SMAP data ranging from May to September 2020, where CYGNSS data provide DDMs and seven features, as shown in Table 3, and SMAP data offer vegetation information and the target label derived from the classification results of inundated versus non-inundated areas classified by SMAP soil moisture [43]. In this study, 50,000 samples were randomly selected from the dataset as the sample set of the DBNN model, and the sample set was divided into training, validation, and testing subsets at a rate of 80%, 15%, and 5%, respectively [15]. These subsets are designed to provide sufficient data to train the network, evaluate its performance, and tune the hyper-parameters. In addition, all the remaining samples in the dataset are predicted by the DBNN model for the inversion of flood monitoring.

The information forward propagation and the error back-propagation algorithms are adopted for the training of the DBNN model. During the forward propagation, the parameters of the neural network are constant, while in the backward propagation, the parameters are automatically updated to minimize the loss function using the Adam optimizer. The loss function of the model is given in Equation (6):

$$L = \frac{1}{M}\sum_{m=1}^{M}\sum_{k=1}^{K} y_m^k \times \log(h_\theta(x_m, k)) \tag{5}$$

where $M$ is the number of training samples in each round; $K$ is the number of classes, set to 2 in this study; $y_m^k$ is the target label, which is derived from the classification results of inundated versus non-inundated areas classified by SMAP soil moisture [43], for training example $m$ for class $k$; $x$ is the input for training example $m$; $h_\theta$ is the neural network model with weights $\theta$. In each round of training, the weights $\theta$ of the DBNN model are updated according to the following formula:

$$\begin{cases} g = \nabla_\theta L \\ m = \beta_1 m + (1-\beta_1) g \\ s = \beta_2 s + (1-\beta_2) g^2 \\ \hat{m} = \frac{m}{1-\beta_1^t} \\ \hat{s} = \frac{s}{1-\beta_2^t} \\ \theta = \theta - \eta\hat{m}/\sqrt{\hat{s}+\epsilon} \end{cases} \tag{6}$$

where $\nabla_\theta$ and $g$ denote the gradient operation and the gradient value of the loss function $L$ on the parameter $\theta$; $\eta$ is the learning step, defaulted to 0.001; $m$ and $s$ denote the first- and second-order moment variables (initialized to be 0), respectively; $\beta_1$ and $\beta_2$ represent the exponential decay coefficients of $m$ and $s$; $\epsilon$ is a small constant, set to $10^{-6}$; $t$ is the number of iterations, which is set to 300 in this experiment.

Specifically, the training process of the proposed DBNN model can be described as follows:

(1) Initialize the parameters of the neural network, and input the training data into the neural network.
(2) Proceed forward propagation and compute the loss function using Formula (6).
(3) Update the parameters in the neural network by means of the back propagation and Adam optimizer.
(4) Complete the training when the variation of the loss function values is less than 0.001 among ten consecutive epochs, or the training times reach the preset number of iterations.

### 3.3. Prediction of DBNN Model

The trained DBNN model can be applied as a classifier for flood monitoring. The flow of flood monitoring is presented in Figure 6a: (1) CYGNSS data are input into the DBNN model to obtain the probability values that the CYGNSS data belong to the submerged area. Specifically, probability values greater than 0.5 are regarded as the submerged areas, while those less than 0.5 are considered as the non-submerged areas. (2) Afterwards, the geolocation of each DDM is realized by using the longitude and latitude of its specular point, and the scatter maps of the predicted results are drawn by combining the prediction results with the location of DDMs. (3) Finally, the scatter maps of the prediction results are gridded into 9 km $\times$ 9 km images.

## 4. Results and Discussion

Flood monitoring is conducive to understanding the dynamic process of flood occurrence and development, which further helps with disaster prevention, relief, and post-disaster reconstruction. In this study, the experimental district was drawn from South Asia, since it had experienced severe flood incidents in the past decade, and especially in 2020. The effectiveness of the DBNN method was first validated by comparing it with the conventional SR and PR methods, and then the spatio-temporal dynamic process of the flood inundation during the 2020 rainy season was investigated using the proposed method.

### 4.1. Effectiveness Validation

In this study, the proposed DBNN model is compared with conventional SR and PR methods, which are frequently employed for flood monitoring. The SR can be calculated following the equations in the literature [20]. Most studies on flood monitoring by SR methods use simple threshold judgments [18,19], where the average value of the surface reflectivity of permanent water bodies throughout the study area is used as the classification threshold [18–20]. In this study, 15 dB is used as the threshold of the SR method to delineate the flood inundation areas. In addition, the PR is computable according to the literature [23] and uses the constant value of 2 as the threshold to distinguish between flood or land. In this study, the CYGNSS data ranging from 1 to 15 October 2020 are used as an example to investigate the inundation extent of the study area. In Figure 7, the inversion and classification results are presented for the flood monitoring in the study area, where the inversion results refer to the continuous values output by the methods, such as the surface reflectivity output of the SR method, and the classification results refer to the extent of flooding and land divided by the thresholds in various methods. In this study, the reference for judging the flood inundation range is obtained by the SMAP soil moisture threshold method due to the lack of measured data of the surface flood extent. Areas with soil moisture greater than 0.4 $cm^3/cm^3$ are classified as the inundated areas, and areas less than 0.4 $cm^3/cm^3$ are regarded as non-inundated areas [43]. Figure 7a shows the continuous spatial distribution of soil moisture, and Figure 7b shows flood inundation extent extracted by the soil moisture threshold method.

In this study, type I error, type II error, and overall accuracy are used to evaluate the classification results, which can be calculated as follows:

$$\begin{aligned} E_1 &= \frac{a}{a+b+c} \\ E_2 &= \frac{b}{a+b+c} \\ C &= \frac{c}{a+b+c} \end{aligned} \tag{7}$$

where $E_1$ and $E_2$ denote the type I error and type II error, respectively. $C$ represents the overall accuracy, $a$ indicates the number of samples that misclassify water bodies as land, $b$ is the number of samples that misclassify land as water bodies, and $c$ is the number of correctly classified samples.

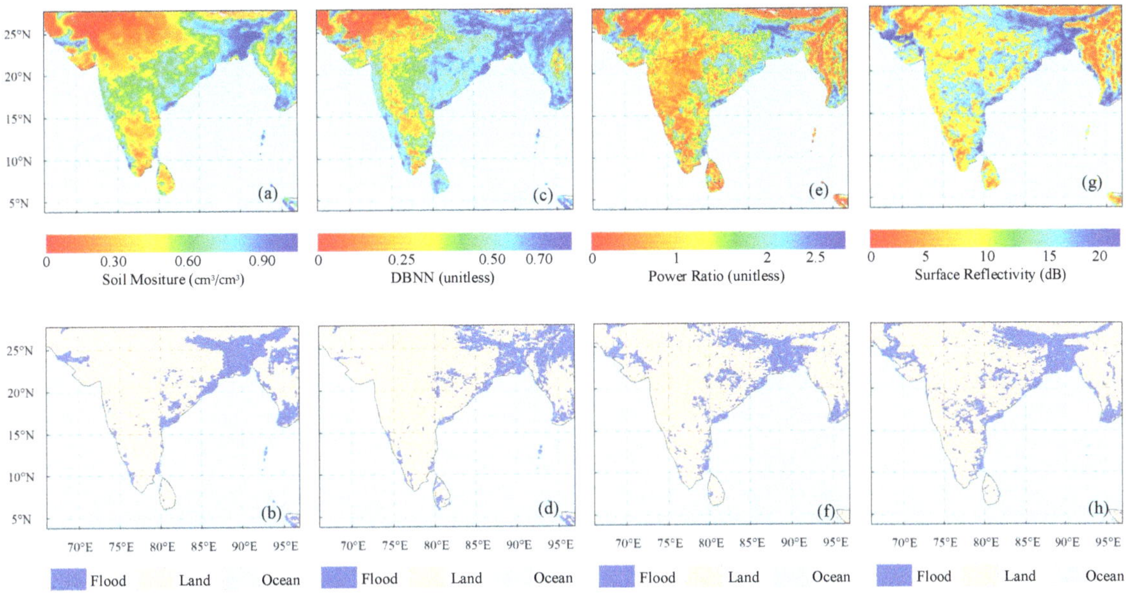


**Figure 7.** The inversion and classification results for the flood monitoring in the study area. (**a**) Soil moisture values provided by SMAP mission and (**b**) flood inundation extent classified by threshold method, respectively. The inversion results of (**c**) DBNN method, (**e**) PR method, and (**g**) SR method. The classification result of (**d**) DBNN method, (**f**) PR method, and (**h**) SR method.

According to flood inundation extent delineated by the SMAP soil moisture threshold method, the overall accuracy, type I error, and type II error of the classification result of SR, PR, and DBNN methods are shown in Figure 8 and Table 4, where the DBNN method has the highest inversion accuracy of 85.4%, and the SR and PR methods have inversion accuracies of 80.17% and 81.34%, respectively. In addition, the DBNN method is additionally superior to SR and PR methods in terms of both type I error and type II error, which can be mainly attributed to the following aspects. Firstly, the DBNN model can fully exploit the underlying abstract features from DDMs, while combining the representative features as inputs of the model, which greatly enhances the utilization of GNSS-R data. Secondly, the DBNN model considers the influence of the vegetation factor on GNSS direct signals and reflected signals. However, it is worth noting that the DBNN model is fed with vegetation information from SMAP, whereas the previous SR and PR methods operate based on the CYGNSS data only.

**Table 4.** Classification error and accuracy of PR and DBNN.

| | Type I Error (%) | Type II Error (%) | Accuracy (%) |
|---|---|---|---|
| SR | 10.76 | 9.07 | 80.17 |
| PR | 9.25 | 9.41 | 81.34 |
| DBNN | 7.29 | 7.17 | 85.54 |

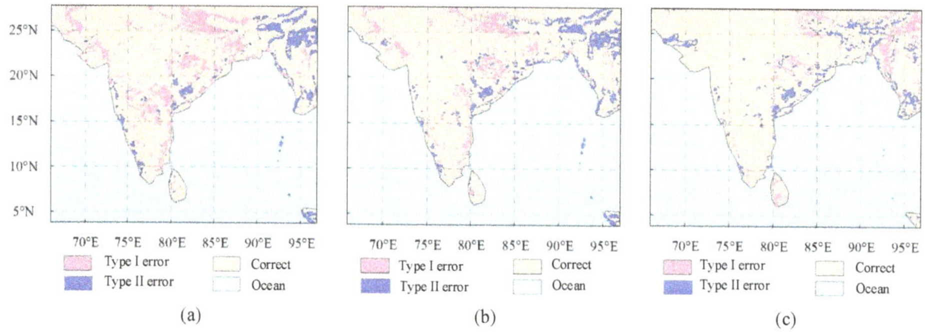


**Figure 8.** (**a**) The difference image between SR classification result and SMAP classification result. (**b**) The difference image between PR classification result and SMAP classification result. (**c**) The difference image between DBNN classification result and SMAP classification result.

*4.2. Spatio-Temporal Analysis of Flood Inundation*

In order to provide support for disaster-resistant activities in the study area, it is crucial to understand the development process of flood events. Therefore, the particularly severe flood incident occurring from May to September 2020 is selected as a flood monitoring case, and the experimental results retrieved by the DBNN model on CYGNSS data are shown in Figures 9 and 10. Before the onset of the 2020 rainy season in the study area, the regions occupied by water bodies are mainly distributed in coastal areas such as Bangladesh, covering about 6.6% of the total land area. From 20 May 2020, the areas inundated by floods gradually expanded to Nepal and the northeast of India with the arrival of frequent rains. The flooded areas accounted for about 17.9% of the total area on 20 June 2020, and reached 28.7% by 20 July 2020. On 20 August 2020, the areas flooded reached their largest, covering around 34.8% of the total area. After 20 August 2020, the flooded areas gradually reduced because of the decrease in rainfalls. Observing the process of flood changes, we could find the increased area, which refers to the difference between the maximum flood area detected during the rainy season and the area of water bodies before the rainy season, accounting for 28.2% of the total study area, mainly in Nepal and the northeastern states of India. In particular, the proportion of the increased flood area in some states is calculated and summarized in Table 5, where Bihar State experienced the most severe flooding event with an increased area accounting for 89.92% of the total area, while Magway State had the smallest increased area, only covering about 5.50%.

**Table 5.** The proportion of the increased flood area to the state area.

| State | Proportion (%) | State | Proportion (%) |
|---|---|---|---|
| West Bengal | 39.83 | Bihar | 89.92 |
| Assam | 16.35 | Narayani Zone | 37.33 |
| Rakhine State | 18.65 | Lumbini Zone | 41.79 |
| Chattogram | 15.88 | Uttar Pradesh | 76.12 |
| Karnali Zone | 25.92 | Odisha | 51.70 |
| Seti Zone | 66.67 | Ayeyarwady | 45.12 |
| Janakpur Zone | 31.84 | Yangon | 67.20 |
| Sagarmatha Zone | 28.44 | Bago | 39.74 |
| Bhojpur | 24.65 | Jharkhand | 65.25 |
| Mechi Zone | 17.12 | Madhya Pradesh | 17.42 |
| Sagaing | 12.26 | Chhattisgarh | 53.82 |
| Rangpur | 19.57 | Magway | 5.50 |

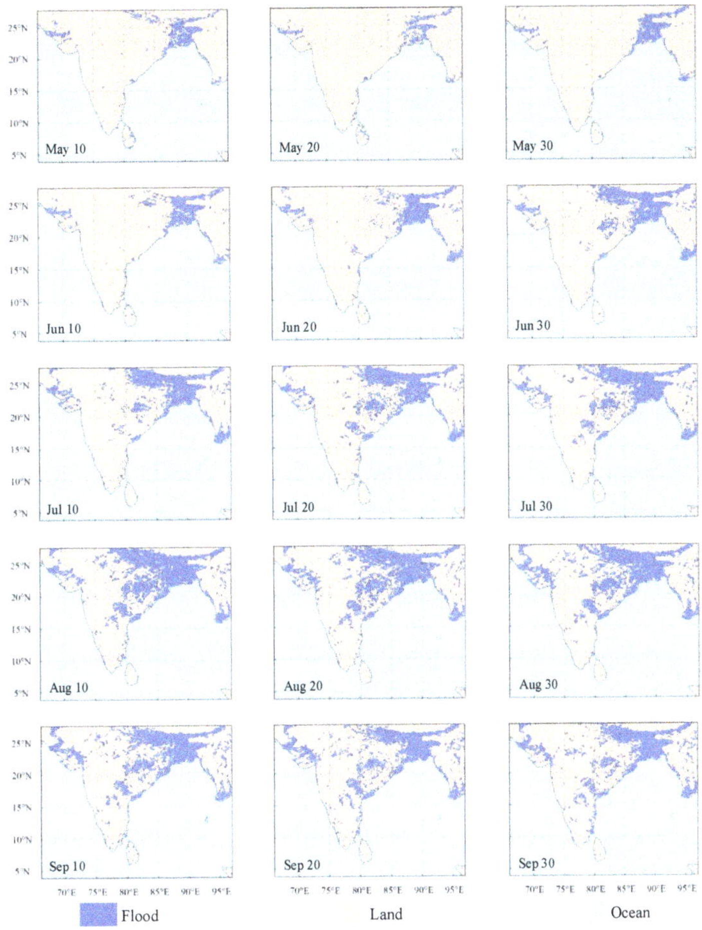


**Figure 9.** The spatial distribution of the study area flood inundation retrieved by DBNN model on CYGNSS. The period spans from May to September 2020.

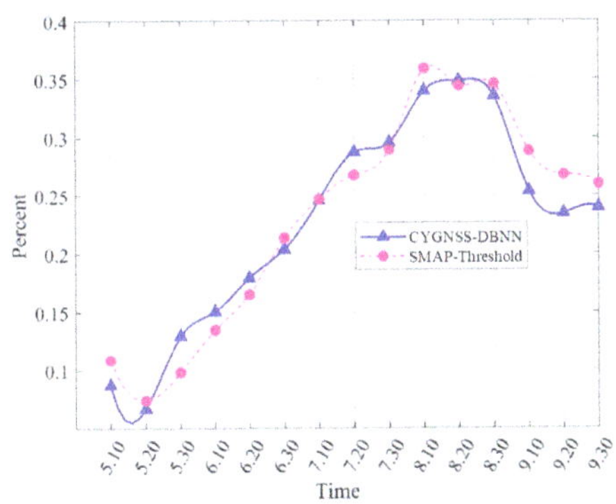


**Figure 10.** The temporal variation process of the flooded area, spanning from May to September 2020. "CYGNSS-DBNN" indicates flood inundation area retrieved by DBNN method based on CYGNSS data, while "SMAP-Threshold" denotes flood inundation area classified by threshold method based on SMAP data.

Through further observation of Figures 9 and 10, it can be found that the flooded areas are mostly located in Bangladesh, Nepal, and northeast of India, which is mainly caused by three factors; i.e., the large quantities of water vapor carried by the southwest monsoons, as well as the influence of topography and water systems. In terms of topography, the region is located at the southern foothills of the Himalayas and on the windward slope of the southwest monsoons. When the southwest monsoon is blocked by the northern mountains, it is forced to lift up and form topographic rain, leading to high rainfall in the study area. In terms of the water system, rainwater from the northern foothills is usually collected in the northeast of the study area, due to the low topography and dense river networks, making the region more vulnerable to the threat of flooding.

### 4.3. Discussion

Compared to SMAP data, the regions with more errors in the inversion results of CYGNSS data are mainly located in high altitude and inland permanent water regions, such as the Western Ghats and the Malwa plateau. The reason for such errors is that CYGNSS uses the DTU10 digital elevation model to calibrate the locations of specular points [23], which does not sufficiently consider the effect of land topography, resulting in high errors in the estimated positions of specular points with altitudes greater than 600 m on land. In addition, the difference in spatial resolution between CYGNSS data and SMAP data can also increase errors in the inversion results. For example, a small area of water may be identified by CYGNSS data, but not by SMAP data with a lower spatial resolution, which will increase the errors in inversion results.

Although the DBNN method could achieve a relatively higher inversion accuracy in flood monitoring, its accuracy would decrease in some special areas, such as flat areas. The DBNN method may misclassify the flat areas as water bodies because both of them have less surface roughness, thus producing similar coherent DDMs. In future research, inversion accuracy in flat areas is expected to be improved by extracting new physical features. In addition, considering that the DBNN method needs to learn an abundance of parameters and takes a large amount of computation to construct the optimal neural network model compared with the traditional methods, more lightweight models for flood monitoring should be anticipated to highly improve their operational efficiency in the future.

## 5. Conclusions

This study proposes a DBNN model for GNSS-R flood monitoring, which is mainly composed of a CNN module and a BP neural network module. The former is adopted to extract the underlying abstract features from DDMs, while the latter takes typical GNSS-R physical features and vegetation information as input. This kind of dual-branch neural network scheme can adequately combine GNSS-R physical features with the abstract features mined by CNN, which helps the DBNN model better utilize GNSS-R data for flood inversion and dynamic monitoring of inundation. Taking the study area in South Asia as an example, the effectiveness of the DBNN method was verified by comparison with the SR and PR methods. Then, the 2020 flood inundation in the study area was retrieved by the DBNN method, and it was found that DBNN had significant flood monitoring capabilities and could track the evolution process of floods over time. This study indicates that although CYGNSS is designed to observe ocean surface wind speed during hurricanes, it also possesses a promising future in flood monitoring applications.

**Author Contributions:** Conceptualization, methodology, D.S. and Q.Z.; software, Q.Z.; validation, writing—original draft preparation, D.S., Q.Z. and B.W.; writing—review and editing, D.S., Q.Z., C.Y. and J.X.; visualization, Q.Z.; funding acquisition, D.S. and B.W. All authors have read and agreed to the published version of the manuscript.

**Funding:** This work was supported by the National Key Research and Development Program of China under Grant 2019YFC1509202, the National Natural Science Foundation of China under

Grant 61371189, 41701513, 41772350, and the Key Research and Development Program of Shandong Province under Grant 2019GGX101033.

**Data Availability Statement:** The CYGNSS datasets used during the study are available from NASA, https://podaac.jpl.nasa.gov (accessed on 10 September 2021). The SMAP datasets analyzed during the current study are available in https://nsidc.org/data/SPL3SMP_E/versions/4 (accessed on 16 October 2021).

**Acknowledgments:** The authors are grateful to the public CYGNSS data from the NASA Physical Oceanography Distributed Active Archive Center and SMAP data provided by National Snow and Ice Data Center.

**Conflicts of Interest:** There is no conflict of interest among the authors of this paper.

## Abbreviations

The following abbreviations are used in this manuscript:

| | |
|---|---|
| BP | Back Propagation |
| CNN | Convolutional Neural Network |
| CYGNSS | Cyclone Global Navigation Satellite System |
| DBNN | Dual-branch neural network |
| DDM | Delay-Doppler map |
| DL | Deep Learning |
| GNSS-R | Global Navigation Satellite System-Reflectometry |
| PR | Power ratio |
| SMAP | Soil Moisture Active Passive satellite |
| SR | Surface reflectivity |

## References

1. Diakakis, M.; Boufidis, N.; Grau, J.M.S.; Andreadakis, E.; Stamos, I. A systematic assessment of the effects of extreme flash floods on transportation infrastructure and circulation: The example of the 2017 Mandra flood. *Int. J. Disaster Risk Reduct.* **2020**, *47*, 101542. [CrossRef]
2. Messner, F.; Meyer, V. Flood damage, vulnerability and risk perception—Challenges for flood damage research. In *Flood Risk Management: Hazards, Vulnerability and Mitigation Measures*; Springer: Dordrecht, The Netherlands, 2006; pp. 149–167.
3. Sharafi, S.; Kamangir, H.; King, S.A.; Safaierad, R. Effects of extreme floods on fluvial changes: The Khorramabad River as case study (western Iran). *Arab. J. Geosci.* **2021**, *14*, 1140. [CrossRef]
4. Brivio, P.; Colombo, R.; Maggi, M.; Tomasoni, R. Integration of remote sensing data and GIS for accurate mapping of flooded areas. *Int. J. Remote Sens.* **2002**, *23*, 429–441. [CrossRef]
5. Drusch, M.; Del Bello, U.; Carlier, S.; Colin, O.; Fernandez, V.; Gascon, F.; Hoersch, B.; Isola, C.; Laberinti, P.; Martimort, P. Sentinel-2: ESA's optical high-resolution mission for GMES operational services. *Remote Sens. Environ.* **2012**, *120*, 25–36. [CrossRef]
6. Platnick, S.; King, M.D.; Ackerman, S.A.; Menzel, W.P.; Baum, B.A.; Riédi, J.C.; Frey, R.A. The MODIS cloud products: Algorithms and examples from Terra. *IEEE Trans. Geosci. Remote Sens.* **2003**, *41*, 459–473. [CrossRef]
7. Yang, W.; Gao, F.; Xu, T.; Wang, N.; Tu, J.; Jing, L.; Kong, Y. Daily Flood Monitoring Based on Spaceborne GNSS-R Data: A Case Study on Henan, China. *Remote Sens.* **2021**, *13*, 4561. [CrossRef]
8. Kuenzer, C.; Guo, H.; Huth, J.; Leinenkugel, P.; Li, X.; Dech, S. Flood mapping and flood dynamics of the Mekong Delta: ENVISAT-ASAR-WSM based time series analyses. *Remote Sens.* **2013**, *5*, 687–715. [CrossRef]
9. Hein, G.W. Status, perspectives and trends of satellite navigation. *Satell. Navig.* **2020**, *1*, 22. [CrossRef] [PubMed]
10. Rodriguez-Alvarez, N.; Camps, A.; Vall-Llossera, M.; Bosch-Lluis, X.; Monerris, A.; Ramos-Perez, I.; Valencia, E.; Marchan-Hernandez, J.F.; Martinez-Fernandez, J.; Baroncini-Turricchia, G. Land geophysical parameters retrieval using the interference pattern GNSS-R technique. *IEEE Trans. Geosci. Remote Sens.* **2010**, *49*, 71–84. [CrossRef]
11. Chew, C.; Shah, R.; Zuffada, C.; Hajj, G.; Masters, D.; Mannucci, A.J. Demonstrating soil moisture remote sensing with observations from the UK TechDemoSat-1 satellite mission. *Geophys. Res. Lett.* **2016**, *43*, 3317–3324. [CrossRef]
12. Foti, G.; Gommenginger, C.; Jales, P.; Unwin, M.; Shaw, A.; Robertson, C.; Rosello, J. Spaceborne GNSS reflectometry for ocean winds: First results from the UK TechDemoSat-1 mission. *Geophys. Res. Lett.* **2015**, *42*, 5435–5441. [CrossRef]
13. Wang, C.; Yu, K.; Qu, F.; Bu, J.; Han, S.; Zhang, K. Spaceborne GNSS-R Wind Speed Retrieval Using Machine Learning Methods. *Remote Sens.* **2022**, *14*, 3507. [CrossRef]
14. Guan, D.; Park, H.; Camps, A.; Wang, Y.; Onrubia, R.; Querol, J.; Pascual, D. Wind direction signatures in GNSS-R observables from space. *Remote Sens.* **2018**, *10*, 198. [CrossRef]
15. Roberts, T.M.; Colwell, I.; Chew, C.; Lowe, S.; Shah, R. A Deep-Learning Approach to Soil Moisture Estimation with GNSS-R. *Remote Sens.* **2022**, *14*, 3299. [CrossRef]

16. Azemati, A.; Melebari, A.; Campbell, J.D.; Walker, J.P.; Moghaddam, M. GNSS-R Soil Moisture Retrieval for Flat Vegetated Surfaces Using a Physics-Based Bistatic Scattering Model and Hybrid Global/Local Optimization. *Remote Sens.* **2022**, *14*, 3129. [CrossRef]
17. Chen, S.; Yan, Q.; Jin, S.; Huang, W.; Chen, T.; Jia, Y.; Liu, S.; Cao, Q. Soil Moisture Retrieval from the CyGNSS Data Based on a Bilinear Regression. *Remote Sens.* **2022**, *14*, 1961. [CrossRef]
18. Chew, C.; Reager, J.T.; Small, E. CYGNSS data map flood inundation during the 2017 Atlantic hurricane season. *Sci. Rep.* **2018**, *8*, 9336. [CrossRef]
19. Wan, W.; Liu, B.; Zeng, Z.; Chen, X.; Wu, G.; Xu, L.; Chen, X.; Hong, Y. Using CYGNSS data to monitor China's flood inundation during typhoon and extreme precipitation events in 2017. *Remote Sens.* **2019**, *11*, 854. [CrossRef]
20. Chew, C.; Small, E. Estimating inundation extent using CYGNSS data: A conceptual modeling study. *Remote Sens. Environ.* **2020**, *246*, 111869. [CrossRef]
21. Clarizia, M.P.; Pierdicca, N.; Costantini, F.; Floury, N. Analysis of CYGNSS data for soil moisture retrieval. *IEEE J. Sel. Top. Appl. Earth Obs. Remote Sens.* **2019**, *12*, 2227–2235. [CrossRef]
22. Unnithan, S.K.; Biswal, B.; Rüdiger, C. Flood inundation mapping by combining GNSS-R signals with topographical information. *Remote Sens.* **2020**, *12*, 3026. [CrossRef]
23. Al-Khaldi, M.M.; Johnson, J.T.; Gleason, S.; Loria, E.; O'Brien, A.J.; Yi, Y. An algorithm for detecting coherence in cyclone global navigation satellite system mission level-1 delay-Doppler maps. *IEEE Trans. Geosci. Remote Sens.* **2020**, *59*, 4454–4463. [CrossRef]
24. Morris, M.; Chew, C.; Reager, J.T.; Shah, R.; Zuffada, C. A novel approach to monitoring wetland dynamics using CYGNSS: Everglades case study. *Remote Sens. Environ.* **2019**, *233*, 111417. [CrossRef]
25. Dong, Z.; Jin, S. Evaluation of the land GNSS-Reflected DDM coherence on soil moisture estimation from CYGNSS data. *Remote Sens.* **2021**, *13*, 570. [CrossRef]
26. Zhang, L.; Zhang, L.; Du, B. Deep learning for remote sensing data: A technical tutorial on the state of the art. *IEEE Geosci. Remote Sens. Mag.* **2016**, *4*, 22–40. [CrossRef]
27. Rumelhart, D.E.; Hinton, G.E.; Williams, R.J. Learning representations by back-propagating errors. *Nature* **1986**, *323*, 533–536. [CrossRef]
28. Entekhabi, D.; Njoku, E.G.; O'Neill, P.E.; Kellogg, K.H.; Crow, W.T.; Edelstein, W.N.; Entin, J.K.; Goodman, S.D.; Jackson, T.J.; Johnson, J. The soil moisture active passive (SMAP) mission. *Proc. IEEE* **2010**, *98*, 704–716. [CrossRef]
29. Chew, C.; Small, E. Soil moisture sensing using spaceborne GNSS reflections: Comparison of CYGNSS reflectivity to SMAP soil moisture. *Geophys. Res. Lett.* **2018**, *45*, 4049–4057. [CrossRef]
30. Gleason, S.; Ruf, C.S.; O'Brien, A.J.; McKague, D.S. The CYGNSS level 1 calibration algorithm and error analysis based on on-orbit measurements. *IEEE J. Sel. Top. Appl. Earth Obs. Remote Sens.* **2018**, *12*, 37–49. [CrossRef]
31. Zavorotny, V.U.; Voronovich, A.G. Scattering of GPS signals from the ocean with wind remote sensing application. *IEEE Trans. Geosci. Remote Sens.* **2000**, *38*, 951–964. [CrossRef]
32. Al-Khaldi, M.M.; Shah, R.; Chew, C.C.; Johnson, J.T.; Gleason, S. Mapping the Dynamics of the South Asian Monsoon Using CYGNSS's Level-1 Signal Coherency. *IEEE J. Sel. Top. Appl. Earth Obs. Remote Sens.* **2020**, *14*, 1111–1119. [CrossRef]
33. Lei, F.; Senyurek, V.; Kurum, M.; Gurbuz, A.C.; Boyd, D.; Moorhead, R.; Crow, W.T.; Eroglu, O. Quasi-global machine learning-based soil moisture estimates at high spatio-temporal scales using CYGNSS and SMAP observations. *Remote Sens. Environ.* **2022**, *276*, 113041. [CrossRef]
34. O'Neill, P.E.; Chan, S.; Njoku, E.G.; Jackson, T.; Bindlish, R.; Chaubell, J. *SMAP Enhanced L3 Radiometer Global Daily 9 km EASE-Grid Soil Moisture*, 4th ed.; NASA National Snow and Ice Data Center Distributed Active Archive Center: Boulder, CO, USA, 2020.
35. Boulder, Colorado USA. NASA National Snow and Ice Data Center Distributed Active Archive Center. Available online: https://nsidc.org/data/spl3smp_e/versions/4 (accessed on 16 October 2021).
36. Chaubell, J.; Yueh, S.; Entekhabi, D.; Peng, J. Resolution enhancement of SMAP radiometer data using the Backus Gilbert optimum interpolation technique. In Proceedings of the 2016 IEEE International Geoscience and Remote Sensing Symposium (IGARSS), Beijing, China, 11–15 July 2016; pp. 284–287.
37. Guo, W.; Du, H.; Cheong, J.W.; Southwell, B.J.; Dempster, A.G. GNSS-R wind speed retrieval of sea surface based on particle swarm optimization algorithm. *IEEE Trans. Geosci. Remote Sens.* **2021**, *60*, 4202414. [CrossRef]
38. Chew, C.; Small, E. Description of the UCAR/CU soil moisture product. *Remote Sens.* **2020**, *12*, 1558. [CrossRef]
39. Yan, Q.; Huang, W. Spaceborne GNSS-R sea ice detection using delay-Doppler maps: First results from the UK TechDemoSat-1 mission. *IEEE J. Sel. Top. Appl. Earth Obs. Remote Sens.* **2016**, *9*, 4795–4801. [CrossRef]
40. Goodfellow, I.; Bengio, Y.; Courville, A. *Deep Learning*; MIT Press: Cambridge, MA, USA, 2016.
41. Gao, B.; Pavel, L. On the properties of the softmax function with application in game theory and reinforcement learning. *arXiv* **2017**, arXiv:1704.00805.
42. Aueb, T.R. One-vs-each approximation to softmax for scalable estimation of probabilities. *Adv. Neural Inf. Process. Syst.* **2016**, *29*.
43. Rahman, M.S.; Di, L.; Shrestha, R.; Eugene, G.Y.; Lin, L.; Zhang, C.; Hu, L.; Tang, J.; Yang, Z. Agriculture flood mapping with Soil Moisture Active Passive (SMAP) data: A case of 2016 Louisiana flood. In Proceedings of the 2017 6th International Conference on Agro-Geoinformatics, Fairfax, VA, USA, 7–10 August 2017; pp. 1–6.

remote sensing

MDPI

*Article*

# Quality Control of CyGNSS Reflectivity for Robust Spatiotemporal Detection of Tropical Wetlands

**Hironori Arai [1,2,3,4,*], Mehrez Zribi [1], Kei Oyoshi [5], Karin Dassas [1], Mireille Huc [1], Shinichi Sobue [5] and Thuy Le Toan [1]**

1 CESBIO (CNRS/CNES/UPS/IRD), 18 av. Edouard Belin, bpi 2801, CEDEX 9, 31401 Toulouse, France
2 Japan Society for the Promotion of Science, Chiyoda, Tokyo 102-0083, Japan
3 International Rice Research Institute (IRRI), Km 2 Pham Van Dong str., Tu Liem District, Hanoi 100000, Vietnam
4 Graduate School of Information Science and Technology, Osaka University, Osaka 565-0871, Japan
5 Earth Observation Research Center, Japan Aerospace Exploration Agency, Tsukuba 305-8505, Japan
* Correspondence: h-arai@ist.osaka-u.ac.jp; Tel.: +84-(0)96-465-5654

**Abstract:** The aim of this study was to develop a robust methodology for evaluating the spatiotemporal dynamics of the inundation status in tropical wetlands with the currently available Global Navigation Satellite System-Reflectometry (GNSS-R) data by proposing a new quality control technique called the "precision index". The methodology was applied over the Mekong Delta, one of the most important rice-production systems comprising aquaculture areas and natural wetlands (e.g., mangrove forests, peatlands). Cyclone Global Navigation Satellite System (CyGNSS) constellation data (August 2018–December 2021) were used to evaluate the spatiotemporal dynamics of the reflectivity Γ over the delta. First, the reflectivity Γ, shape and size of each specular footprint and the precision index were calibrated at each specular point and reprojected to a 0.0045° resolution (approximately equivalent to 500 m) grid at a daily temporal resolution (Lv. 2 product); then, the results were obtained considering bias-causing factors (e.g., the velocity/effective scattering area/incidence angle). The Lv. 2 product was temporally integrated every 15 days with a Kalman smoother (+/− 14 days temporal localization with Gaussian kernel: 1σ = 5 days). By applying the smoother, the regional-annual dynamics over the delta could be clearly visualized. The behaviors of the GNSS-R reflectivity and the Advanced Land Observing Satellite-2 Phased-Array type L-band Synthetic Aperture Radar-2 quadruple polarimetric scatter signals were compared and found to be nonlinearly correlated due to the influence of the incidence angle and the effective scattering area.

**Keywords:** CyGNSS; GNSS-R; inundation; wetland; Mekong Delta

**Citation:** Arai, H.; Zribi, M.; Oyoshi, K.; Dassas, K.; Huc, M.; Sobue, S.; Toan, T.L. Quality Control of CyGNSS Reflectivity for Robust Spatiotemporal Detection of Tropical Wetlands. *Remote Sens.* **2022**, *14*, 5903. https://doi.org/10.3390/rs14225903

Academic Editors: Hugo Carreno-Luengo, Dallas Masters and Chun-Liang Lin

Received: 28 September 2022
Accepted: 18 November 2022
Published: 21 November 2022



## 1. Introduction

Global Navigation Satellite System Reflectometry (GNSS-R) data have the potential to regionalize methane ($CH_4$) emissions from land surface images by detecting their inundation status. Methane is an important greenhouse gas (GHG); its global warming potential over a 100-year horizon is 28 times higher than that of carbon dioxide ($CO_2$) [1]. In 2011, the $CH_4$ concentration was 1803 ppb, 150% higher than the preindustrial level, and a predominantly biogenic post-2006 increase has also been reported [2]. Concurrently, atmospheric methane's $\delta^{13}C_{CH4}$ value has trended towards lighter ($^{13}C$-depleted) values, implying a significant shift in the balance between the sources and sinks of $CH_4$ [3] and a greater contribution of biogenic $CH_4$ emission sources rather than fuel combustion to this rapid $CH_4$ concentration increase [2]. Several hypotheses have been postulated for the cause of this isotopic shift, and these hypotheses can be summarized as one or a combination of the following: (i) a change in the oxidative capacity of the atmosphere [4]; (ii) changes in the relative strengths of anthropogenic sources, such as land-use changes on tropical wetlands to agriculture or waste and fossil fuel emissions with an overall net effect of increasing emissions (e.g., [5]); and (iii) an increase in natural sources such as wetlands, potentially as a feedback effect from regional climatic change (e.g., [3]). Large gaps still exist between top-down and bottom-up $CH_4$ total global emissions calculations, with much of the uncertainty associated with the emissions of wetlands and other natural emissions categories [6,7], particularly in tropical wetlands [7–9].

Because $CH_4$ is emitted from inundated soil, which is spatiotemporally heterogeneous and has a flux pattern characterized by non-Gaussian/nonlinear behaviors [7], the appropriate evaluation of the $CH_4$ flux requires the monitoring of the inundation status with spatiotemporally high-resolution techniques [10]. GNSS-R data became a popular input source in microwave remote sensing techniques following the deployment of the Cyclone Global Navigation Satellite System (CyGNSS), an eight-microsatellite constellation data system [11]. Every single CyGNSS microsatellite has two left-hand circular polarization (LHCP) down-looking antennas pointing to the Earth's surface with an inclination angle of approximately 28 degrees on either side of the satellite ground track [12].

The data can be used to globally detect the land surface inundation status almost daily with high-spatial-resolution L-band microwave signals (with estimated spatial resolutions of approximately 500–7000 m [13]) compared to common passive L-band microwave radiometers. A few studies have reported that the use of CyGNSS-based inundation maps for land surface methane emission simulations improved the representation of the $CH_4$ emission status compared to the results obtained using common wetland maps (e.g., simulating a greater amount of $CH_4$ emissions by detecting inundation under clouds/vegetation with GNSS-R data [14]).

There are several studies on the detection of inundation over wetlands with GNSS-R data e.g., [15–19]. However, the results in most studies remain spatiotemporally sparse. In most cases, the spatiotemporal interpolation is conducted with monthly observation datasets, or spatially interpolated with optical observation sensors e.g., [15,19]. Due to the limitations of L-band fine-spatial-resolution microwave remote sensing data like GNSS-R, there are only a few studies conducting the cross-validation of GNSS-R and L-band SARs observations [15]. Furthermore, from the perspective of the application of this study, most of the time, this sort of fine-spatial-resolution, satellite-derived wetland/inundation observation is downsampled or spatially thinned (a.k.a., superobservations) before being used in advanced simulation modeling approaches accompanied with high computation costs (e.g., coarse-spatial-resolution ensemble simulations or the use of superobservations to deal with observation error covariance in data assimilation tasks) by degrading the spatial resolution or thinning the observations (e.g., [14,20]). Due to the local heterogeneity of the inundation status and the non-Gaussian/nonlinear characteristics of the spatiotemporal $CH_4$ emission distribution at the local scale [7–9], the deterioration of the spatial resolution of data can introduce large discrepancies to the emission values obtained between the top-down approach and bottom-up approach [6–9,20]. Therefore, the regionalization of $CH_4$ emissions based on high-spatial-resolution L-band microwave data as a bottom-up approach still remains important [7–9,17]. Since most studies have used GNSS-R data for regional-scale simulations at a relatively coarse spatial resolution compared to remote sensing observations (e.g., 0.01°-resolution CyGNSS-based watermasks are downsampled to a 0.5° resolution to match the WetCHARTs simulation grid [14]), few studies have paid attention to the differences among each specular point's footprint size (i.e., the glistening area). To rasterize each piece of specular point-scale vector data without downsampling for use in local-scale simulations, one must consider the difference among each specular point's footprint size to use these signals in fine-spatial-resolution, local-scale simulations (e.g., 10–50 m resolution irrigation models [10]). This information would also be essential for determining the spatial localization scale to ensure efficient data assimilation by determining the spatial localization scale at each specular point and adequately addressing the spatial observation error covariance.

More fundamentally, the amount of data of a certain quality provided by the GNSS-R microsatellite constellation is still limited, and the observations are prone to being conducted sparsely in space; in addition, the incidence angle varies widely among specular points, which is known to cause biases in the microwave reflectivity observations (unlike other spatially continuous microwave remote sensing observations, such as those obtained from passive microwave radiometers or synthetic aperture radars). The local incidence angles of the Phased-Array L-band Synthetic Aperture Radar-2 (PALSAR-2) ScanSAR instruments vary from 25 to 50 degrees, while CyGNSS incidence angles vary from 0 to 70 degrees [13,17]). To prepare inundation maps based on GNSS-R data for future applications or to be assimilated into simulation models, the spatiotemporal interpolation step needs to be processed before the data can be used in applications. Therefore, an adaptive quality control method that considers the size of each specular point and depends on each specular point vector and incidence angle but does not require ad hoc parameter tuning or region-specific empirical parameterization with external data, such as the normalized differential vegetation index (NDVI) or digital elevation model (DEM) data, is essential for this robust interpolation preprocessing step. To realize this globally consistent rasterization at a fine spatial resolution, the authors have developed a precision index calibration scheme implemented while processing the raw specular vector data to rasterize data while considering the differences in incidence angles and specular points' sizes/shapes/velocities. To compensate for the spatially sparse distribution of GNSS-R specular data, the temporal Kalman smoother is applied by using the precision index as the reciprocal observation error number in each 15-day cycle over the Mekong Delta as a demonstration; this case study area consists of double-/triple-rice-cropping systems, aquacultural ponds, mangroves and peatlands. Cross-validation with the PALSAR-2 quadruple polarimetric data (3–6 m resolution) product is also conducted, and the results are validated with ground inundation observation datasets [7–9,17]. The goal of this study is to demonstrate the usefulness of this quality control method by applying it to fine-spatiotemporal-resolution analyses over the Mekong Delta [i.e., (I) comparing it with a common change detection algorithm with the daily temporal resolution, (II) applying it with a 500 m rasterization with a 15-day temporal resolution, (III) and surveying the consistency with 3–6 m spatial-resolution L-band SAR backscatter intensities].

## 2. Materials and Methods

This study consists of (1) the introduction of the "precision index" for use in the quality control assessment of GNSS-R data; (2) a daily rasterization demonstration over the Mekong Delta based on the precision index obtained from the Lv. 2 product; (3) a demonstration of the temporal Kalman smoother over the Mekong Delta using the precision index as the reciprocal number of observation errors (Lv. 3 product); and (4) cross-validation with PALSAR-2 quadruple polarimetric data that have been preprocessed with a polarimetric decomposition method and ground observation data. A flowchart presenting the methodology of this study is illustrated in Figure 1.

### 2.1. Sites along with the Collection of Field Data

We prepared ground observation datasets obtained at six sites (A–E) located in six different districts: Site A, in Thot Not, Can Tho (10°10′N, 105°33′E); Site B, in Chau Thanh (10°16′N, 105°08′E); Site C, in Cho Moi (10°25′N, 105°27′E); Site D, in Thoai Son (10°16′N, 105°08′E); and Site E, in Tri Ton, An Giang (10°23′N, 105°06′E) [7,8,10,21–27] Figure S1. The soils at sites A–C are classified as silty clay fluvisol (a type of alluvial soil; [17]), while the soils at sites D and E are classified as sulfuric humaquepts (a type of alluvial soil [17]).

In Can Tho and An Giang, 50 farmers' rice paddies (30 in site A, five each in sites B–E) were chosen as regions of interest (ROIs). At the center of each ROI, field water level data were collected for the supervised classification of the satellite remote sensing data with a water level gauge (daily, 10:00 AM–12:00 PM at site A) or with a HOBO CO-U20L-04 water level data logger (Onset Computer Corporation, United States; collected every 4 h at sites B–E). At the same time, we collected information about the history of field operations (e.g., fertilization and land preparation/sowing/harvesting dates) at each ROI throughout the observation period. The numbers of ROIs in the inundated/non-inundated rice paddies are described in the cited literature [10].

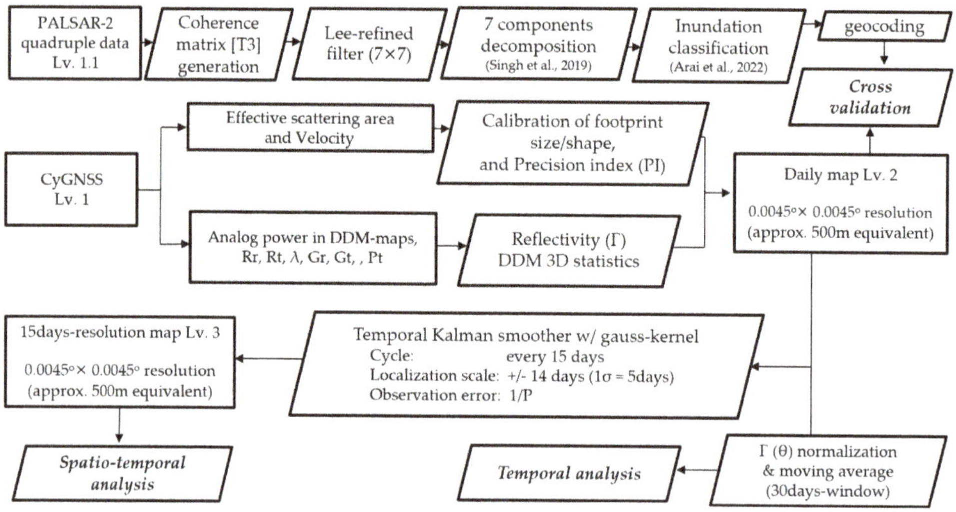


**Figure 1.** Flowchart of the methodology of this study outlining the data, processing and analysis steps.

*2.2. CyGNSS GNSS-R Datasets and Their Preprocessing Methods*

All CyGNSS [10] Lv.1 Version 3.0 data observed from the first observation (August 2018) until December 2021 were downloaded from https://podaac.jpl.nasa.gov/dataset/CYGNSS_L1_V3.0 (accessed on 15 January 2022). The reflectivity (Γ) data were calibrated [28] using Equation (1):

$$\Gamma(\theta) = \frac{(4\pi)^2 (P_{DDM} - N)(R_r + R_t)^2}{\lambda^2 G_r G_t P_t} \quad (1)$$

where $P_{DDM}$ is the maximum value of the analog power in the delay/Doppler maps (DDM), $N$ is the noise floor related to the *DDM*, *Rr* is the receiver–specular point (SP) distance, *Rt* is the transmitter–SP distance, $\lambda$ is the wavelength, $\theta$ is the incidence angle, *Gr* is the receiver antenna gain in the direction of the SP and *GtPt* is the transmitter equivalent isotropically radiated power (EIRP). The noise floor is computed as the mean value of the DDM subset, where the signal is absent (located above the characteristic horseshoe shape of the DDMs). The effect of the scattering area with the highest analog power in the DDM maps was used as the size of the specular point. Since the CyGNSS GPS signal integration time is fixed at 1 s, the footprint shape was inversely computed using the integration time, the velocity of the SP and the effective scattering area.

Our precision index model's design was inspired by the spatial localization technique of common data assimilation methods such as the local ensemble transform Kalman filter or local particle filters [10]. The precision index (PI) was calibrated using the following equation in the grid covered by the SP footprint, as shown in Equation (2):

$$PI = \frac{\cos(\theta) \times GS}{sqrt(ESA) \times \exp(3.0 \times (DistSP/SemiDSP)^2)} \propto \text{ObsEr} \quad (2)$$

where $\theta$ is the incidence angle, *GS* is the grid spacing, *ESA* is the effective scattering area, *DistSP* is the distance from the center of the specular point, *SemiDSP* is the semidiameter of the ellipsoidal-shaped specular point, and *ObsEr* is the observation error (Figure 2).

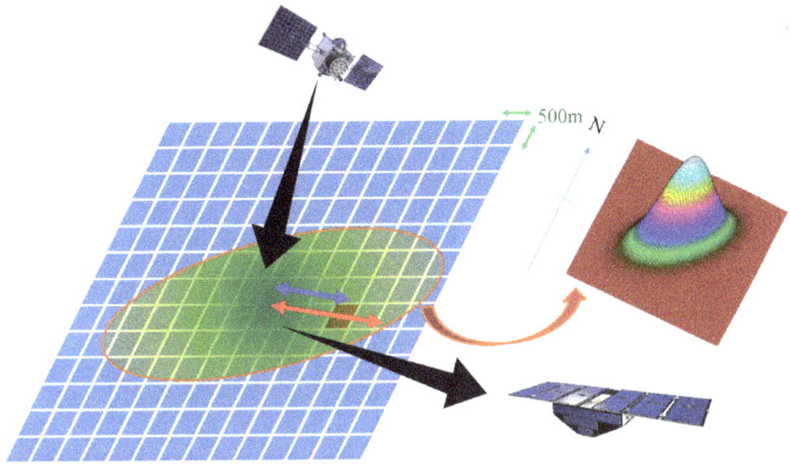


**Figure 2.** Illustration of the precision index. The light blue tiles are rasterization grid cells. The yellow circular area is the effective scattering area. The red tile in the yellow/green circle effective scattering area is the corresponding grid. The green, blue and red arrows are equivalent to the *GS*, *DistSP* and *SemiDSP* terms in Equation (2).

Each specular point in the CyGNSS data format contains analog power in a 17 × 11 array of DDM bins [17 rows for Delay with a 0.25-chip resolution, 11 columns for Doppler with a 500-Hz resolution]. We also analyzed the analog power in the DDM by regarding the power as a probability density of a 3-dimensional histogram (representing skewness and kurtosis) as described in Equation (3) and Figure 3 by targeting 5 × 5 arrays surrounding the element containing the maximum analog power over the DDM at each specular point. If the kurtosis value was greater than 0.01, the precision index ($P$) zeroed out before its use to omit the noise derived from the specular effects over highly rough land surfaces.

$$
\begin{gathered}
SKW\_a = \sum_{i=-2}^{2}\sum_{i=-2}^{2}\left(\frac{(P_{ij}-P_{00})/(TP\times 25)}{1/3+\sqrt{2}\left|(DPL\_P_{ij}-DPL\_P_{00})\right|}\right)^{3} \\
SKW\_b = \sum_{i=-2}^{2}\sum_{i=-2}^{2}\left(\frac{(P_{ij}-P_{00})/(TP\times 25)}{1/3+\sqrt{2}\left[-(DPL\_P_{ij}-DPL\_P_{00})+(DLY\_P_{ij}-DLY\_P_{00})\right]}\right)^{3} \\
SKW\_c = \sum_{i=-2}^{2}\sum_{i=-2}^{2}\left(\frac{(P_{ij}-P_{00})/(TP\times 25)}{1/3+\sqrt{2}\left[(DPL\_P_{ij}-DPL\_P_{00})+(DLY\_P_{ij}-DLY\_P_{00})\right]}\right)^{3} \\
DDM\_SKW = MAX\{abs(SKW\_a), abs(SKW\_b), abs(SKW\_c)\} \\
DDM\_KTS = \sum_{i=-2}^{2}\sum_{i=-2}^{2}\left(\frac{(P_{ij}-P_{00})/(TP\times 25)}{1/3+\sqrt{2\times\left[(DPL\_P_{ij}-DPL\_P_{00})^{2}+(DLY\_P_{ij}-DLY\_P_{00})^{2}\right]}}\right)^{4}
\end{gathered}
\tag{3}
$$

where *SKW* is skewness, KTS is kurtosis, $P_{00}$ is the maximum analog power value (W) on the DDM arrays of each specular point ["00" indicates the index of the element containing the maximum analog power among all arrays in the *DDM*; i.e., $P_{00}$ in Equation (3) is equivalent to $P_{DDM}$ in Equation (1)], $i$ and $j$ are array indexes over the *DDM* surrounding the maximum analog power element ($i$ is the Delay row index and $j$ is the Doppler column index), *TP* indicates the sum of the power analog values of all arrays in the *DDM*, and *DPL* and *DLY* indicate the doppler–delay index of the target element (i.e., $P_{ij}$ or $P_{00}$) over the *DDM* array.

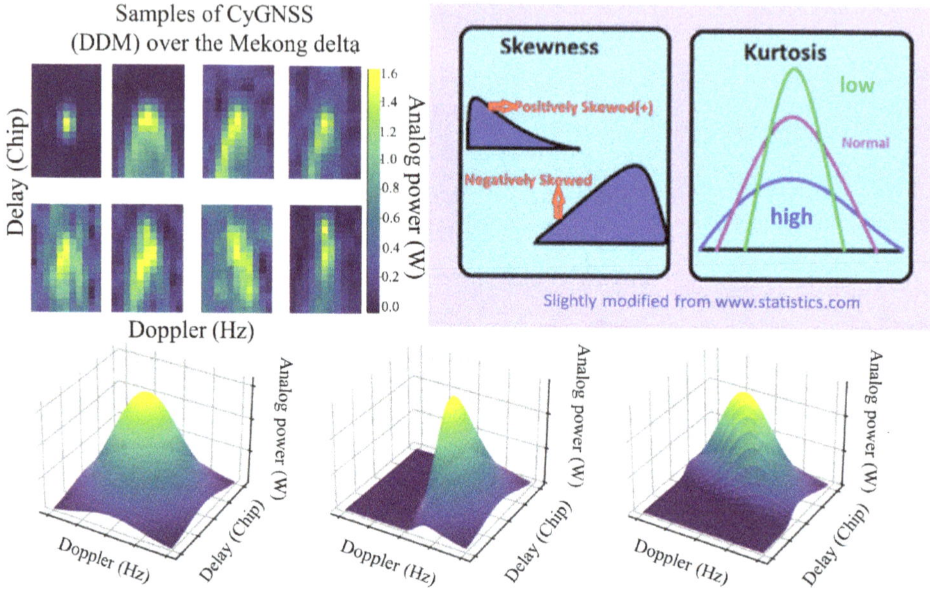


**Figure 3.** Illustration of DDM 3D statistics (skewness and kurtosis). The calibration is conducted by assuming that the analog power is equivalent to the probability density of the DDM 3D histogram (Delay, Doppler and Analogue power).

After calibrating the reflectivity (Γ) and PI on a latitude/longitude map (with a 500 m resolution and a daily temporal resolution) (Lv. 2, Figure 4), the data were applied to a temporal Kalman smoother on each 15-day cycle (temporal localization scale: 14 days. 1σ = 5 days) to obtain the Lv. 3 product (Figure 5) for the subsequent spatiotemporal analysis. The Lv. 2 data were also applied for the temporal analysis (with a slight modification to the change detection algorithm described in [29]) just after being applied in the $\Gamma(\theta)$-normalization task with Equation (4) and in a 30-day moving average filter; then, the results were compared with the ALOS-2/PALSAR-2 products reported in [10]. To generate the Lv. 3 products, this study simply used a linear Kalman filter (i.e., the time-evolution of the model was assumed to be negligible). Γ-reflectivity was treated as both the states and measurements.

$\Gamma_{normalized} = \frac{\Gamma - \Gamma_{\min}}{\Gamma_{\max} - \Gamma_{\min}}$ referring to a paper [29]

$$\Gamma(\theta)_{normalized} = \frac{\Gamma(\theta) - \Gamma(\theta)_{\min}}{\Gamma(\theta)_{\max} - \Gamma(\theta)_{\min}} \quad (4)$$

where $\Gamma(\theta)_{\max/\min}$ is the temporal maximum/minimum value of the corresponding incidence angle bin. Due to the data quantity limitations of specular points obtained during the 2018–2022 period, we calibrated the incidence angle bins prepared for every 5° interval (i.e., 0–5°, 5–10°, 10–15°, 15–20°, 20–25°, 25–30°, 30–35°, 35–40°, 40–45°, 45–50°, 50–55°, 55–60°, 60–65° and 65–70° bins) in each grid.

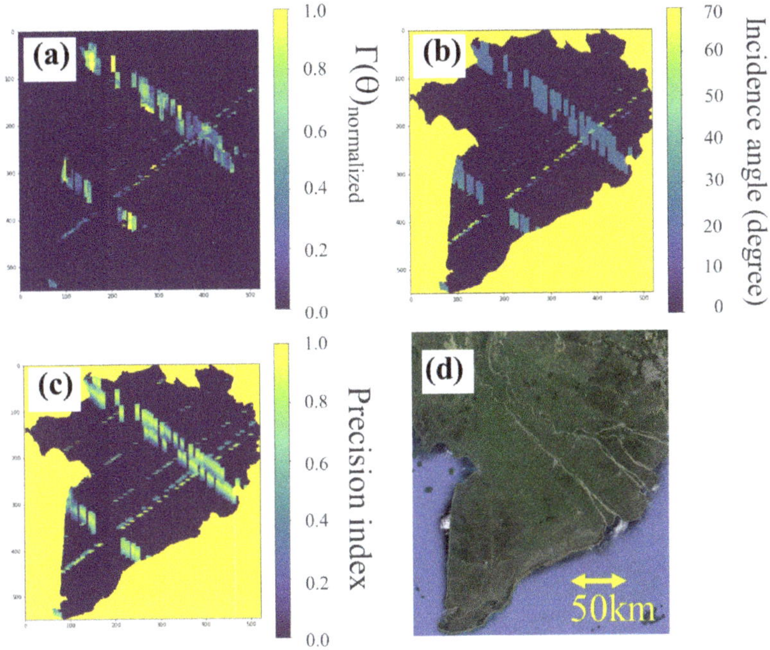


**Figure 4.** A sample of the Lv. 2 daily product (**a**) $\Gamma(\theta)_{\text{normalized}}$; (**b**) incidence angle; (**c**) precision index clipped at the Mekong Delta ((**d**) Optical image) on 4 January 2021.

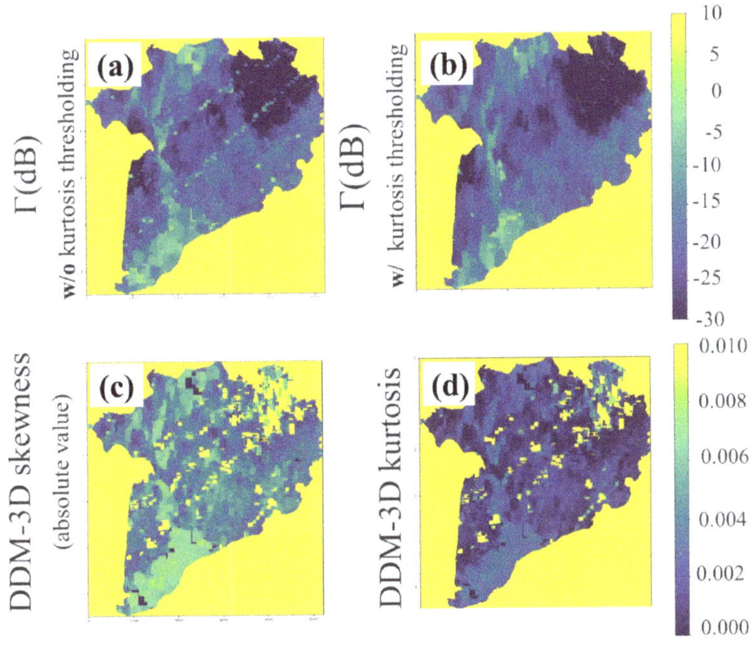


**Figure 5.** A sample of the Lv. 3 15−day−cycle Kalman smoother product based on the precision index [Γ(dB) without or with applying the precision index (**a**,**b**), zeroed out based on the kurtosis threshold and DDM 3D statistics such as skewness (**c**) and kurtosis (**d**)] clipped at Mekong Delta on 1 August 2018.

After calibrating the reflectivity (Γ) and PI on a latitude/longitude map (with a 500 m resolution and a daily temporal resolution) (Lv. 2, Figure 4), the data were applied to a temporal Kalman smoother on each 15-day cycle (temporal localization scale: 14 days. $1\sigma$ = 5 days) to obtain the Lv. 3 product (Figure 5) for the subsequent spatiotemporal analysis. The Lv. 2 data were also applied for the temporal analysis (with a slight modification to the change detection algorithm described in [29]) just after being applied in the $\Gamma(\theta)$-normalization task with Equation (4) and in a 30-day moving average filter; then, the results were compared with the ALOS-2/PALSAR-2 products reported in [10].

### 2.3. PALSAR-2 Datasets, Corresponding Preprocessing Methods and Cross-Validation Scheme with CyGNSS Data

PALSAR-2's quadruple observation datasets (Lv. 1.1; 40–50 km observation widths, 70 km observation length; 307 scenes; August 2018–December 2021, Table S1) containing observations of the Mekong Delta were prepared after the radiometric and polarimetric calibration factors of the PALSAR-2 standard product were updated (on 24 March 2017 [30]). The high-spatial-resolution (4.3 m azimuthal resolution and 5.1 m range resolution at a 37° incidence angle) quadruple data were decomposed to characterize the microwave scattering pattern in inundated paddy soils and non-inundated paddy soils at different rice growth stages. The phase and polarimetry data in PALSAR-2's quadruple observation datasets were converted into a coherency matrix; a refined Lee filter (7 × 7 window) was applied to ease speckle noise; and the data were then decomposed with Singh 7 components [31]. The digital number of the HH/HV/VH/VV microwave data was used in the backscatter reflectivity calibration expressed in Equation (5):

$$\sigma^0 = 10 \cdot \mathrm{Log}_{10} < I^2 + Q^2 > -\ 105.0 \quad (5)$$

where $\sigma^0$ is the backscattering coefficient, $I$ is the value of the imaginary component and $Q$ is the value of the quadrature component of the digital numbers. The value of −105 is the calibration factor noted in the literature [30]. An inundation detection classification task (i.e., to determine whether the field water level was higher than the soil surface or not) was conducted with a support vector obtained in the previous supervised classification study [9] during ground observation collection (a total of 624 ROIs considering different rice growth stages), as mentioned above in Section 2.1. The backward geocoding of the abovementioned products was conducted by the Newton–Raphson method with ellipsoidal height data (DEM: Shuttle Radar Topography Mission 3 (SRTM3) version 4 and the EGM2008 geoid model) and the ALOS-2 orbital data (3D-spline-interpolated on every azimuth line).

The cross-validation was conducted with the PALSAR-2 preprocessed quadruple data and the CyGNSS specular points Lv. 2 data product following the calibration described in Section 2.2; these data were observed over the same locations as the PALSAR-2 geocoded images within ±3 days of the PALSAR-2 observation date. First, the PALSAR-2 data were spatially downsampled to a 500 m resolution, and then the precision index of each corresponding specular point was calibrated over the geocoded PALSAR-2 images. Finally, each weighted mean of PALSAR-2 signals (e.g., the 7-component scattering intensities, $\sigma^0$, and the spatial inundation rate) was further weighted based on the precision index derived value over the PALSAR-2 image, and the results were compared with the CyGNSS reflectivity Γ data.

These SAR data processes were necessary for the robust validation to compensate for the footprint size difference of the inundation status that was observed between the ground point observations and the GNSS-R data that were detected from space.

## 3. Results

### 3.1. Spatiotemporal Dynamics Evaluation over the Mekong Delta by CyGNSS GNSS-R Measurements

The annual/seasonal dynamics of Γ (i.e., high values in the rainy season from June to October, and low values in the dry season from February to May) could clearly be visualized in the CyGNSS GNSS-R product, as shown in Figure 6. By improving the change detection algorithm by considering the difference in the local incidence angle among each grid cell (i.e., from Γ-normalization to Γ(θ)-normalization), two peaks with high Γ values could be detected annually.

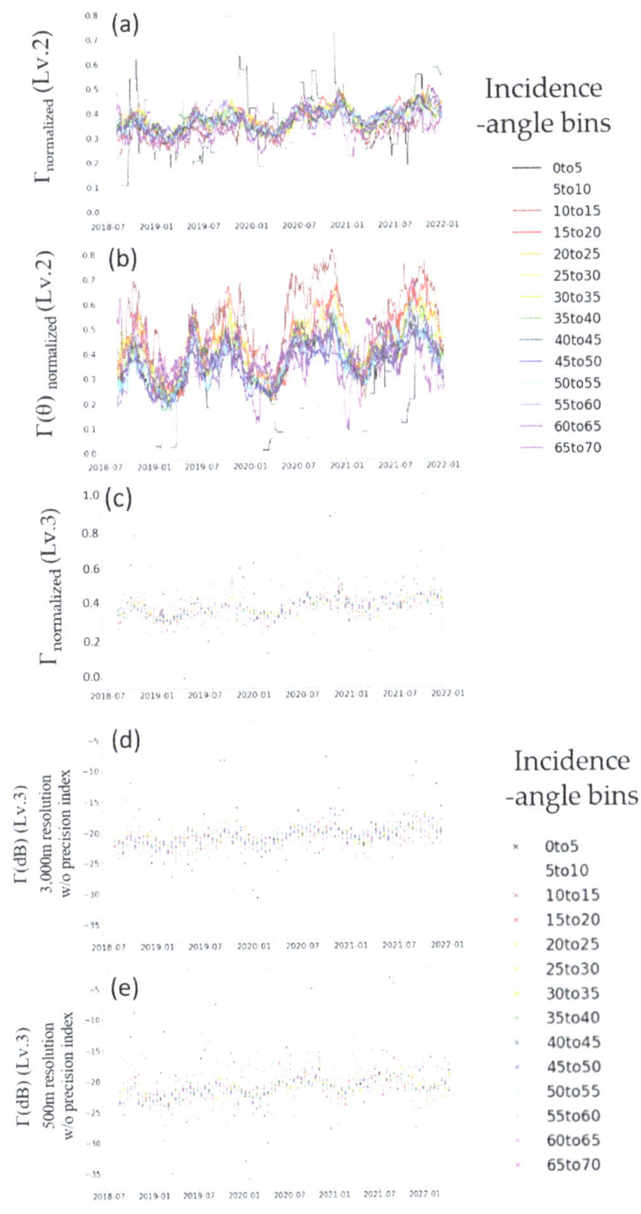


**Figure 6.** Temporal dynamics of the Lv. 2 daily product with a 30-day moving average ((**a**): $\Gamma_{\text{normalized}}$ and (**b**): $\Gamma(\theta)_{\text{normalized}}$) and the Lv. 3 15-day-cycle Kalman smoother product [(**c**): $\Gamma_{\text{normalized}}$ (500 m resolution with the precision index), (**d**): Γ (3000 km resolution without the precision index) and e: Γ (500 m resolution without the precision index)]. Each line/plot denotes spatially averaged values over the delta.

For the Lv. 2 product, a moving average was required to see the seasonal dynamics. However, this seasonal pattern was clearly illustrated in the Lv. 3 product even if the change detection algorithm was not applied (Figure 6e). Particularly for high incidence angles (>55°), the non-normalized Γ series shows a wide distribution among relatively high dB values (−20>) during the rainy season. This indicated that our proposed precision index worked adequately as a Kalman smoother weight-mean processing tool and enabled robust spatiotemporal comparisons. Compared with the result that was obtained from the 3000 m grid spacing rasterization result without the precision index (Figure 6d), the 500 m grid spacing rasterization that was enabled by using the precision index displayed the seasonal contrast more clearly (Figure 6e).

The Γ normalization step applied to each incidence angle [i.e., $\Gamma(\theta)_{normalized}$] significantly improved the sensitivity of the results to the temporal dynamics of the incidence angle by increasing the dynamic range [0.2–0.4 for Γ and 0.1–0.7 for $\Gamma(\theta)$]. $\Gamma(\theta)$ values with lower incidence angles tended to show a greater dynamic range than values with higher incidence angles (Figure 6b).

The Lv. 3 product's spatial distribution snapshot maps showed relatively strong Γ values in the northwest triple-rice-cropping region (a.k.a., Dong Thap and An Giang provinces, Figure 7). Irrespective of seasonal differences, the northeastern non-rice-cropping upland zone showed low Γ reflectivity (Figure 7). These results were consistent with the L-band SAR data-based rice paddy distribution map and rice floodability map (Figure 8 [21]). The southwestern coastal wetland zone (comprising mangrove forests, fishponds and peatlands) showed continuously high Γ values throughout the year. High Γ (dB) noise occasionally remained in the specular data in the fine-spatial-resolution (i.e., low effective scattering area) Lv. 3 product (Figure 5). However, the noise was accompanied by high DDM 3D skewness/kurtosis values because the noise was derived from the locally high land surface roughness.

### 3.2. Cross-Validation with PALSAR-2 Quadruple Observation Products

The relationship between CyGNSS reflectivity Γ values and PALSAR-2 backscatter σ0 values was differentiated depending on the specular point incidence angles and effective scattering area (Tables S2 and S3). Positive relationships between the Γ values and $\sigma^0$ values were found with 0–10° incidence angles (Figures S2–S6a,b; Table S2). For specular points obtained at 10–70°, the correlations became negative, with a few exceptions observed for a fine specular point group (i.e., for incidence angles of 30–35°, the square root value of the effective scattering area is smaller than 6 km, Table S2, Figure 9). The spatial inundation rates and Γ values showed mostly positive correlations among the groups with incidence angles of 10–50°. In contrast, negative correlations tended to be dominant for low-end incidence angle groups of 0–10° and high-end incidence angle groups of 50–70° (Table S2, Figure S2). In such high-/low-end incidence angle groups, the double bounce factor tended to show the most significant co-relationship with the Γ values among the 7-component scatterings (odd/double/volume scatterings listed in Table S1 and shown in Figures S3–S5. The remaining component analysis results are not shown in this paper since the correlations were weaker than those of the odd/double/volume scatterings. In contrast, for the middle-incidence-angle groups (10–50°), the volume diffusion results tended to show the most significant correlations with the Γ values (Table S2, Figures S3–S5). Among the PALSAR-2 HH/HV/VV backscatters, HV tended to show the most significant correlations with the Γ values (Table S2, Figure S6).

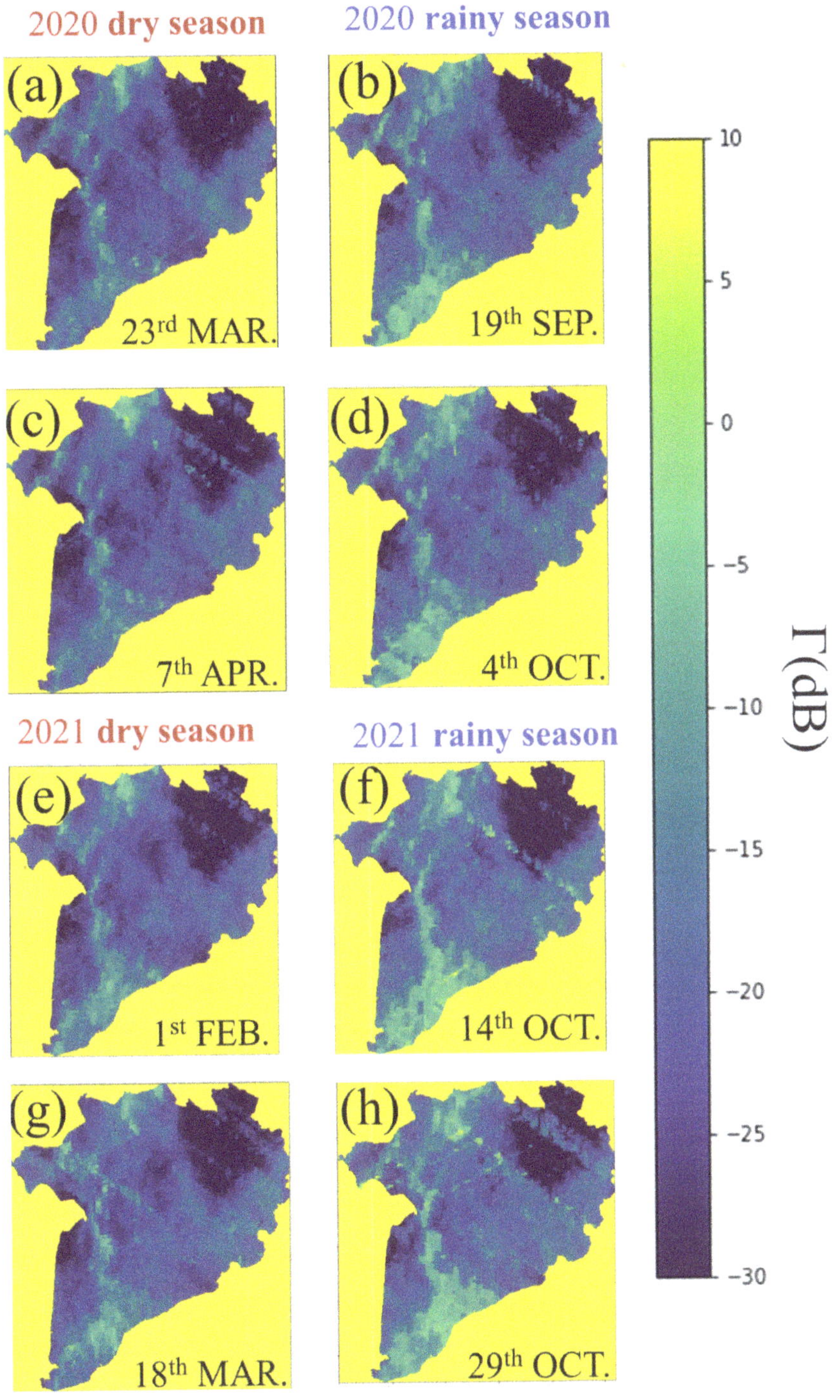


**Figure 7.** Samples of nonnormalized Γ values in 2020 (**a**–**d**) and 2021 (**e**–**h**). The left-hand side scenes are snapshots obtained in dry seasons (**a**,**c**,**e**,**g**). The right-hand side scenes are snapshots obtained in rainy seasons (**b**,**d**,**f**,**h**).

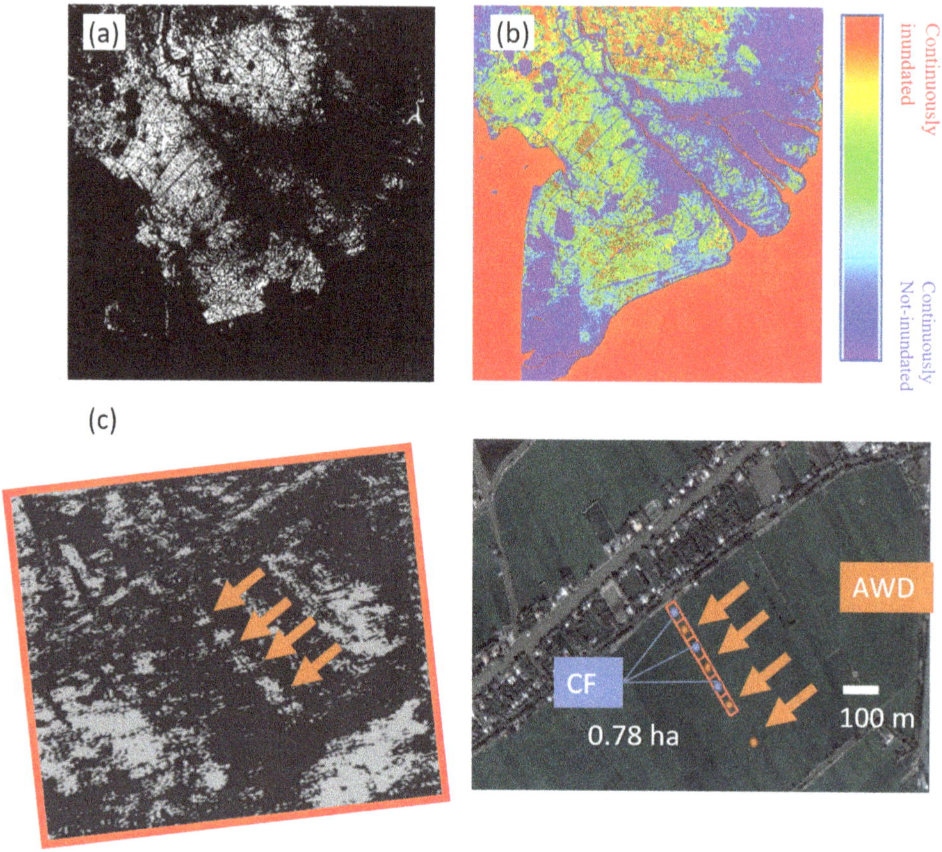


**Figure 8.** PALSAR-2 data-based rice map ((**a**); white pixels indicate rice paddies), PALSAR-2 data-based rice floodability map (**b**) and inundation detection snapshot obtained by PALSAR-2 above one of the study sites [Thot not, Can Tho city, Vietnam, on 6 May 2016 (69 days after sowing)] with the corresponding aerial photo (**c**); CF: Continuously inundated paddy; AWD: Alternate wetting and drying paddy; the temporal water level dynamics of these blocks are presented in the referenced literature [21,26,27]).

The CyGNSS reflectivity Γ and PALSAR-2 backscatter data series showed a highly nonlinear relationship, and this was one of the causes of the low Pearson correlation coefficients (Table S2). Particularly for the relationship between the Γ values and the PALSAR-2-based spatial inundation percentages, three domains with unique characteristics were found (Figure S2g,m). First, for the specular points whose Γ values are approximately smaller than −20 dB, relatively high inundation percentages were found (Figure S2g,m; domain shown by the green arrow). In such a domain, the Γ values tended to show a linearly positive correlation with the inundation percentages. Second, for specular points with Γ values between approximately −20 dB and 0 dB, specular points with 0% spatial inundation percentages were detected (Figure S2g,m; domain shown by the red arrow). In this domain, the Γ values tended to correspond to upwardly convex negative nonlinear correlations. Finally, for the specular points with Γ values greater than approximately 0 dB (Figure S2g,m; domain shown by the blue arrow), relatively high inundation percentages were detected. In this domain, the Γ values tended to show upwardly convex positive nonlinear correlations.

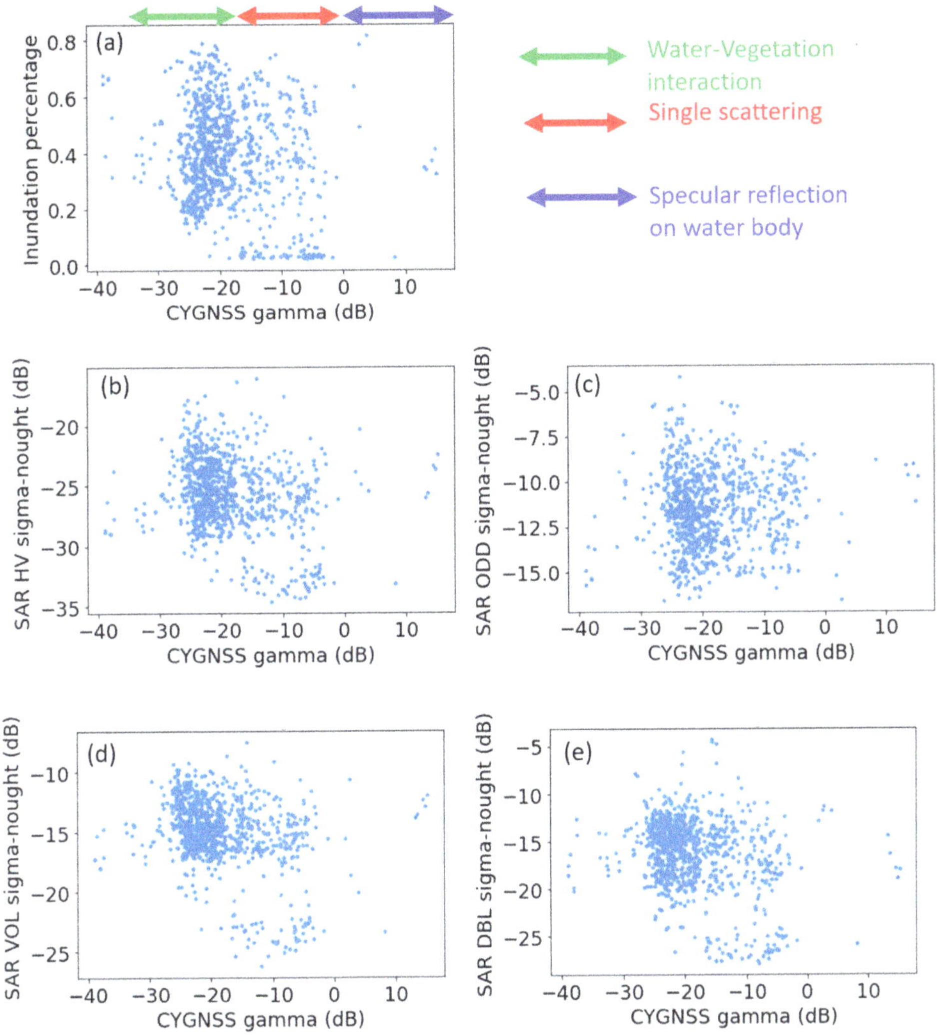


**Figure 9.** Two-dimensional scatterplots between the CyGNSS reflectivity Γ (dB) and PALSAR-2 based spatial inundation percentage (**a**) and PALSAR-2 back scatters $\sigma^0$ (dB) values ((**b**) HV, (**c**) odd scattering, (**d**) volume diffusion, (**e**) double bounce) at specular points with 30–35° incidence angles. The statistical analysis results representing these relationships are described in Table S2.

Since the relationship between the CyGNSS reflectivity Γ and PALSAR-2 backscattering σ0 values was also highly nonlinear (Figures S3–S6), a quadratic polynomial fitting analysis was carried out to survey the direction of convexity (downwardly convex, linear, or upwardly convex; Table S3). Although the relationship was mostly downwardly convex for the groups with incidence angles of 0–60°, an upwardly convex nonlinear relationship became dominant for groups with incidence angles of 5–15° and 60–70° (Table S3, Figures S3–S6). In contrast, for the middle-incidence-angle groups (15–60°), downwardly convex nonlinear relationships represented the majority.

## 4. Discussion

### *4.1. Performance of the Precision Index in the Rasterization Process without Sacrificing Spatial Resolution*

The specular point vector data derived from the Lv. 1 product were rasterized to form the Lv. 2 product, as shown in Figure 4. All the specular points depicted from southwest to northeast had relatively fine spatial resolutions due to the low effective-scattering-area values. In contrast, all the specular points depicted from northwest to southeast had relatively coarse spatial resolutions due to the relatively large effective scattering areas. It is known that the effective scattering area and footprint size/shape are mainly controlled by the Delay and Doppler effects [32]. This indicates that the relationship between the velocity (particularly in the advancing direction) of the GNSS receiver and the transmitter is the main factor controlling the spatial resolution of each specular point rather than the difference in the incidence angle or land surface roughness over lowlands such as the Mekong Delta (which has an elevation approximately 2 m above the sea surface) [21]. In this context, further GNSS-R receivers are expected to flexibly choose/adjust their transmitters to continuously receive only fine-spatial-resolution GNSS signals. The future use of geostationary GNSS transmitters or quasi-zenith-satellite-system-boarded transmitters (QZSSs) is also expected to be selected occasionally in specific regions.

The precision index developed in this study was designed to be maximized at the centers of the specular points, as the maximum analog power was detected at the center of the DDM (i.e., the neutral Delay/Doppler position), as shown in Figure 3. Since high-delay specular points are occasionally found in the Mekong Delta, hollow-ring-shaped Gaussian kernels might be appropriate for such unique specular points [32]. To further improve the index, such spatial localization regarding the *dst.centerSP/semidiameter.SP* ratio in the denominator of Equation (2) should be implemented for the further development of specular points with relatively high-delay chips. Considering the unique specular points for which the maximum analog power is located in the non-neutral Doppler bin, performing spatial localization while considering the specular advancing direction and Doppler effect would also be desirable, although such specular points were rarely found over the delta in this study.

The temporally Kalman-smoothed product (Lv. 3) clearly visualized the spatial pattern throughout the year, even without spatial interpolation/filtering/smoothing. This indicated that spatial inundation mapping can be accomplished even without performing a bias correction, depending on ad hoc parameter tuning to deal with incidence angle differences or even without depending on external NDVI data to deal with vegetation interactions. Although noise associated with relatively high $\Gamma$ values is occasionally detected with small effective scattering specular points (Figure 5a), such specular noise was seemingly found to be accompanied by high DDM 3D skewness/kurtosis values (Figure 5c,d) and could thus be denoised naturally, as shown in Figure 5b.

### *4.2. Spatiotemporal Dynamics or Inundation Detection by CyGNSS*

The $\Gamma(\theta)$ normalization results obtained for the Lv. 2 product indicated two peaks annually. The first peak was generally detected in the latter half of the dry season from April to June (Figure 6b), and the second peak was detected in the latter half of the rainy season from August to October. These findings indicate that the inundation status over the entire Mekong Delta is primarily controlled by double-/triple-rice-cropping irrigation activities. Approximately 57.4% of the rice-cropping area in 2012 was estimated to be triple-cropped [33]. Interestingly, the northwestern region where the most intensive triple rice cropping is conducted (i.e., the An Giang and Dong Thap Districts) showed significantly greater reflectivity $\Gamma$ values in the rainy season than in the dry season. Interestingly, the southwestern coastal wetlands consisted of mangroves and peatlands surrounded by acid-sulfate soils [21]. The spatially high reflectivity values found in such coastal regions, even in the dry season, might have been the result of aquacultural activities, including prawn-rice cropping rotations [34]. Because the delta receives greater attention for being exposed to salinity intrusions exacerbated by rising sea levels [35], increased upstream dam construction [36] and groundwater depletion [37], further long-term observations over the delta are necessary for future assessments of the freshwater inundation status and the salinity intrusion succession status.

One of the novel features of our work that is presented in this paper is that our methodology realized the generation of spatiotemporally continuous data sets with a finer resolution (500 m spatial resolution, 15-day temporal resolution) than commonly used methods (that mostly have 3 km and 30-day resolutions, e.g., [13]), even though we did not use any spatiotemporal interpolation methods. Simple gridding without considering the size/shape of specular points cannot spatially rasterize the continuous CyGNSS GNSS-R data even with a lower resolution due to the data quantity limitation [13].

### 4.3. Comparison with Quadruple Polarimetric L-Band SAR Backscattering Signals

Statistically significant Pearson correlations were confirmed through the precision-index-based comparison between the CyGNSS reflectivity $\Gamma$ and the PALSAR-2 backscattering intensity $\sigma^0$ or the spatial inundation percentage. We defined the inundation status based on PALSAR-2's 3–6 m resolution quadruple polarimetric data and ground truth observations [10]. To compensate for the spatial footprint size difference between the GNSS-R data and the inundation status observations with a finer spatiotemporal resolution in this study, we employed the product based on SAR data. There was still a discrepancy between the CyGNSS observations and the SAR-based inundation status product due to the heterogeneity surrounding the rice paddies over the Mekong Delta (e.g., buildings, forests, dykes), which was contaminated in the GNSS-R specular observations. The relationship was highly nonlinear, and its convexity was highly dependent on incidence angle differences. However, correlations were still found between these different microwave remote sensing methodologies even with the different observation resolutions over the heterogenous ground objects in this study.

As with other error-causing factors, notably, there are various factors causing geometric errors. For example, to propose a methodology that was independent from external data in this study, the ellipsoidal height that was derived from DEM and geoidal height information was not used for the rasterization process of the specular points. Most importantly, the grid-based rasterization of specular points was conducted by assuming that the velocity at each specular point was constant throughout each integration time (i.e., the acceleration of each specular point was assumed to be 0). It is still expected that the cross-validation performance could be better improved by rasterizing each specular point without the velocity-constant assumption. Regarding the geometric error correction, we also conducted a tuning experiment of the Gaussian function parameter of the precision index model [i.e., a value of 3.0 was used in this study, as described in Equation (2)] without downsampling. However, the tuning of the model parameter did not significantly differentiate the validation performance with the PALSAR-2 product (data not shown). Hence, rasterization with the consideration of acceleration was more important for tuning this model parameter. Without the acceleration information, the model tuning did not reliably improve the validation performance. For the current data interpretation, we also have to note the temporal differences between the observation times/dates of the CyGNSS and PALSAR-2 products. Due to the quantitative limitation of available specular points in this study, a low effective scattering area specular point group occasionally showed the opposite correlation with the PALSAR-2 backscatters (i.e., 30–35° incidence angles, 0–6 km square root values of the effective scattering area, Table S2). One of the causes of this result is the limited availability of quantitative specular points from CyGNSS over the Mekong Delta and the limited observable swath data of the quadruple PALSAR-2 observations (only 40–50 km widths were used to avoid incidence-angle-difference-derived biases in the polarimetric decomposition analysis). These data quantity limitations might have only partially caused the local optimization of the nonlinear function. Further observations are expected to enable the global optimization of the nonlinear function when estimating the spatial inundation percentage or backscatter intensity from CyGNSS specular reflectivity data. The most importantly, we need to reshape this gaussian function along with the specular points velocity vector (i.e., shifting the gaussian center considering the doppler frequency, and reshape the skewness regarding its delay time in DDM information [32]).

Unlike the specular points with incidence angles wider than 10°, a positive relationship was found between the CyGNSS reflectivity Γ and PALSAR-2 backscattering intensity $\sigma^0$ series for specular points with incidence angles narrower than 10°. Since the difference between the microwave-energy-advancing vector directions of the backscatters and specular reflection values decreased as the incidence angle decreased, these positive relationships could have been found for specular points with such low incidence angles. This indicates that the reflectivity is highly dependent on the dielectric properties, particularly for low-incidence-angle specular points. Because such specular points tend to have low effective scattering areas (i.e., fine spatial resolutions), the incidence angle bias correction on such low-incidence-angle specular points is necessary to enable high-quality information on land surface properties to be derived. For most specular points with incidence angles ranging from 15 to 60°, the CyGNSS reflectivity Γ and PALSAR-2 backscatter intensity $\sigma^0$ tended to show downwardly convex relationships (Table S3). This indicated that wetlands on relatively dry ground with a relatively low dielectric constant do not activate multi-time scattering (e.g., the double/triple bounce effect). Hence, the negative correlations between Γ values and $\sigma^0$ values tended to appear to be simply controlled by the specular reflection or single scattering effect. However, wetlands on wet ground, which have high dielectric constants at a certain level, also enhance multi-time scattering to emit relatively strong power levels not only oriented forwards but also backward. The specular points with incidence angles wider than 60° tended to show upwardly convex negative relationships between the CyGNSS reflectivity Γ and PALSAR-2 backscattering intensity $\sigma^0$ series. These findings indicated that if the incidence angle was wider than a certain level, the ground-volume interactions between inundated soil and wetland vegetation would be more prone to occur than if the specular points had lower incidence angles.

The three domains classified in the 2D scatter plots between the CyGNSS reflectivity Γ values and PALSAR-2-based spatial inundation percentages indicated a microwave scattering status difference among each domain (Figure S2g,m). The specular points in the first domain with Γ values lower than approximately −20 dB (Figure S2g,m; domain shown with the green arrow) tended to reflect relatively high odd/double bounce values. This finding indicated that the ground-volume interaction between inundated soil and the land-covering vegetation in wetlands plays a dominant role in the scattering process in this domain. Because positive correlations tended to appear between the CyGNSS reflectivity Γ and spatial inundation percentage series in this domain, this domain would be more sensitive to inundation than to soil moisture. In the second domain, where the Γ values were between approximately −20 dB and 0 dB, specular points with a 0% spatial inundation percentage were detected (Figure S2g,m; domain shown with the red arrow). In this domain, the Γ values mostly showed negative correlations with the spatial inundation rate and backscattering intensities. This indicates that the multi-time scattering effect would not have played a major role in this domain. Instead, the single scattering effect would have played a major role in such dry ground areas with relatively low dielectric constants. The negative correlations also indicate the possibility that the soil moisture and the vegetation water content may have greater roles than the spatial inundation percentage in such non-inundated wetland ROIs. In the third domain, where the specular points had Γ values greater than approximately 0 dB (Figure S2g,m; domain shown with the blue arrow), the Γ values tended to become significantly high, although the PALSAR-2 backscatter intensity values (including the odd scattering and double bounce values) tended to be low. These results indicate that the contribution of multi-time scattering was negligible and that the surface roughness in this domain would also be low. The presence of a water body without vegetation would have enabled such strong specular reflection conditions under weak backscattering effects. Consistently, Arai et al. [7,8,10] also reported three similar domains from HH/HV backscatter 2D distribution plots. Thus, this might be a common characteristic of L-band active microwave scattering signals collected over tropical wetlands.

For further development, the application of a precision index to a finer-spatial-resolution GNSS-R product (e.g., the CyGNSS interferometric coherence ratio product [38]) would be desirable to improve the spatial resolution of the resulting reflectivity Γ product. Since the differentiation of multi-time scattering processes using the phase information of scattered microwaves is mandatory to improve the inundation detection performance, the Stokes vector-based pseudo-3-component decomposition approach [39] or multi-polarimetric reflectivity/phase information (e.g., HydroGNSS) also need to be addressed for use with the GNSS-R data. To prepare for a robust comparison between SAR data and such polarimetric GNSS-R data, further improvement must be made to the precision index model.

In this study, effective scattering area was employed as the footprint size for the following two reasons. The L-band SAR polarimetric decomposition study of the rice paddies revealed that the SAR backscattering intensity is mainly controlled by ground vegetation and is sensitive to both canopy structure and ground inundation status and coherence was mostly low, impeding the possibility of using polinsar approach [10,21,23]. From this study, we also detected that most of the rice paddies whose L-band SAR backscattering intensity is relatively high showed low GNSS-R reflectivity (Figure 9). This indicated that the GNSS-R signal over the lowland wetlands/rice paddy is sensitive to ground-vegetation interaction and that the reflective property is incoherent rather than coherent. In subsequent studies study, the First Fresnel zone [40] should be considered as the footprint size, particularly for non-vegetated wetlands or paddies with immature rice paddies whose number of days since sowing is shorter than three weeks.

Regarding the nonlinear relationships between the CyGNSS reflectivity Γ and PALSAR-2 backscattering intensity $\sigma^0$ and between the CyGNSS reflectivity Γ and the inundation percentages as affected by incidence angle differences, a model parameterization scheme with an improved precision index model is desirable if both SAR data and GNSS-R data are to be used cooperatively to overcome their observation scale differences.

## 5. Conclusions

For the operational use of GNSS-R data for sustainable tropical wetland management, a simple quality control method was proposed in this study. Even without ad hoc parameter tuning, the proposed simple model comprising the "precision index" and DDM 3D statistics showed a fine performance in visualizing the spatiotemporal dynamics of wetlands at a fine spatiotemporal resolution (500 m spatial resolution, 15-day temporal resolution). Even without using a common change detection algorithm, the precision-index-model-based approach showed temporal dynamics similar to those obtained using a change detection algorithm. By considering the incidence angle difference, we also succeeded in improving the sensitivity and dynamic range of the change detection results. As a result, we now are able to detect two annual inundation peaks over the Mekong Delta, indicating that the multi-cropping rice system dominating this region plays a major role in controlling the inundation status of the delta. The DDM 3D statistics approach was applied to successfully denoise the locally abnormal specular points by adaptively detecting specular points collected over rough land surfaces. The comparison with L-band microwave SAR data based on the precision index showed a reasonable mutual correlation and provided knowledge of how the microwave scattering pattern is affected by the incidence angle over tropical wetlands. Further study is required with a shorter-integration-time coherence product or a polarimetric decomposition product (e.g., stokes vector) containing GNSS-R data with 1st-/2nd-order specular point velocity (e.g., acceleration) derivatives to enable more precise comparisons with L-band SAR data.

**Supplementary Materials:** The following supporting information can be downloaded at: https://www.mdpi.com/article/10.3390/rs14225903/s1.

**Author Contributions:** H.A., M.Z., K.O., S.S. and T.L.T. conceived and designed the experiments; H.A., K.D. and M.H. performed the experiments; H.A., M.Z. and K.D. analyzed the data; H.A., M.Z. and M.H. preprocessed the base datasets; all authors supported the data interpretation and model design; H.A., M.Z. and T.L.T. wrote the paper; and all authors read the paper and provided revision suggestions. All authors have read and agreed to the published version of the manuscript.

**Funding:** This research received no external funding.

**Data Availability Statement:** Not applicable.

**Acknowledgments:** A portion of the field survey activities described in this work was supported by the Japan Society for the Promotion of Science (JSPS) Grants-in-aid for Scientific Research & Scholarship received from JSPS as research execution expenses and by JSPS KAKENHI Grant Numbers 15J00001 and 16J02509. The remote sensing study was financially supported by the Japan Aerospace Exploration Agency (JAXA) and by the JSPS KAKENHI Project (Area No. 16J02509) and JSPS Overseas Research Fellowships (No. 201960100). ALOS-2/PALSAR-2 data are provided under the "Agreement between ESA and JAXA on the cooperation for the use of synthetic aperture radar satellites in Earth science and applications".

**Conflicts of Interest:** The authors declare no conflict of interest.

## References

1. Masson-Delmotte, V.; Zhai, P.; Pirani, A.; Connors, S.L.; Péan, C.; Berger, S.; Huang, M.; Yelekci, O.; Yu, R.; Zhou, B.; et al. Climate change 2021: The physical science basis. In *Contribution of Working Group I to the Sixth Assessment Report of the Intergovernmental Panel on Climate Change*; Cambridge University Press: Cambridge, UK, 2021.
2. Schaefer, H.; Fletcher, S.E.M.; Veidt, C.; Lassey, K.R.; Brailsford, G.W.; Bromley, T.M.; Dlugokencky, E.J.; Michel, S.E.; Miller, J.B.; Levin, I.; et al. A 21st-century shift from fossil-fuel to biogenic methane emissions indicated by 13 CH4. *Science* **2016**, *352*, 80–84. [CrossRef] [PubMed]
3. Nisbet, E.G.; Manning, M.R.; Dlugokencky, E.J.; Fisher, R.E.; Lowry, D.; Michel, S.E.; Lund Myhre, C.; Platt, S.M.; Allen, G.; Bousquet, P.; et al. Very strong atmospheric methane growth in the 4 years 2014–2017: Implications for the Paris Agreement. *Glob. Biogeochem. Cycles* **2019**, *33*, 318–342. [CrossRef]
4. Rigby, M.; Montzka, S.A.; Prinn, R.G.; White, J.W.C.; Young, D.; O'Doherty, S.; Lunt, M.F.; Ganesan, A.L.; Manning, A.J.; Simmonds, P.G.; et al. Role of atmospheric oxidation in recent methane growth. *Proc. Natl. Acad. Sci. USA* **2017**, *114*, 5373–5377. [CrossRef] [PubMed]
5. Jackson, R.B.; Saunois, M.; Bousquet, P.; Canadelle, J.G.; Poulter, B.; Stavert, A.R.; Bergamaschi, P.; Niwa, Y.; Segers, A.; Tsuruta, A. Increasing anthropogenic methane emissions arise equally from agricultural and fossil fuel sources. *Environ. Res. Lett.* **2020**, *15*, 071002. [CrossRef]
6. Saunois, M.; Stavert, A.R.; Poulter, B.; Bousquet, P.; Canadell, J.G.; Jackson, R.B.; Raymond, P.A.; Dlugokencky, E.J.; Houweling, S.; Patra, P.K.; et al. The global methane budget 2000–2017. *Earth Syst. Sci. Data* **2020**, *12*, 1561–1623. [CrossRef]
7. Arai, H.; Takeuchi, W.; Oyoshi, K.; Nguyen, L.D.; Fumoto, T.; Inubushi, K.; Le Toan, T. Pixel-Based Evaluation of and Related Greenhouse Gas in the Integrating SAR Data and ground observation. In *Remote Sensing of Agriculture and Land Cover/Land Use Changes in South and Southeast Asian Countries*; Springer: Cham, Switzerland, 2021; pp. 251–266.
8. Arai, H.; Takeuchi, W.; Oyoshi, K.; Nguyen, L.D.; Tachibana, T.; Uozumi, R.; Terasaki, K.; Miyoshi, T.; Yashiro, H.; Inubushi, K. Low cost and transparent MRV system of GHG emissions based on satellite remote sensing data: Case study on CH4 emission from the Mekong delta. In Monitoring of Global Environment and Disaster Risk Assessment from Space: The IIS Forum Proceedings; Volume 27 (National Diet Library, Japan) pp. 3–10. Available online: https://cir.nii.ac.jp/crid/1520853833371853952 (accessed on 10 February 2022).
9. France, J.L.; Fisher, R.E.; Lowry, D.; Allen, G.; Andrade, M.F.; Bauguitte, S.J.B.; Bower, K.; Broderick, T.J.; Daly, M.C.; Forster, G.; et al. δ13C methane source signatures from tropical wetland and rice field emissions. *Phil. Trans. R. Soc. A* **2021**, *380*, 20200449.
10. Arai, H.; Le Toan, T.; Takeuchi, W.; Oyoshi, K.; Fumoto, T.; Inubushi, K. Evaluating irrigation status in Mekong Delta through polarimetric L-band SAR data assimilation. *Remote Sens. Environ.* **2022**, *279*, 113139. [CrossRef]
11. Ruf, C.S.; Atlas, R.; Chang, P.S.; Clarizia, M.P.; Garrison, J.L.; Gleason, S.; Katzberg, S.J.; Jelenak, Z.; Johnson, J.T.; Majumdar, S.J.; et al. New Ocean Winds Satellite Mission to Probe Hurricanes and Tropical Convection. *Bull. Am. Meteorol. Soc.* **2016**, *97*, 385–395. [CrossRef]
12. Carreno-Luengo, H.; Luzi, G.; Crosetto, M. Impact of the Elevation Angle on CYGNSS-R Bistatic Reflectivity as a function of the Effective Surface Roughness Over Land Surfaces. *Remote Sens.* **2018**, *10*, 1749. [CrossRef]
13. Stilla, D.; Zribi, M.; Pierdicca, N.; Baghdadi, N.; Huc, M. Desert Roughness Retrieval Using CYGNSS GNSS-R Data. *Remote Sens.* **2020**, *12*, 743. [CrossRef]

14. Gerlein-Safdi, C.; Bloom, A.A.; Plant, G.; Kort, E.A.; Ruf, C.S. Improving representation of tropical wetland methane emissions with CYGNSS inundation maps. *Glob. Biogeochem. Cycles* **2021**, *35*, e2020GB006890. [CrossRef]
15. Rodriguez-Alvarez, N.; Podest, E.; Jensen, K.; McDonald, K.C. Classifying inundation in a tropical wetlands complex with GNSS-R. *Remote Sens.* **2019**, *11*, 1053. [CrossRef]
16. Zuffada, C.; Chew, C.; Nghiem, S.V. Global navigation satellite system reflectometry (GNSS-R) algorithms for wetland observations. In Proceedings of the 2017 IEEE International Geoscience and Remote Sensing Symposium (IGARSS), Fort Worth, TX, USA, 23–28 July 2017; pp. 1126–1129.
17. Chew, C.; Small, E. Estimating inundation extent using CYGNSS data: A conceptual modeling study. *Remote Sens. Environ.* **2020**, *246*, 111869. [CrossRef]
18. Setti, P.D.T.; Tabibi, S.; Van Dam, T. CYGNSS GNSS-R Data for Inundation Monitoring in the Brazilian Pantanal Wetland. In Proceedings of the IGARSS 2022-2022 IEEE International Geoscience and Remote Sensing Symposium, Kuala Lumpur, Malaysia, 17–22 July 2022; pp. 5531–5534.
19. Yang, W.; Gao, F.; Xu, T.; Wang, N.; Tu, J.; Jing, L.; Kong, Y. Daily Flood Monitoring Based on Spaceborne GNSS-R Data: A Case Study on Henan, China. *Remote Sens.* **2021**, *13*, 4561. [CrossRef]
20. Zhang, G.; Xiao, X.; Dong, J.; Zhang, Y.; Xin, F.; Qin, Y.; Doughty, R.B.; Moor, B., III. Reply to: "Correlation between paddy rice growth and satellite-observed methane column abundance does not imply causation". *Nat. Commun.* **2021**, *12*, 1189. [CrossRef]
21. Arai, H.; Takeuchi, W.; Oyoshi, K.; Nguyen, L.D.; Inubushi, K. Estimation of Methane Emissions from Rice Paddies in the Mekong Delta Based on Land Surface Dynamics Characterization with Remote Sensing. *Remote Sens.* **2018**, *10*, 1438. [CrossRef]
22. Arai, H. The Anthropogenic Greenhouse Gas Emission from Tropical High Carbon Reservoirs. Chiba University Library Online Public Library Catalog. 2015. Available online: http://opac.ll.chiba-u.jp/da/curator/900119174/HMA_0068.pdf (accessed on 11 April 2018). (In Japanese).
23. Arai, H.; Takeuchi, W.; Oyoshi, K.; Nguyen, L.D.; Tachiba, T.; Inubushi, K. Regional evaluation on greenhouse gas-mitigation & yield-increase performance of a water-saving irrigation practice's dissemination in rice paddies in the Mekong Delta. *Monit. Glob. Environ. Disaster Risk Assess. Space IIS Forum Proc.* **2018**, *26*, 43–50.
24. Van Hong, N.P.; Nga, T.T.; Arai, H.; Hosen, Y.; Chiem, N.H.; Inubushi, K. Rice straw management by farmers in a triple rice production system in the Mekong Delta, Viet Nam. *Trop. Agr. Develop.* **2014**, *58*, 155–162.
25. Arai, H.; Hosen, Y.; Van Hong, N.P.; Chiem, N.H.; Inubushi, K. Greenhouse gas emissions derived from rice straw burning and straw-mushroom cultivation in a triple rice cropping system in the Mekong Delta. *Soil Sci. Plant Nutr.* **2015**, *61*, 719–735. [CrossRef]
26. Arai, H.; Hosen, Y.; Chiem, N.H.; Inubushi, K. Alternate wetting and drying enhanced the yield of a triple-cropping rice paddy of the Mekong Delta. *Soil Sci. Plant Nutr.* **2021**, *67*, 493–506. [CrossRef]
27. Arai, H. Increased rice yield and reduced greenhouse gas emissions through alternate wetting and drying in a triple-cropped rice field in the Mekong Delta. *Sci. Total Environ.* **2022**, *842*, 156958. [CrossRef] [PubMed]
28. Motte, E.; Zribi, M.; Fanise, P.; Egido, A.; Darrozes, J.; Al-Yaari, A.; Baghdadi, N.; Baup, F.; Dayau, S.; Fieuzal, R.; et al. GLORI: A GNSS-R Dual Polarization Airborne Instrument for Land Surface Monitoring. *Sensors* **2016**, *16*, 732. [CrossRef] [PubMed]
29. Pierdicca, N.; Comite, D.; Camps, A.; Carreno-Luengo, H.; Cenci, L.; Clarizia, M.P.; Costantini, F.; Dente, L.; Guerriero, L.; Mollfulleda, A.; et al. Potential of spaceborne GNSS reflectometry for soil moisture, biomass and freeze-thaw monitoring: Summary of an ESA-funded study. *IEEE Geosci. Remote Sens. Mag.* **2021**, *10*, 8–38. [CrossRef]
30. Shimada, M.; Japan Aerospace Exploration Agency-Earth Observation Research Center. ALOS-2 characteristics, CAL/VAL results and operational status. In Proceedings of the ALOS Kyoto & Carbon Initiative 21th Science Team Meeting, Kyoto, Japan, 3–5 December 2014; Available online: http://www.eorc.jaxa.jp/ALOS/kyoto/dec2014_kc21/pdf/3-02_KC21_ALOS-2_Shimada-JAXA.pdf (accessed on 11 April 2018). (In Japanese).
31. Singh, G.; Malik, R.; Mohanty, S.; Rathore, V.S.; Yamada, K.; Umemura, M.; Yamaguchi, Y. Seven-component scattering power decomposition of POLSAR coherency matrix. *IEEE Trans. Geosci. Remote Sens.* **2019**, *57*, 8371–8382. [CrossRef]
32. Collett, I.W. Applying GNSS Reflectometry-Based Stare Processing to Modeling and Remote Sensing of Wind-Driven Ocean Surface Roughness. Ph.D. Thesis, University of Colorado at Boulder, Boulder, CO, USA, 2021.
33. Son, N.T.; Chen, C.F.; Chen, C.R.; Duc, H.N.; Chang, L.Y. A phenology-based classification of time-series MODIS data for rice crop monitorin in Mekong Delta, Vietnam. *Remote Sens.* **2013**, *6*, 135–156. [CrossRef]
34. Loc, H.H.; Lixian, M.L.; Park, E.; Dung, T.D.; Shrestha, S.; Yoon, Y.J. How the saline water intrusion has reshaped the agricultural landscape of the Vietnamese Mekong Delta, a review. *Sci. Total Environ.* **2021**, *794*, 148651. [CrossRef]
35. Le Toan, T.; Huu, N.; Simioni, M.; Phan, H.; Arai, H.; Mermoz, S.; Bouvet, A.; Eccher, I.d.; Diallo, Y.; Duong, T.H.; et al. Agriculture in Viet Nam under the impact of climate change. In Climate Change in Viet Nam. Impacts and Adaptation. A COP26 Assessment Report of the GEMMES Viet Nam Project; 2021, HAL, France. Available online: https://hal.inrae.fr/hal-03456472/file/2021_Simioni_Rapport_AFD_Climate%20change_GEMMES-pages-191-228.pdf (accessed on 1 November 2022).
36. Kondolf, G.M.; Schmitt, R.J.P.; Carling, P.A.; Goichot, M.; Keskinen, M.; Arias, M.E.; Bizzi, S.; Castelletti, A.; Cocharane, A.; Darby, S.E.; et al. Save the Mekong Delta from drowning. *Science* **2022**, *376*, 583–585. [CrossRef]
37. Minderhoud, P.S.J.; Erkens, G.; Pham, V.H.; Bui, V.T.; Erban, L.; Kooi, H.; Stouthamer, E. Impacts of 25 years of groundwater extraction on subsidence in the Mekong delta, Vietnam. *Environ. Res. Lett.* **2017**, *12*, 064006. [CrossRef]

38. Carreno-Luengo, H.; Warnock, A.; Ruf, C.S. The CYGNSS coherent End-to-end simulator: Development and results. *IEEE Trans. Geosci. Remote Sens.* **2022**, *60*, 7441–7444.
39. Bosch-Lluis, X.; Munoz-Martin, J.F.; Rodriguez-Alvarez, N.; Oudrhiri, K. Towards GNSS-R hybrid compact polarimetry: Introducing the stokes parameters for SMAP-R dataset. *IEEE Trans. Geosci. Remote Sens.* **2022**, *60*, 7425–7428.
40. Camps, A. Spatial resolution in GNSS-R under coherent scattering. *IEEE Trans. Geosci. Remote Sens.* **2019**, *17*, 32–36. [CrossRef]

remote sensing

MDPI

Article

# Enhancing Spatial Resolution of GNSS-R Soil Moisture Retrieval through XGBoost Algorithm-Based Downscaling Approach: A Case Study in the Southern United States

**Qidi Luo, Yueji Liang *, Yue Guo, Xingyong Liang, Chao Ren, Weiting Yue, Binglin Zhu and Xueyu Jiang**

College of Geomatics and Geoinformation, Guilin University of Technology, Guilin 541004, China; luoqidi@glut.edu.cn (Q.L.); guoyue@nnct.edu.cn (Y.G.); liangxy@glut.edu.cn (X.L.); renchao@glut.edu.cn (C.R.); yueweiting@glut.edu.cn (W.Y.); zhubinglin@glut.edu.cn (B.Z.); xiaojiang@glut.edu.cn (X.J.)
* Correspondence: lyjayq@glut.edu.cn; Tel.: +86-158-783-57721

**Abstract:** The retrieval of soil moisture (SM) using the Global Navigation Satellite System-Reflectometry (GNSS-R) technique has become a prominent topic in recent years. Although prior research has reached a spatial resolution of up to 9 km through the Cyclone Global Navigation Satellite System (CYGNSS), it is insufficient to meet the requirements of higher spatial resolutions for hydrological or agricultural applications. In this paper, we present an SM downscaling method that fuses CYGNSS and SMAP SM. This method aims to construct a dataset of CYGNSS observables, auxiliary variables, and SMAP SM (36 km) products. It then establishes their nonlinear relationship at the same scale and finally builds a downscale retrieval model of SM using the eXtreme Gradient Boosting (XGBoost) algorithm. Focusing on the southern United States, the results indicate that the SM downscaling method exhibits robust performance during both the training and testing processes, enabling the generation of a CYGNSS SM product with a 1 day/3 km resolution. Compared to existing methods, the spatial resolution is increased threefold. Furthermore, in situ sites are utilized to validate the downscaled SM, and spatial correlation analysis is conducted using MODIS EVI and MODIS ET products. The CYGNSS SM obtained by the downscaling model exhibits favorable correlations. The high temporal and spatial resolution characteristics of GNSS-R are fully leveraged through the downscaled method proposed. Furthermore, this work provides a new perspective for enhancing the spatial resolution of SM retrieval using the GNSS-R technique.

**Keywords:** GNSS-R; CYGNSS; SMAP; downscaled; soil moisture

**Citation:** Luo, Q.; Liang, Y.; Guo, Y.; Liang, X.; Ren, C.; Yue, W.; Zhu, B.; Jiang, X. Enhancing Spatial Resolution of GNSS-R Soil Moisture Retrieval through XGBoost Algorithm-Based Downscaling Approach: A Case Study in the Southern United States. *Remote Sens.* **2023**, *15*, 4576. https://doi.org/10.3390/rs15184576

Academic Editors: Hugo Carreno-Luengo and Chun-Liang Lin

Received: 14 July 2023
Revised: 29 August 2023
Accepted: 14 September 2023
Published: 17 September 2023



## 1. Introduction

Soil moisture (SM) plays a pivotal role in many natural phenomena and processes. For instance, it directly affects crop growth and can be a significant factor in natural disasters such as land degradation, floods, and landslides [1]. These issues have profound impacts, including on food security and the stability of ecological environments, making accurate and real-time monitoring of SM particularly important. However, traditional SM detection methods have notable limitations. These methods primarily rely on direct measurements from ground detectors or meteorological stations, which means they require substantial human and material resources and are time-consuming [2]. Moreover, due to the limitations of these methods, they cannot achieve large-scale, efficient, and low-cost SM retrieval. For vast areas and complex terrains, their detection performance is severely limited. Fortunately, the advent of remote sensing technology provides a new avenue to address this issue. Remote sensing technology can use satellites or drones to monitor the ground from the air, avoiding the difficulties of ground detection and thus achieving large-scale SM retrieval [3,4]. In fact, the European Space Agency (ESA) and the National Aeronautics and Space Administration (NASA) have launched the Soil Moisture and Ocean Salinity (SMOS) satellite [5] and the Soil Moisture Active Passive (SMAP) mission [6] for

SM retrieval. Both missions can achieve global SM retrieval with a spatial resolution of about 40 km, and they can revisit the globe every 2–3 days. However, while remote sensing technology and related satellite missions such as SMOS and SMAP provide global SM retrieval capabilities, the resolution of these products is relatively low, making them more suitable for large-scale applications. For medium- and small-scale applications that require more detailed observations, such as irrigation management in farmland or flood warnings in specific areas, these methods may not meet the needs.

The technique of Global Navigation Satellite System-Reflectometry (GNSS-R) represents a novel type in the field of remote sensing. Its internal L-band signal source is adequate and exhibits high penetration capabilities for vegetation, soil, snow, etc. It is capable of all-weather, all-day observation and has excellent potential for SM retrieval [7–9]. GNSS-R receivers are generally installed on the ground or on aircraft. Although they have an excellent detection accuracy, the monitoring range limits its ability to achieve a wide range of SM retrieval [10]. CYGNSS was successfully launched in 2016, with a revisit cycle of 2.8 (median) and 7.2 (average) hours [11], providing ample data for SM retrieval by GNSS-R technique. Thus, using the GNSS-R technique to retrieve SM has become a hot research topic in recent years. Chew et al. [12] showed that there is a strong linear relationship between the surface reflectance of CYGNSS and SMAP SM, and a global SM product with a resolution of 36 km was produced through linear method. Ruf [13] proposed that SMAP SM can be supplemented by using the relative signal-to-noise ratio (rSNR) of CYGNSS to SM retrieval. Al-Khaldi et al. [14] considered that vegetation and surface roughness would affect SM. They proposed a method for CYGNSS SM retrieval through time series. A global SM product of $0.2^\circ \times 0.2^\circ$ was finally generated. Considering that the terrain, vegetation, and surface roughness have an impact on the GNSS signal, the relationship between the signal and SM is relatively complex and nonlinear. Machine learning has been frequently used in the study of CYGNSS SM retrieval because of its great advantages in handling nonlinear situations. Eroglu et al. [15] combined CYGNSS observables with in situ sites observations, Vegetation Water Content (VWC), Normalized Vegetation Index (NDVI), and topography features. Finally, a daily SM product with a resolution of 9 km was generated using the Artificial Neural Network (ANN). Senyurek et al. [16] obtained the daily SM of the United States with a resolution of 36 km using CYGNSS and in situ site observations based on machine learning algorithms. The results showed that the prediction effect of Random Forest (RF) was the best, with an RMSE of 0.052. Jia et al. [17] pre-classified land cover types and used the eXtreme Gradient Boosting (XGBoost) method for SM retrieval. Compared with the accuracy of SM retrieval without pre-classification, there was an improvement, with an RMSE of 0.052.

However, the SM products obtained from the aforementioned microwave remote sensing data have a coarse resolution, which limits their utility in medium- and small-scale hydrological and agricultural applications. Zhan et al. [18] first introduced an empirical polynomial for downscaling, marking an initial exploration of effective strategies to address this issue. Subsequently, Chauhan et al. [19] improved upon Zhan's method, enhancing its performance. In this empirical polynomial downscaling method, high-resolution SM is expressed as a polynomial function of surface temperature, plant index, and surface reflectance derived from brightness temperature data. This innovative method provides a fresh perspective for tackling the downscaling of SM. Piles et al. [20] further optimized this downscaling polynomial fitting method. Their improvement replaced surface reflectance in the polynomial equation with coarse-resolution brightness temperature data, making the method more flexible and efficient in handling practical problems. Moreover, this polynomial fitting downscaling method has been widely applied in the downscaling of various SM products, such as SMOS and AMSR-E, and also in various high-resolution remote sensing image products, such as MODIS and MSG-SEVIRI. This has been confirmed by many scholars [21–26]. Their research further validates the practicality and broad application value of this method. In order to retrieve daily SM at a 9 km resolution, Das et al. [27] downscaled the coarse-resolution (approximately 40 km) SMAP L-band brightness temperature data using the high-resolution (1–3 km) L-band Synthetic Aperture Radar (SAR) backscatter observations. Based on artificial intelligence techniques including Support Vector Machines, Artificial Neural Networks, and Associated Vector Machines, Srivastava et al. [28] fused MODIS surface temperature with SMOS SM and enhanced the spatial resolution of SMOS SM by using downscaling methods. The factor used to represent the high-resolution state of SM plays a crucial role in determining the accuracy of the downscaled SM. The downscaled SM has higher accuracy compared to the original coarse-resolution SMOS and AMSR-E SM, with the $R$ rising from 0.27 to 0.96 [29]. Compared to the observed data, the accuracy of the downscaled SM has improved relative to the products of SMOS and AMSR-E [30]. This means that downscaling methods could be attempted to provide high-resolution SM for products such as SMAP, SMOS, AMSR-E, and NASA-USDA.

The aforementioned research demonstrates both the significant advantages of using GNSS-R technique for SM retrieval and the notable effects of using downscaling methods to enhance the spatial resolution of SM products. However, no studies have yet used the downscaling method to improve the spatial resolution of GNSS-R technique. At present, the spatial resolution achieved by SM retrieval based on spaceborne GNSS-R is limited (up to 9 km). Spatial downscaling of microwave SM is a crucial strategy. It addresses the pressing need for higher spatial resolution SM data, which is essential for local hydrological or agricultural applications. Therefore, this paper proposes a method for constructing a SM downscaling model. This method aims to fuse the CYGNSS observables and auxiliary variables with SMAP SM (36 km) products, forming a nonlinear relationship at the same scale. Finally, a downscaling model will be built based on the XGBoost algorithm to retrieve SM with a spatial resolution of 3 km. In the end, the SM retrieval using GNSS-R technique is successfully spatially downscaled, improving the spatial resolution of SM retrieval.

## 2. Materials and Methods

### 2.1. Study Area

The study area is located in the southern United States, characterized by diverse terrains and rich ecosystems. The region experiences a subtropical humid climate, with an average annual precipitation of approximately 834.45 mm, contributing to the area's rich biodiversity and thriving ecosystems. Geographically, the study area exhibits significant variations in altitude, ranging from −88 m to 4277 m, with an average altitude of 1778 m. The terrain generally features higher elevations in the west and lower in the east. This variation in terrain provides excellent conditions for studying the relationship between SM and environmental factors such as terrain and climate. Ecologically, the primary land cover types in the study area are grasslands and tropical savannas, collectively accounting for 55% of the total area. Additionally, a considerable portion of the western region is covered by open shrublands, making up 13% of the total area. The elevation and land cover types of the study area are shown in Figure 1.

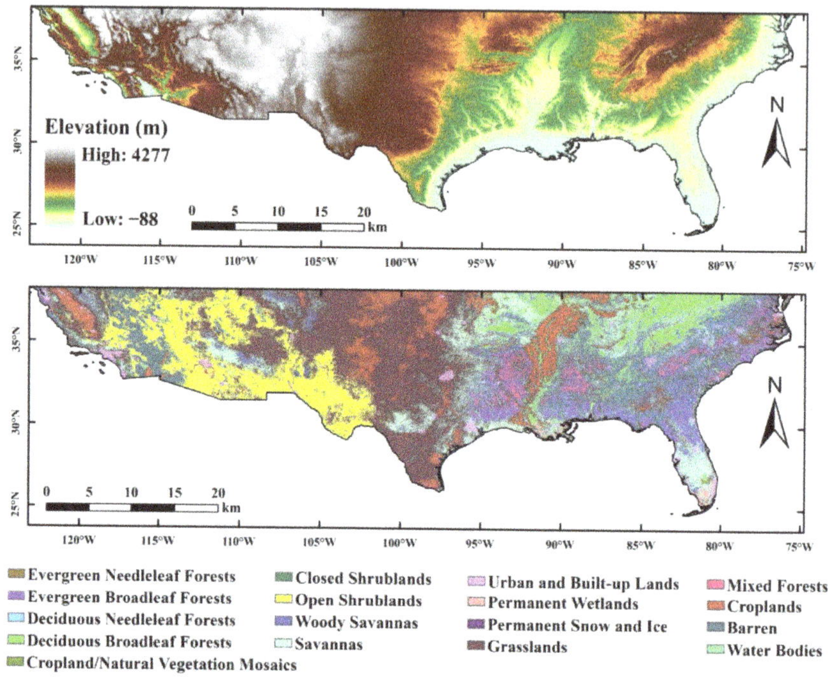


**Figure 1.** DEM and land cover type map of the study area.

### 2.2. *Cyclone Global Navigation Satellite System*

As a component of NASA's Earth System Science Pathfinder project, the Cyclone Global Navigation Satellite System (CYGNSS) was launched on 15 December 2016. The observatories are composed of eight microsatellites. They offer almost uninterrupted coverage of the Earth due to their orbit inclination of approximately 35° to the equator. This positioning results in an average revisit time of 7 h and a median revisit time of 3 h. This inclination allows CYGNSS to cover an observational range from 38°N to 38°S. Therefore, we selected the southern part of the United States as the study area (CYGNSS observables cannot cover the entire US).

The objective of this study is to retrieve SM within a specific region. To achieve this, we utilized the CYGNSS Level-1 (L1) version 2.1 product, with data sourced from the Physical Oceanography Distributed Active Archive Center (PO.DAAC, https://podaac.jpl.nasa.gov/, accessed on 1 April 2023). The primary goal of CYGNSS is to enhance understanding and prediction of tropical cyclone intensity by leveraging signals from the Global Navigation Satellite System (GNSS). The core component of this system is the Delay Doppler Mapping Instrument (DDMI), whose main task is to generate Delay Doppler Maps (DDMs) [31]. DDMs represent the received surface power of each observed specular reflection point through a series of time delays and Doppler frequencies, measured on a bin-by-bin basis. In other words, they provide a two-dimensional representation of the reflection characteristics of GNSS signals. These characteristics are influenced by factors such as SM and vegetation cover, and can therefore be used to infer SM. It is important to note that the DDMI initially measures in uncalibrated "counts", which have a linear relationship with the total signal power it processes. The total signal power includes thermal radiation from the Earth and the DDMI itself, as well as GPS signals scattered from the land surface. However, during the Level-1A calibration process, each bin in the DDM converts these raw counts into watts, allowing for a more intuitive understanding and analysis of the data. The CYGNSS observables used in this paper cover the period from 1 January to 31 December 2019.

The surface reflectivity can be estimated through a variety of methods with various coherence and incoherence assumptions using the observables in the L1 data [15,32,33]. In water accumulation areas such as lakes, rivers, and wetlands, low surface roughness leads to dominant coherent scattering in forward scattering. Even with higher SM, coherent forward scattering remains strong due to water. However, GPS signals interacting with vegetation introduce some incoherent components. Higher SM regions show a stronger signal intensity due to a relatively higher SNR compared to lower SM areas. Thus, in this paper, we adopted the approach proposed by Rodriguez-Alvarez et al. [32] to calculate reflectivity, under the assumption that the observed GNSS-R signal is predominantly made up of coherent reflections. This involves using the BRCS (denoted as '*brcs*' in CYGNSS L1) and the range terms to calculate the reflectivity ($\Gamma_{RL}(\theta_i)$) as:

$$\Gamma_{RL}(\theta_i) = (\frac{4\pi}{\lambda})^2 \frac{P_{RL}^{coh}(rst + rsr)^2}{P_t G_t G_r} \quad (1)$$

where $P_{RL}^{coh}$ represents the dual base radar coherent receive power. The subscripts $R$ and $L$ stand for the right circularly polarized GNSS transmit antenna and the left circularly polarized GNSS-R antenna, respectively. The GNSS signal wavelength is denoted by $\lambda$. $rst$ and $rsr$ refer to the distances from the specular reflection point to the GNSS transmitter and the GNSS-R receiver, respectively. $P_t$ signifies the peak power of the transmitting GNSS signal. $G_t$ and $G_r$ are the gains of the transmitting and receiving antennas, respectively. Lastly, $\Gamma_{RL}(\theta_i)$ is the surface reflectance at an incidence angle of $\theta_i$.

Leading Edge Slope (LES) and Trailing Edge Slope (TES) are indicators associated with coherent or incoherent scattering conditions. An increase in the incoherent reflection component within the reflected signal typically results in a corresponding increase in the absolute values of both LES and TES. Following the methodologies presented by Carreno-Luengo et al. [34] and Rodriguez-Alvarez et al. [32], LES and TES can be calculated as follows:

$$\text{LES} = \frac{\Gamma_m - \Gamma_{m-3}}{3\Delta} \quad (2)$$

$$\text{TES} = \frac{\Gamma_{m+3} - \Gamma_m}{3\Delta} \quad (3)$$

where $\Gamma_m$ represents the peak reflectivity at the reflection point, $\Gamma_{m-3}$ is the reflectivity at the third point before the reflection point, $\Gamma_{m+3}$ is the reflectivity at the third point after the reflection point, and $\Delta$ stands for the delay resolution of the Doppler delay map, which is 0.2552 chips.

DDM_SNR is one of the most basic variables in CYGNSS observables. When the value of SM increases in the same area, the difference between the corresponding values of DDM_SNR also increases. Therefore, DDM_SNR is added to the model as a factor affecting the SM retrieval. For SM retrieval in the machine learning framework, the derived reflectivity, together with LES, TES, and DDM_SNR, are used as the input layer characteristics of CYGNSS observables.

### *2.3. Soil Moisture Active Passive Data*

The reference products utilized in this study primarily originate from the SMAP satellite, launched by NASA in 2015. The primary mission of this satellite is to monitor global surface SM and freeze-thaw states, aiming to gain a deeper understanding and knowledge of the Earth's surface water cycle, climate change, ecosystem dynamics, and the impact of human activities. The SMAP satellite employs an L-band radiometer for its observations, a device capable of penetrating clouds and most vegetation to directly measure microwave radiation from the ground, thereby inferring SM and freeze-thaw states. The SMAP satellite revisits each location every 2–3 days, offering a very short global coverage cycle. Notably, the SMAP satellite carries out two types of observations: ascending

(6:00 a.m.) and descending (6:00 p.m.) [35]. This design allows for comparisons and analyses at different times for the same location, providing more comprehensive information.

The data used in this study is obtained from the National Snow and Ice Data Center (NSIDC, https://nsidc.org/, accessed on 15 April 2023). We selected the SMAP Level-3 (L3) Radiometer Global Daily 36 km EASE-Grid Soil Moisture (Version 8, SPL3SMP) as the reference product. This product offers daily estimates of global land surface conditions. The data derived from SMAP's L-band are resampled to a global, cylindrical, 36-km Equal-Area Scalable Earth Grid. The data period is from 1 January to 31 August 2019, providing ample samples for our study.

The original format of the SMAP product is HDF5. In this study, we use the HEG tool (HDFEOS To GeoTIFF Conversion Tool) to convert it into an easily processed Geotiff data format.

### 2.4. International Soil Moisture Network

In this paper, the in situ SM observations from the ISMN sites [36] are used to validate the CYGNSS SM data predicted by the downscaling model. Globally, the ISMN has set up more than 50 SM monitoring networks that are either operational or experimental. These networks provide a unified in situ SM database on a global scale, with a standardized data format and pre-processing quality flags [37]. The majority of sites that offer time and space co-located with CYGNSS observables are located in North America. Consequently, we selected 78 available sites within the spatial coverage of CYGNSS for our study (Figure 2). These sites primarily belong to the Soil Climate Analysis Network (SCAN), the U.S. Climate Reference Network (USCRN), and the Snow Telemetry Network (SNOTEL). The hourly SM data from the ISMN was processed by filtering it with the provided quality mark (marked with a "G" for "good") and subsequently converting it into daily averages. The surface SM data utilized was at a depth of 5 cm, aligning with the penetration depth of L-band microwave signals. For a comprehensive overview of ISMN, readers can refer to [36,38]. The ISMN dataset can be accessed publicly (http://ismn.geo.tuwien.ac.at, accessed on 20 April 2023).

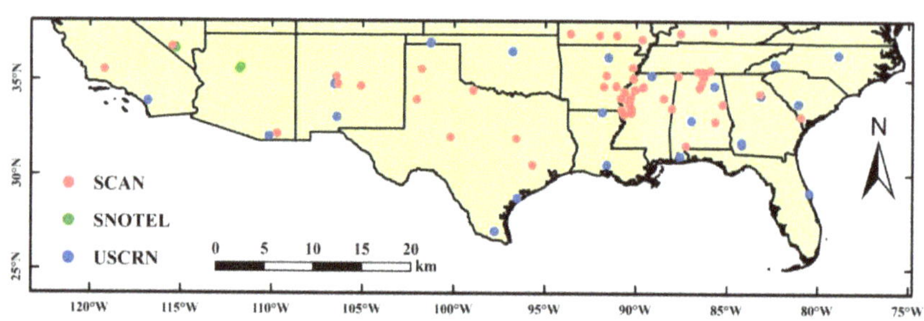


**Figure 2.** Geographic distribution of International Soil Moisture Network (ISMN) sites in the study area.

### 2.5. Auxiliary Data

According to existing research, SM is influenced by additional variables in addition to rainfall, including elevation, land cover type, annual accumulated days, Normalized Difference Vegetation Index (NDVI), and latitude and longitude information of satellite sampling points [31,39]. These factors are typically used as auxiliary variables in downscaling methods [40–43].

Topography, as a significant non-living factor, greatly influences the variability of soil hydrothermal resources. The differences in elevation directly impact the spatial redistribution of solar radiation and rainfall. Therefore, in our downscaling model, we incorporated altitude as the topographic variable. The source of altitude data is the Shuttle Radar Topography Mission (SRTM) [44]. The influence on SM varies with different types of land cover, as they have different effects on the storage and release of moisture. The land cover type data used in this paper is based on the International Geosphere Biosphere Programme (IGBP) [45] land cover map derived from the MODIS. NDVI is widely used to assess vegetation growth, drought conditions, and ecological environments. Since NDVI exhibits a high sensitivity to factors such as vegetative cover and SM content, it is also used for retrieving SM and vegetation covering [46]. The NDVI product is calculated from the daily 250 m product provided by MODIS (MOD09GQ). The precipitation plays a significant role in vegetation growth and has a strong impact on SM. Precipitation affects SM as it comes into contact with the soil, and there is a positive correlation between SM and precipitation. Therefore, we include precipitation as an input variable in the model. The daily average precipitation in the study area is obtained through the Climate Hazards Group InfraRed Precipitation with Station data (CHIRPS) project. Table 1 summarizes the fundamental characteristics of the auxiliary variables used in this paper.

**Table 1.** Overview of data utilized for the downscaling process in this paper.

| Datasets | Variables | Temporal Resolution (Day) | Spatial Resolution | Time |
|---|---|---|---|---|
| MOD09GQ | NDVI | 1 | 250 m | 1 January–31 December 2019 |
| MCD12Q1 | Land cover | - | 500 m | 2019 |
| SRTM | DEM | - | 30 m | 2019 |
| CHIRPS | Precipitation | 1 | 0.5° | 1 January–31 December 2019 |
| - | Lon, Lat | 1 | - | 1 January–31 December 2019 |
| - | Doy | 1 | - | 1 January–31 December 2019 |

## 3. Soil Moisture Downscaling Framework

### *3.1. Random Forest (RF)*

Ho et al. [47] first proposed the concept of Random Forest (RF) in 1998; then, Breiman et al. [48] systematically developed it in 2001. RF is a collective model constructed on the foundation of decision trees. It is implemented through the Bagging concept of ensemble learning, aiming to solve the problem of overfitting that is common in single decision tree algorithms. The decision tree is a fundamental component. Due to its significant advantages in handling high-dimensional feature data and large datasets, RF is widely used in multivariate regression problems. Compared with ordinary decision trees, RF makes some improvements in the process of building decision trees. During the generation of a regular decision tree, the optimal feature among all sample features on the node is used. However, the RF algorithm randomly selects a certain number of attribute features when generating a decision tree, and then picks the optimal feature from these randomly selected features to construct the decision tree. The decision trees built using RF have different structures. They will not lead to overfitting due to the addition of more trees, but instead produce a limited value of generalization error. This approach not only reduces fitting errors but also avoids repetitive learning, which helps to enhance the predictive performance of the final model and improve its generalization ability.

The process of the random forest algorithm is: (1) Performing $n$ random samplings on the training dataset, each time taking $m$ samples, to obtain a subset of data $S_n = \{(x_1, y_1), (x_2, y_2), (x_3, y_3), \cdots, (x_m, y_m)\}$ containing $m$ samples. (2) Using these sub-datasets to train $n$ weak prediction models $f_n(x)$ separately. (3) When training decision tree model nodes, a subset of feature samples is selected from all samples. Then, the optimal feature for splitting the decision tree is chosen from this randomly selected subset of feature samples. (4) The results of the various weak prediction models are consolidated according to the specific problem at hand. For regression functions, the final output is the arithmetic average of all the weak prediction models.

### 3.2. eXtreme Gradient Boosting (XGBoost)

The eXtreme Gradient Boosting (XGBoost) method, proposed by Chen et al. [49], is also an ensemble learning method based on gradient boosting machines. Similar to RF, XGBoost is a learner based on Classification and Regression Trees (CART). It implements ensemble learning of multiple CART trees by optimizing the traditional Gradient Boosting Decision Tree (GBDT). It can be used to solve various machine learning problems, including classification and regression. While each tree in the RF algorithm is trained in parallel, the decision trees in XGBoost are not mutually independent. The construction process of the XGBoost model is as follows: First, an initial tree is built using the training set for model training, which results in residuals between the model's predicted and actual values. Then, during each iteration, a tree is added to fit the residuals from the model's previous prediction until the model's learning process is terminated. Ultimately, this forms an iterative residual tree collection, an ensemble of numerous tree models. The predicted value can be calculated as follows:

$$\hat{y}_i = \sum_{k=1}^{K} f_k(x_i) \tag{4}$$

where $\hat{y}_i$ represents the final model prediction value, $K$ represents all the built CART trees, $x_i$ represents the features of the $i$ sample, and $f_k(x_i)$ represents the prediction value of the $k$ tree. The objective function calculation formula for XGBoost is shown in Equation (5):

$$O_{bj} = \sum_{i=1}^{m} l(\hat{y}_i, y_i) + \sum_{k=1}^{K} \Omega(f_k) \tag{5}$$

where $m$ represents the total amount of sample data imported into the $k$ tree. The first term is the loss function, which measures the error between the true value $y_i$ and the predicted value $\hat{y}_i$. The second term is the regularization term, used to control the model's complexity and prevent overfitting. The complexity of each tree is defined as:

$$\Omega(f) = \gamma T + \frac{1}{2}\lambda||w||^2 \tag{6}$$

where $\gamma$ represents the difficulty of node splitting, $T$ represents the number of leaf nodes, $\lambda$ is the L2 regularization coefficient to prevent overfitting, and $w$ is the modulus of the leaf node vector.

### 3.3. Light Gradient Boosting Machine (LGBM)

Developed by Microsoft Research in 2017, Light Gradient Boosting Machine (LGBM) stands as one of the most effective and advanced machine learning algorithms [50]. LGBM has evolved from the boosting regression algorithms. It employs a histogram-based algorithm, storing continuous features into discrete bins. The use of a histogram-based method accelerates the training speed and reduces memory usage. Additionally, LGBM utilizes the leaf-wise tree growth algorithm. The growth process involves choosing the leaf with the highest delta loss. This contrasts with many boosting algorithms (such as XGBoost) that use a level-wise approach. Although a level-based approach ensures a consistent number of leaves at each level, the leaf-wise strategy leads to a different number of leaves at each respective level. This approach helps LGBM achieve lower loss. The main process of the LGBM algorithm is shown in Equation (7):

$$F_n(x) = \alpha_0 f_0(x) + \alpha_1 f_1(x) + \cdots + \alpha_n f_n(x) \tag{7}$$

where the classifier begins with $n$ decision trees, and the weight assigned to the training samples is $\frac{1}{n}$. The weak classifier $f(x)$ and its weight $\alpha$ are determined. The process continues, with the classifier adjusting the weights until it arrives at the final classifier, denoted as $F_n(x)$. In summary, the main goal of the LGBM algorithm is to improve training efficiency and accuracy through feature parallelization and a histogram-based decision tree algorithm. It also uses gradient boosting methods to continuously optimize the model, thereby achieving better classification and regression results.

### 3.4. Genetic Algorithm, Back Propagation (GA-BP)

The Back Propagation (BP) neural network, a classic model in ANN (Artificial Neural Networks), was first proposed by Hecht-Nielsen et al. [51]. This network comprises an input layer, hidden layers, and an output layer, with neurons connecting each layer. The output of a neuron depends on its input values, activation function, and threshold. The BP neural network consists of two steps: forward propagation of information and backward propagation of errors. Although the BP neural network has excellent self-learning, adaptability, and self-organization capabilities and can effectively handle non-linear problems, it has some limitations: Firstly, in order to reduce error and improve accuracy, an appropriate number of neurons in the hidden layer need to be selected. However, there is a lack of a clear method for this selection. Secondly, the BP neural network randomly generates initial weights and thresholds. This results in adaptive and global approximation processes that are time-consuming, thereby slowing the network's convergence rate. Lastly, the use of gradient descent by the BP neural network can often lead to it becoming trapped in local minima.

The Genetic Algorithm (GA) is a global optimization probabilistic search method based on the principles of biological inheritance and evolution [52]. The GA mainly includes three operations: (1) Selection operation: The probability of an individual entering the next generation population is determined based on the fitness value. The higher the fitness, the greater the chance of inheritance. (2) Crossover operation: This is a key part of the algorithm. Two individuals are selected from the population, and a portion of their genes are exchanged to produce more optimal individuals in the new generation. (3) Mutation operation: An individual is randomly selected from the population, and a mutation is performed at a certain locus of its chromosome to produce a more optimal individual. Combining crossover and mutation operations can achieve optimal search performance. The GA has the characteristics of global search and parallel computation, but it lacks learning ability. The application of GA can optimize the BP neural network. This combines the GA's global search traits with the BP's learning and non-linear mapping abilities. As a result, the network's output accuracy improves.

### 3.5. Performance Metrics and Evaluation

The performance of models and the retrieval accuracy of downscaled SM are evaluated in this paper using three indicators: Mean Absolute Error (MAE), Root Mean Square Error (RMSE), and correlation coefficient ($R$). MAE is calculated as the average of the absolute differences between each observation and the mean. This method circumvents the issue of error cancellation, thereby providing a more accurate representation of the actual prediction error magnitude. RMSE is commonly used as a standard to measure the prediction results of machine learning models. The $R$ can be used to measure the degree of correlation between two variables. The calculation formulas for the three indicators are as follows:

$$\mathrm{MAE} = \frac{1}{n}\sum_{i=1}^{n}|X_i - Y_i| \tag{8}$$

$$\mathrm{RMSE} = \in \sqrt{\frac{1}{n}\sum_{i=1}^{n}(X_i - Y_i)^2} \tag{9}$$

$$R(X,Y) = \frac{Cov(X,Y)}{\sqrt{Var[X]Var[Y]}} \tag{10}$$

where $n$ represents the amount of data used for modeling, $X$ is the reference value of SM, and $Y$ is the retrieved value of SM. These three values are crucial for us to evaluate the prediction accuracy of the model. Among them, $X$ is the known true value, and $Y$ is the value predicted by our models. To more accurately evaluate the performance of the model, we introduce several key statistical indicators. Among them, $Cov(X, Y)$ represents the covariance of $X$ and $Y$, which describes the degree of joint variation of $X$ and $Y$. At the same time, $Var[X]$ is the variance of $X$, and $Var[Y]$ is the variance of $Y$. These two indicators describe the range of variation of $X$ and $Y$, respectively. These three indicators jointly evaluate the performance and prediction accuracy of the model. Covariance describes the correlation between the model's predicted values and the true values, while variance shows the dispersion of the data. The changes in these data directly affect the predictive ability of the model.

### 3.6. Downscaling Process

This study employed four ML techniques, including RF, XGBoost, LGBM, and GA-BP, to downscale CYNGSS SM retrieval to 3 km, respectively. Each of the proposed downscaling methods operates on a common principle: they create a statistical link between CYGNSS, geospatial variables (such as elevation and land cover type), land-surface variables (such as NDVI and precipitation), and SMAP SM at a coarse resolution of 36 km. In addition, SMAP SM was used as the reference value of SM, predicted by the downscaling model. Finally, the output covariates of the input variables were linked by using the following equation:

$$SM = f(\rho_1, \rho_2, \rho_3, \ldots, \rho_n) + \varepsilon \quad (11)$$

where SM represents the downscaled SM data, which is determined by the regression function of the machine learning models (RF, XGBoost, LGBM, and GA-BP), $\varepsilon$ is the model retrieval error, $\rho_1, \rho_2, \rho_3, \ldots, \rho_n$ represent the input covariates (i.e., SNR, SR, LES, TES, NDVI, DEM, land cover type, and precipitation). The total number of predictors is represented by $n$. The steps of the downscaling method mentioned above can be briefly summarized as follows:

1. Aggregation: The training procedure is carried out on the 36-km grid of SMAP SM. To maintain consistency with the spatial resolution of SMAP SM, the high-resolution CYGNSS observables and auxiliary variables (i.e., predictive factors) are aggregated to a 36 km scale using a simple arithmetic averaging method. It is worth noting that the theoretical resolution of the CYGNSS dataset used in this study is $7 \times 0.5$ km, while the SMAP product resolution is 36 km. This means that multiple CYGNSS observation points inevitably exist within the same SMAP grid. To address this, we average the CYGNSS sample points within the SMAP grid during the spatial matching process. Figure 3 presents the daily count of CYGNSS sampling points within the study area, as well as the counts after matching with SMAP. Specifically, from January to August, the counts correspond to the match with SMAP's 36-km grid; from September to December, the counts result from matching with the resampled 3-km grid of SMAP.
2. Model building: The aggregated data is divided into 70% for training and 30% for testing. SMAP SM is used as the response variable, and CYGNSS observables and auxiliary variables are used as input variables to train the four models: RF, XGBoost, LGBM, and GA-BP.
3. Resampling: The CYGNSS observables and auxiliary variables are resampled to a high resolution of 3 km using the nearest neighbor method. Subsequently, a spatial connection is established between them.
4. Model application: The resampled 3-km high-resolution CYGNSS observables and auxiliary variables were input to the downscaling models to obtain the downscaled CYNGSS SM with a spatial resolution of 3 km.
5. Model evaluation: Once the most optimal downscaling models were determined based on the lowest RMSE as a benchmark, these models were then evaluated for accuracy using the testing set. Then, the 3-km downscaled SM obtained was validated

using in situ data. Spatial analysis of the downscaled CYGNSS SM is conducted using MODIS EVI and MODIS EV products.

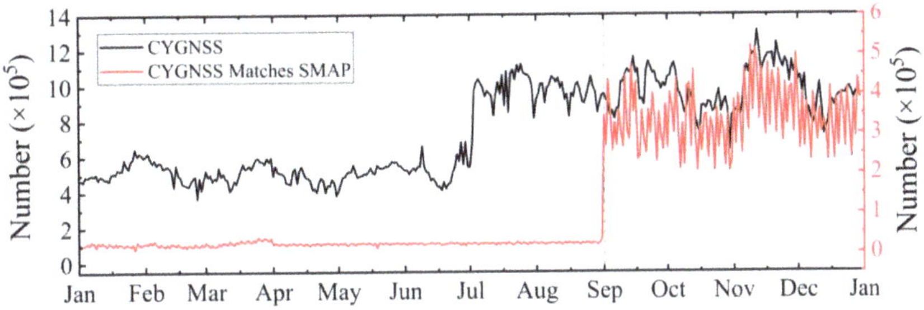


**Figure 3.** Daily sampling counts of CYGNSS and their corresponding matched counts with SMAP.

The experimental process is based on the assumption that the spatial scale relationship among SMAP SM, CYGNSS observables, and auxiliary variables maintains consistency. In other words, the relationship models established at a coarse resolution are still applicable at a high resolution [39,53,54]. The above experimental process is shown in Figure 4.

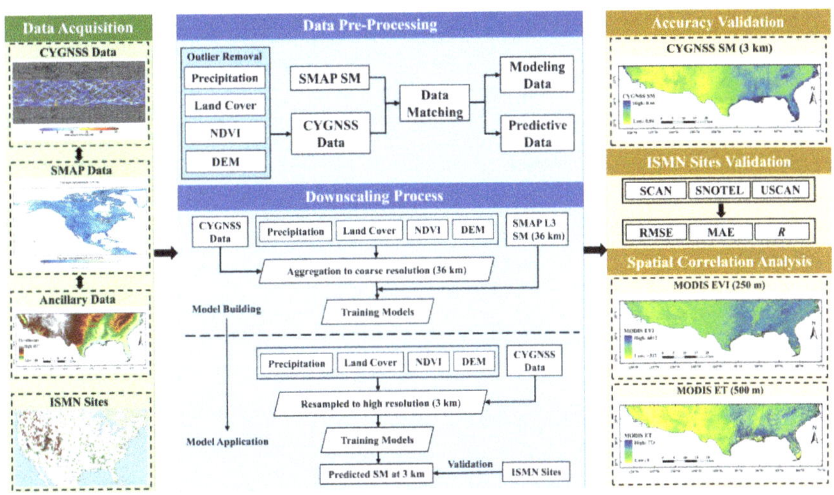


**Figure 4.** Flow chart of downscaling procedure.

## 4. Results

### 4.1. Models Evaluation

Following the methodologies outlined in Section 3, this paper constructs four CYGNSS SM downscaling models and adjusts the hyperparameters for the RF, XGBoost, LGBM, and GA-BP models. Hyperparameters are parameters given in advance in neural networks or machine learning to control the learning process of the model. The appropriate selection of hyperparameters is crucial for the predictive performance of the model and can also prevent the occurrence of overfitting or underfitting. The common hyperparametric methods are Grid search, Random search, and Bayesian optimization. This paper uses the Grid Search method for hyperparameter adjustment. Although Grid Search requires a longer runtime compared to the other two hyperparameter selection methods, it is a more exhaustive search method that ensures the best hyperparameter combination is found within the given parameter range. The final hyperparameter adjustment results are shown in Table 2.

**Table 2.** Results of hyperparameter adjustment for the four models.

| Model | Hyperparameters |
|---|---|
| GA-BP | popu = 50, learning_rate = 0.001, epochs = 100, n_hidden layer = 10 |
| RF | n_estimators = 100, max_depth = 6, max_leaf_nodes = None, min_samples_leaf = 1, min_samples_split = 2 |
| XGBoost | booster = tree, max_depth = 8, min_child_weight = 1, leaing_rate = 0.25, n_estimators = 100, subsample = 0.9, colsaple_bytree = 0.6, gamma = 0 |
| LGBM | learning_rate = 0.09, n_estimators = 100, min_samples_gain = 0.1, max_depth = 6, num_leaves = 50, subsample = 0.8, colsample bytree = 0.8 |

Through this process, we found the best combination of hyperparameters for each model to ensure that the CYGNSS SM downscaling models have high predictive performance. In the subsequent analysis, we will use these optimal hyperparameter combinations to train the model and evaluate its performance. To preliminarily assess the performance of the four models, this paper uses the method of ten-fold cross-validation for comparative analysis. The dataset uses SMAP SM (36 km) as the reference value, and four CYGNSS parameters, including SR, SNR, LES, and TES, as well as the auxiliary variables described in Section 2. This paper selects the coarse resolution data (36 km) from January to August 2019 to construct the downscaling model, yielding a total of 303,354 samples. For the prediction dataset, we use high-resolution data (3 km) from September to December, which provides a total of 4,123,129 samples. The ten-fold cross-validation accuracy of the four models and the running time of the models are shown in Table 3 and Figure 5.

**Table 3.** Summary of the overall accuracy of the ten-fold cross-validation of the four models.

| Name | Time | Model Training | | | Model Testing | | |
|---|---|---|---|---|---|---|---|
| | $t$/s | RMSE $cm^3/cm^3$ | MAE $cm^3/cm^3$ | $R$ - | RMSE $cm^3/cm^3$ | MAE $cm^3/cm^3$ | $R$ - |
| GA-BP | 46.16 | 0.071 | 0.055 | 0.834 | 0.072 | 0.055 | 0.831 |
| RF | 5712.91 | 0.011 | 0.007 | 0.996 | 0.031 | 0.021 | 0.969 |
| XGBoost | 25.90 | 0.037 | 0.027 | 0.958 | 0.038 | 0.028 | 0.955 |
| LGBM | 6.85 | 0.045 | 0.033 | 0.937 | 0.045 | 0.034 | 0.935 |

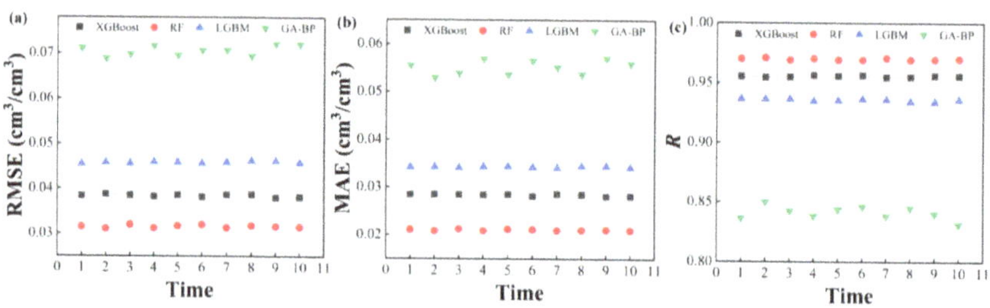


**Figure 5.** Summary of the accuracy of the ten-fold cross-validation of the four models. (**a**) RMSE; (**b**) MAE; (**c**) $R$.

Table 3 and Figure 5 present the accuracy of the ten-fold cross-validation of the four models and their execution times. In terms of execution time, the RF model took the longest, reaching 5712.91 s. This could be due to the fact that the RF model needs to generate a large number of decision trees during the training process and carry out complex voting and averaging operations, thus taking a longer time. The execution time of the GA-BP model was 46.16 s, significantly shorter than the RF model. However, the predictive performance of the GA-BP model was not satisfactory, with an RMSE of 0.072, an MAE of 0.055, and an $R$ of 0.831. These metrics indicate that the GA-BP model has relatively low accuracy and stability in predicting SM. The XGBoost model had a shorter execution time of 25.9 s. Its predictive performance was relatively good, with an RMSE of 0.038, an MAE of 0.028, and an $R$ of 0.955. These metrics indicate that the XGBoost model has high accuracy and stability in predicting SM. The LGBM model had the shortest execution time, only 6.85 s, although its predictive performance was not as good as the RF and XGBoost models, with an RMSE of 0.045, an MAE of 0.034, and an $R$ of 0.935. Nevertheless, these metrics still indicate that the LGBM model has acceptable predictive performance. Therefore, considering both execution time and predictive performance, the LGBM model performs best in terms of time efficiency, but its predictive performance is slightly worse than the RF and XGBoost models. Although the RF model takes the longest time, it has the best predictive performance. The XGBoost model performs well in both execution time and predictive performance. Although the GA-BP model has a shorter execution time, it has the worst predictive performance.

Figure 6 presents the performance of the downscaling models XGBoost, RF, LGBM, and GA-BP, which were constructed based on CYGNSS and auxiliary variables. The CYGNSS SM predictions at a coarser resolution (36 km) were compared with the SMAP SM predictions at the same resolution using the scatter plots for each model. It can be seen that the XGBoost and RF models perform well, exhibiting strong consistency between the CYGNSS SM and SMAP SM in both the training and testing set. The $R$ for the training set is 0.95 and 0.99, respectively, while, for the testing set, they are both 0.95. However, the GA-BP model shows less satisfactory retrieval results, with an $R$ of 0.84 for both the training and testing set. When comparing the RMSE of the models mentioned, XGBoost and RF models clearly outperform, with an RMSE of 0.038 and 0.012 for the training set, and 0.039 and 0.033 for the testing set. In contrast, the GA-BP and LGBM models show a higher RMSE, with 0.069 and 0.045 for the training set, and 0.070 for both in the testing set. The RF model exhibits the lowest MAE, with 0.008 for the training set and 0.022 for the testing set. However, the GA-BP and LGBM models show a higher MAE, with 0.054 and 0.033 for the training set, and 0.054 and 0.055 for the testing set, respectively.

The results indicate that the downscaling models built on RF and XGBoost outperform the models constructed using LGBM and GA-BP. Overall, the RF and XGBoost downscaling models demonstrate superior correlation and less error compared to the other models. This may be due to the robustness and unpredictable nature of the RF and XGBoost algorithms. When dealing with numerous variables at once, these techniques are intended to avoid overfitting. However, compared to the RF model, the XGBoost model achieves high accuracy with less time. Therefore, we mainly focus on downscaled SM from XGBoost model in the following sections.

Figure 7 presents the importance scores for various variables in retrieving outcomes with the XGBoost model. Among all input variables, the greatest impact on both the 36 km and 3 km resolution is seen with land cover and DEM. In particular, NDVI has a more substantial influence at the 36 km resolution, while its effect diminishes at the 3 km resolution. Conversely, influence of DDM_SNR is relatively low at the 36 km scale, but shows an increase at the 3 km resolution.

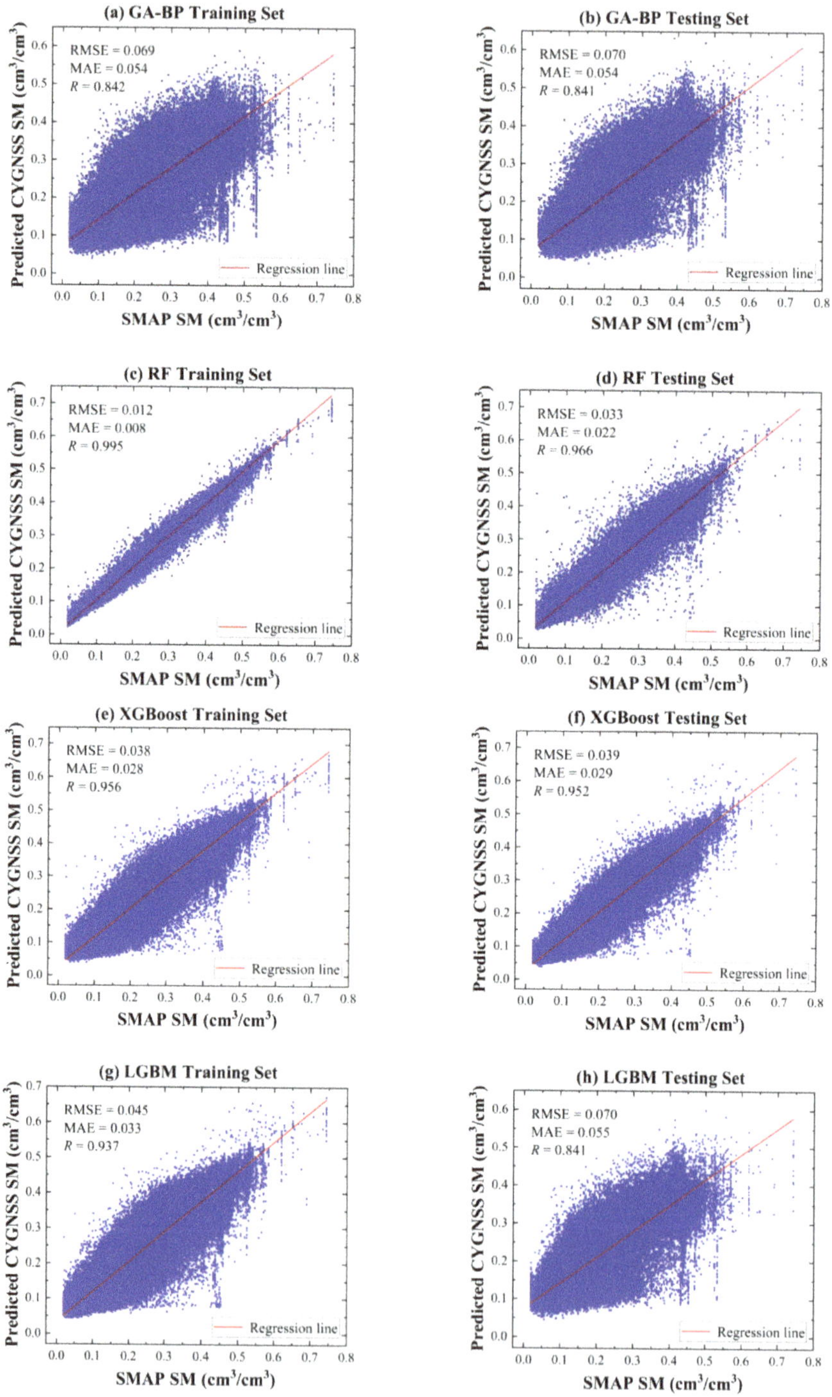


**Figure 6.** The retrieval accuracy of the four models in the training and testing datasets.

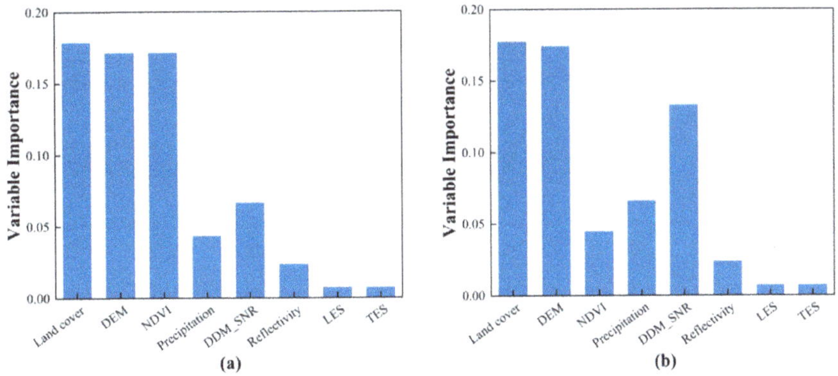


**Figure 7.** Variables' importance scores at (**a**) 36 km and (**b**) 3 km of XGBoost model.

### 4.2. Assessing the Accuracy of Downscaled Soil Moisture Using In Situ Observations

In the study area, we utilized the spatial coverage of CYGNSS and selected 78 sites from the ISMN with ground-based observation data as our research subjects from 1 September to 31 December 2019. These sites mainly come from SCAN, USCRN, and SNOTEL. Given the utilization of SMAP SM as a reference for the downscaling model in this study, it is essential to ensure the credibility of the assessment between downscaled SM and in situ SM observations. To achieve this, we initiated accuracy statistics for SMAP SM and in situ SM observations. Furthermore, to conduct a comprehensive time series analysis, we randomly selected four in situ sites for comparative evaluation with SMAP SM.

Table 4 delineates the comparison between in situ SM observations and corresponding SMAP SM. Analysis reveals that, out of all the in situ sites, 48 exhibit an MAE below 0.6, while 58 showcase an RMSE below 0.7. Additionally, 50 sites demonstrate an $R$ exceeding 0.7. The respective average values for MAE, RMSE, and $R$ stand at 0.051, 0.062, and 0.813. Overall, the majority of in situ sites exhibit commendable accuracy, thereby validating the reliability of the downscaling model constructed with SMAP SM as a reference. Figure 8 further illustrates the time series comparison of in situ SM observations and SMAP SM for randomly selected four in situ sites, with the time frame matching the dates of the downscaling model's prediction set. Of note is the 2–3-day revisit period of the SMAP satellite, which inhibits the guarantee of simultaneous coverage for each in situ site within the study area. Despite this limitation, the temporal variation of in situ SM observations (the blue line) closely aligns with that of SMAP SM (the red line). This alignment underscores SMAP SM's capability to capture the temporal dynamics of in situ SM, thereby validating the rationale for utilizing in situ SM observations in the downscaled SM assessment. To provide a quantitative assessment of the downscaled SM from the XGBoost model, Table 5 includes the accuracy statistical data for the downscaled SM and in situ SM observations.

According to the data analysis results in Table 5, for the total of 78 sites we studied, 62% of the sites, or about 49 sites, have an $R$ greater than 0.600. This value is quite high, indicating that the downscaling model for these sites have good predictive performance. Similarly, we have 54% of sites, about 43, with an RMSE less than 0.070, which also indicates that these sites have a small retrieval error. Further, 53% of sites, or about 42 sites, have an MAE less than 0.060, indicating that our model has a high accuracy of retrieval for these sites. Overall, the average $R$, RMSE, and MAE of all sites are 0.712, 0.065, and 0.058, respectively, demonstrating the excellent performance of our model overall. However, we also found that the type of land cover may affect the accuracy of site validation. To gain a deeper understanding of this issue, we conducted further analysis. After analysis, we concluded that the downscaled SM for most sites using the XGBoost model is reliable compared to the in situ SM observations. However, the validation accuracy of a few sites is relatively poor. To better understand the accuracy of the downscaling model, we also considered the type of land cover at the site location, as shown in Figure 9. This means that the type of land cover at the site location may affect the accuracy of the model. Through further research and model adjustment, we hope to better predict and understand this impact to optimize our model accuracy.

**Table 4.** Accuracy statistics for SMAP SM and in situ SM observations.

| Evaluation Index | Ranges | Number of In Situ Site | Average Value |
|---|---|---|---|
| MAE | <0.04 | 38 | 0.051 |
| | 0.04–0.06 | 10 | |
| | 0.06–0.08 | 20 | |
| | >0.08 | 10 | |
| RMSE | <0.06 | 48 | 0.062 |
| | 0.06–0.07 | 10 | |
| | 0.07–0.08 | 11 | |
| | >0.08 | 9 | |
| *R* | <0.60 | 22 | 0.813 |
| | 0.60–0.70 | 6 | |
| | 0.70–0.80 | 10 | |
| | >0.80 | 40 | |

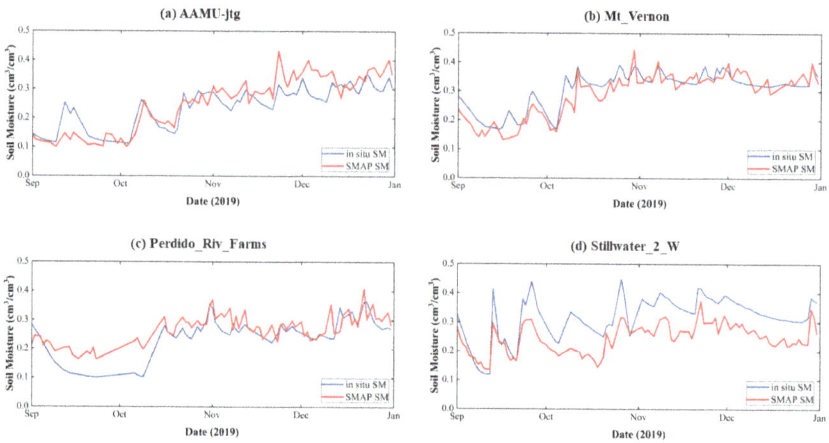


**Figure 8.** Time series of the SMAP SM and the in situ SM observations at the four sites.

**Table 5.** Accuracy statistics for downscaled SM and in situ SM observations.

| Evaluation Index | Ranges | Number of In Situ Site | Average Value |
|---|---|---|---|
| MAE | <0.04 | 24 | 0.058 |
| | 0.04–0.06 | 18 | |
| | 0.06–0.08 | 19 | |
| | >0.08 | 17 | |
| RMSE | <0.06 | 37 | 0.065 |
| | 0.06–0.07 | 5 | |
| | 0.07–0.08 | 11 | |
| | >0.08 | 25 | |
| *R* | <0.60 | 29 | 0.712 |
| | 0.60–0.70 | 9 | |
| | 0.70–0.80 | 7 | |
| | >0.80 | 33 | |

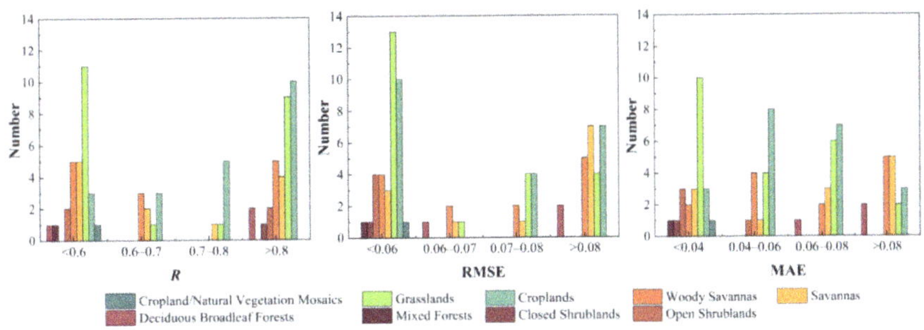


**Figure 9.** Precision statistics of in situ observations of different land cover types.

As seen in Figure 9, in the 78 in situ sites, 22 are located in grassland areas. Among these grassland sites, 14 have an $R$ value less than 0.600, 13 have an RMSE less than 0.060, and 14 have an MAE less than 0.06. For the 21 sites situated in farmland areas, 18 have an $R$ value less than 0.600, 10 have an RMSE less than 0.060, and 11 have an MAE less than 0.060. Of the 12 savanna sites, 7 have an $R$ value less than 0.600, 3 have an RMSE less than 0.060, and 4 have an MAE less than 0.060. In the woody savannas, there are 13 sites, with 8 having an $R$ value less than 0.600, 6 with an RMSE less than 0.060, and 4 with an MAE less than 0.060. Lastly, in the open shrublands, there are 4 sites. Two of these have an $R$ value less than 0.600, all 4 have an RMSE less than 0.060, and 3 have an MAE less than 0.060. For the land cover types of deciduous broadleaf forests, mixed forests, closed shrublands, and cropland/natural vegetation mosaics, the number of sites are 3, 1, 1, and 1, respectively. Correspondingly, the sites with an $R$ greater than 0.600 are 2, 0, 1, and 0. Sites with an RMSE less than 0.060 are found to be 0, 1, 1, and 1, while those with an MAE less than 0.060 are also 0, 1, 1, and 1.

This study primarily investigates the accuracy of downscaled SM models obtained through the application of the XGBoost model across nine different land cover types. The results indicate that sites located in grasslands and farmlands exhibit higher accuracy. This may be attributed to the fact that SM retrieval based on GNSS-R technology tends to be more accurate in flat areas than in areas with significant surface undulations or tree cover. Additionally, grasslands and farmlands are common land use types; hence, we have more sites for observation and validation. Conversely, the other seven land cover types have fewer sites, leading to a lack of sufficient validation data, which could be a significant factor affecting accuracy. Furthermore, we believe that other potential factors might influence the accuracy of SM retrieval. For instance, the varying soil properties and complexities across different regions could impact model performance. Highly heterogeneous soils or areas with significant rock content could lead to inaccurate predictions. Changes in precipitation and meteorological conditions might also affect the accuracy of the model. Prolonged droughts or consistent rainfall could potentially lead to decreased model performance during specific periods. The quality of GNSS-R technology data, the calibration process, and observational errors could impact model accuracy to some extent. Additionally, if there are changes in land use or land cover types in the study area during the observation period, this could affect the training and validation data, consequently influencing the model's accuracy. However, despite lower accuracy in some areas, it is evident from our results (Figure 9 and Table 5) that the downscaled SM model constructed in this study generally achieves satisfactory results. This suggests that our method exhibits adaptability and robustness, providing high accuracy in most scenarios.

### 4.3. Graphical Assessment of Spatial Distribution of Downscaled Soil Moisture

After the downscaled SM was validated using in situ sites, we proceeded to conduct a spatial analysis to evaluate the effectiveness of the downscaling approaches. The downscaled SM from the XGBoost model was visually compared with high-resolution MODIS EVI and MODIS ET products. The Enhanced Vegetation Index (EVI) is an indicator used to assess and monitor the health and growth status of vegetation [55]. When SM is low, vegetation may be constrained by water availability, leading to slowed or stressed plant growth. This may manifest as lower EVI values. Conversely, when SM is high, plants may have ample water supply, promoting growth and resulting in higher greenness and elevated EVI. Evapotranspiration (ET) refers to the sum of evaporation from the land surface and transpiration from plants [56]. When SM is high, there is ample water supply in the soil, and plant roots can absorb sufficient water for transpiration, thereby promoting the ET process. Higher SM typically results in higher ET. Conversely, when SM is low, the water supply in the soil decreases and plants face water limitations, leading to reduced plant transpiration. Lower SM typically results in lower ET. Therefore, examining the variations in EVI and ET within the study area can indirectly reflect changes in SM.

The following sections compare the relationships among the downscaled SM, EVI, and ET for four periods: 9–14, 10–16, 11–17, and 12–19. We processed the downscaled SM from the XGBoost model using simple Kriging interpolation, and then conducted a spatial analysis with MODIS EVI and MODIS ET products.

As seen in Figure 10, the EVI values in the central and eastern of the study area are relatively high, while those in the northwest and southwest are lower. This is related to the vegetation cover in the study area and is consistent with the geographical characteristics of the study area described in Section 2.1. Compared with the downscaled SM and EVI at the same time, we can observe that areas in the study region with higher SM also have higher EVI values, such as the sides of the Central Valley in the middle and the Appalachian Mountains in the east. Conversely, areas with lower SM also have lower EVI values, such as in the western regions of Oklahoma and Salt Lake City. It can be proved that there is a correlation between SM and EVI. As seen in Figure 10(a-1), on September 14, the SM values in the Homochitto National Forest, the Sabine National Forest in the south-central study area, and the southeastern region are relatively high. Comparing this with Figure 10(a-3) at the same time, we can see that the ET values in these areas are also high. The same pattern can be observed when comparing Figure 10(b-1) with Figure 10(b-3), Figure 10(c-1) with Figure 10(c-3), and Figure 10(d-1) with Figure 10(d-3). SM, as one of the main sources of water for ET, may lead to higher ET in areas with higher SM. However, it is worth noting that there are some discrepancies in the spatial distribution of downscaled SM, EVI, and ET in some areas in the south-central study area (areas within the red box in Figure 10). Because the downscaling model is established at a 36 km grid scale, some extreme values are smoothed during the spatial aggregation process. As a result, the training samples selected in the model construction process are all smooth data, with fewer extreme values. This is not unique to our study, as all existing downscaling methods necessitate calibration with coarse-resolution data initially, making the aggregation of high-resolution predictors inevitable [26]. The result is that the downscaled SM has some mistakes.

Overall, the spatial distribution and temporal variation of the downscaled SM product generated in this paper are relatively consistent with EVI and ET, both of which have a certain correlation with SM. Therefore, this indirectly verifies the accuracy of the downscaled SM for the retrieval of SM in the study area.

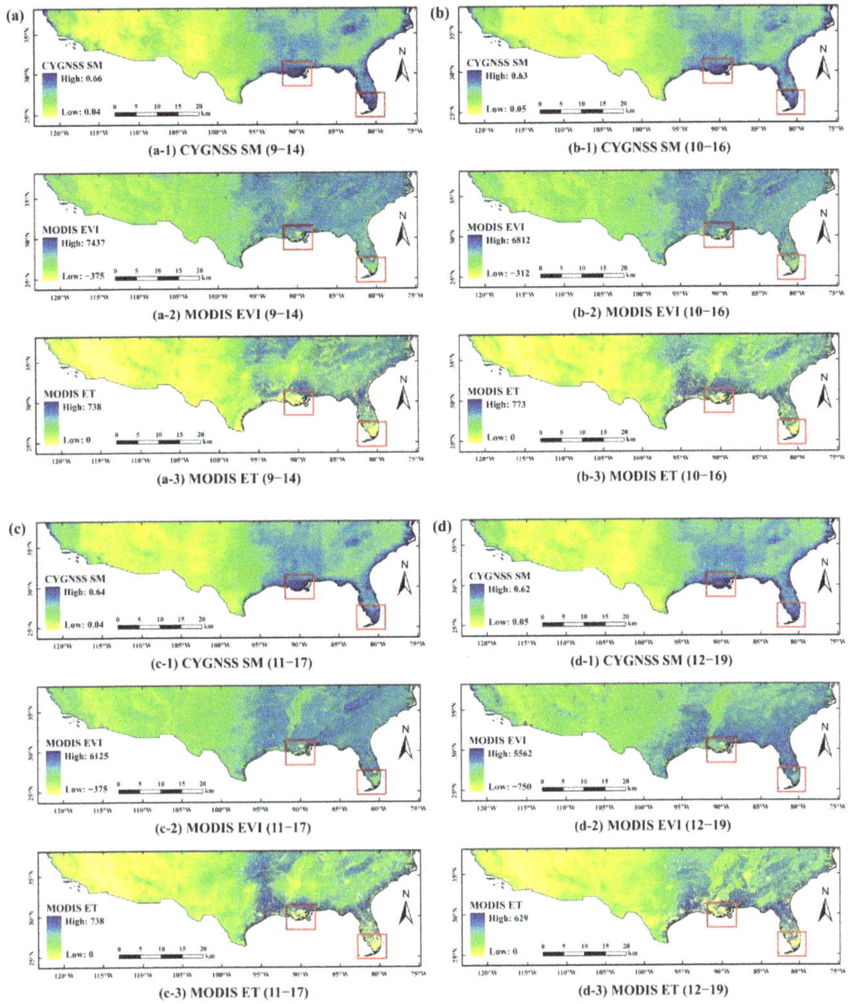


**Figure 10.** Distributions of the downscaled CYGNSS SM, MODIS EVI, and MODIS ET on 14 September 2019, 16 October 2019, 17 November 2019, and 19 December 2019.

## 5. Discussion

A key advantage of this study is the downscaling of CYGNSS based on the XGBoost model using L-band passive microwave SM (i.e., SMAP SM) and auxiliary variables. Instead, most previous studies downscale satellite SM products (AMSR-E, SMOS, and SMAP, etc.) based on optical data [39,53,54,57]. Another advantage is that it improves upon previous research that used CYGNSS to retrieve SM with a maximum spatial resolution of 9 km. Through the method of downscaling, this study has increased the spatial resolution of SM retrieval using CYNGSS to 3 km. Furthermore, the downscaled SM can more finely represent the spatial distribution changes in SM, offering substantial potential for applications such as irrigation planning in agriculture.

The noteworthy limitations of this study may present opportunities for improving the spatial downscaling of satellite SM outputs with coarse resolution. First, CYGNSS observables are collected at pseudo-random positions, with irregular spatial and temporal resolution. This is different from conventional remote sensing technologies, which have repeatable swaths and consistent local collection times. As a result, mapping CYGNSS observables regularly in space presents a challenge in terms of assigning appropriate spatial grid sizes. The spatial resolution of CYGNSS observations can vary greatly, ranging from the first Fresnel zone with coherent reflections (0.5 km) to the scintillation zone with incoherent reflections (i.e., dozens of km). Traditional methods of mapping using regular spatial grids and integral time-step cannot fully account for this complexity in the spatiotemporal resolution of CYGNSS signals [58]. However, a transformation procedure is needed in order to match CYGNSS observables with other remote sensing and modeling data. This conversion process can introduce inaccuracies into the reflectivity, which could have implications, not just for this study, but for all similar research endeavors as well [12,17,59,60].

Second, during the model building process, the input CYGNSS observables and auxiliary variable are aggregated from high resolution to 36-km coarse resolution using a simple arithmetic averaging method. Furthermore, the SMAP SM encapsulates an average representation of SM, which is spread across a spatial resolution of 36 km. The average SM represents the SM of the whole region, and most of the information is ignored due to the coarse resolution. Hence, the training samples chosen during the model building process are smooth data with minimal extreme values. Models built from these samples invariably influence the downscaled SM. The scale discrepancy between the input data for model training and the SMAP products somewhat constrains the selection of suitable data during the regression model construction process. If a large amount of training data is necessary, choosing a research area that is large enough to assure the collection of enough training samples becomes crucial. During the application of the downscaling model, due to the increased heterogeneity and richer data representation at a 3 km resolution, there might be extreme values that were not encountered during model training. This corresponded with the results of Wakigari et al. [54]. Therefore, in practical applications, the downscaled SM has some inevitable errors. These errors are not randomly generated, but are closely related to the variance of SM in our training samples. In other words, the greater the degree of variation or dispersion of SM in the training samples, the greater the retrieval error may be. This is because a large variance means that the SM values in the dataset have greater changes, which may lead to more errors in the model's predictions. At the same time, the results of the downscaling process are significantly affected by the number and representativeness of the training samples. A sufficient number of training samples can provide more comprehensive information, helping the model to better learn and understand the characteristics of the data. The degree of representativeness of the samples directly affects the generalization ability of the model. If the samples can fully represent the characteristics of the entire data, then the model's retrieval results on unknown data will be more accurate.

Third, we used in situ SM observations, which are direct measurements from specific locations. However, these data may introduce some uncertainties when validating our downscaled SM model, mainly due to scale discrepancies. In our model, the downscaled SM represents an average SM value over a 3 km $\times$ 3 km area, which is a broader spatial average than in situ measurements. However, due to geographical conditions and human activities, there may be significant variations in SM within this area. For instance, if a location is under irrigation, it could lead to the recorded SM value at this point being much higher than the area's average. This scale discrepancy could pose some issues during the validation phase. For example, if a site is located in an irrigation area, its SM measurement might be significantly higher than the average SM of the area, leading to a large deviation between the SM measurement and the model retrieval at this site during model validation. This deviation is not a problem with the model, but is caused by spatial scale differences.

Fourth, the input optical data NDVI is inevitably affected by clouds during the model construction and model application. This also leads to the downscaled SM exhibiting optical properties. Factors such as cloud cover can impact the downscaled SM, leading to the occurrence of null values [39]. The presence of clouds may influence the availability of downscaled SM at the corresponding location. Furthermore, the 3 km resolution may present challenges, potentially leading to the presence of missing values in the downscaled SM due to its inability to cover all processed pixels. To address this issue, we adopted an approach similar to that described by Wei Shangguan et al. [53] to fill these gaps. Specifically, we performed Kriging interpolation on the 3-km downscaled CYGNSS SM. However, it is important to acknowledge that the utilization of interpolation unavoidably introduces certain errors. Thus, some inconsistencies will present themselves during the validation stage.

Fifth, we utilized only four auxiliary variables, namely rainfall, land cover type, DEM, and NDVI. It is crucial to adequately consider the spatial scale of CYGNSS observation data, SM reference data, and auxiliary variables for the accuracy of SM retrieval. Factors such as soil type (sand, loam, clay, etc.) affect soil water absorption and the capacity to minimize water loss, as well as surface temperature variations and water evaporation caused by wind speed. As discussed by Volkan Senyurek [16], soil texture features are considered to have the greatest impact on retrieval SM among auxiliary inputs. In summary, while the method proposed in this paper has achieved a commendable accuracy in SM retrieval, there is room for improvement by considering a wider range of auxiliary factors. This has the potential to further enhance the accuracy of SM retrieval using GNSS-R.

Sixth, in assessing the spatial distribution of downscaled SM, this paper has not yet considered the influence of a variety of factors on plant growth and ET. These factors include light conditions, temperature, soil texture, and carbon dioxide concentration, all of which may have an impact on the accuracy of MODIS EVI and MODIS ET products, as changes in these factors may result in changes in vegetation activity and ET. There are limitations in using these products to assess the spatial distribution of downscaled SM. Considering that these factors may add to the complexity of the assessment, future research could attempt to integrate these factors to obtain more accurate downscaled SM estimates.

Seventh, in the process of evaluating the spatial distribution of downscaled SM, we employed the Kriging interpolation method. However, the Kriging interpolation method might not be optimal, as it measures SM with limited physical significance and could result in spatial heterogeneity. Therefore, in future research, it is essential to compare different interpolation methods and investigate their impact on downscaled SM. Selecting the most suitable interpolation method will facilitate the assessment of the spatial distribution of downscaled SM.

Finally, there are some limitations concerning the geographical scope of our study area and the duration of the data utilized. When validating the downscaled SM using in situ sites, we observed that, aside from grasslands and farmlands, the availability of in situ sites for other land cover types was limited. This paucity of data can hinder a comprehensive validation. By expanding the study area, the number of in situ sites for other land cover types would increase, thereby augmenting the validation dataset and enhancing the accuracy and reliability of our model performance assessment. In this study, the SM downscaling model was constructed using data from January to August 2019, while data from September to December was used for SM retrieval. This may introduce seasonal biases into the constructed downscaling model, leading to certain inaccuracies. Extending the data period for a year or even longer could mitigate such seasonal effects, thus boosting the reliability of the downscaling approach.

## 6. Conclusions

In this paper, we propose a downscaling method for CYGNSS SM based on the XGBoost algorithm, using high-resolution CYGNSS observables and auxiliary variables as input data to improve the spatial resolution of GNSS-R technique retrieval of SM to 3 km. The method selects common downscaled variables such as DEM, land cover, NDVI, and rainfall. We enhance and improve the polynomial-based downscaling regression model by incorporating parameters of SR, SNR, LES, and TES from CYGNSS. Experiments were conducted using data covering the southern United States, and the results were validated by 78 in situ sites. The results show that the downscaled SM achieves $R$, RMSE, and MAE of 0.712, 0.068, and 0.058, respectively, compared with the in situ SM observations. Spatial analysis using MODIS EVI and MODIS ET products shows that the spatial distribution and temporal variation of the downscaled CYGNSS SM products are more consistent with the EVI and ET products. The feasibility of the method is proved. Additionally, we discuss a number of problems that came up throughout the downscaling and validation process.

Overall, the findings of this study offer valuable insights for enhancing SM downscaling methods. This is crucial for advancing high-resolution SM retrieval. In future research, it will be key to develop gap-filling methods to address missing remote sensing data and refine the downscaling model. Additionally, researchers could consider using satellite SM products from various sources (e.g., SMAP, SMOS, AMSR-E, NASA-USDA, etc.) as reference values for downscaling models. This could aid in determining the most efficient downscaled SM products that are best suited to the particular conditions of the selected study area. Additionally, future research could consider a gradual downscaling approach (for instance, downsizing from 36 km to 9 km, followed by a reduction from 9 km to 3 km), as opposed to an immediate downscaling from 36 km to 3 km.

**Author Contributions:** Q.L.: Methodology, Writing—original draft, Writing—review and editing. Y.L.: Methodology, Writing—original draft, Writing—review and editing. Y.G.: Visualization, Validation. X.L.: Supervision. C.R.: Project administration. W.Y.: Software. B.Z.: Data curation. X.J.: Investigation. All authors have read and agreed to the published version of the manuscript.

**Funding:** This work was supported by the National Natural Science Foundation of China (No. 41901409); the Natural Science Foundation of Guangxi (No. 2021GXNSFBA220046); the National Natural Science Foundation of China (No. 42064003).

**Data Availability Statement:** CYGNSS data can be downloaded from the Physical Oceanography Distributed Active Archive Center (PO.DAAC, https://podaac.jpl.nasa.gov, accessed on 1 April 2023). SMAP data can be obtained from the National Snow and Ice Data Center (NSIDC, https://nsidc.org, accessed on 15 April 2023). The in situ soil moisture data can be accessed publicly (http://ismn.geo.tuwien.ac.at, accessed on 20 April 2023).

**Acknowledgments:** The authors are grateful to NASA EOSDIS Physical Oceanography Distributed Active Archive Center (DAAC), Jet Propulsion Laboratory, Pasadena, CA, USA, for making the CYGNSS data available, at https://www.esrl.noaa.gov/psd/ (accessed on 1 April 2023).

**Conflicts of Interest:** The authors declare no conflict of interest.

## References


1. McColl, K.A.; Alemohammad, S.H.; Akbar, R.; Konings, A.G.; Yueh, S.; Entekhabi, D. The global distribution and dynamics of surface soil moisture. *Nat. Geosci.* **2017**, *10*, 100–104. [CrossRef]
2. Yan, Q.; Gao, M.; Zafar, Z.; Li, A. Ground based GNSS-R observations for soil moisture. *Chin. J. Geophys.* **2011**, *54*, 2735–2744.
3. Peng, J.; Loew, A.; Merlin, O.; Verhoest, N.E. A review of spatial downscaling of satellite remotely sensed soil moisture. *Rev. Geophys.* **2017**, *55*, 341–366. [CrossRef]
4. Crow, W.T.; Wood, E.F.; Dubayah, R. Potential for downscaling soil moisture maps derived from spaceborne imaging radar data. *J. Geophys. Res. Atmos.* **2000**, *105*, 2203–2212. [CrossRef]
5. Kerr, Y.H.; Waldteufel, P.; Richaume, P.; Wigneron, J.P.; Ferrazzoli, P.; Mahmoodi, A.; Al Bitar, A.; Cabot, F.; Gruhier, C.; Juglea, S.E. The SMOS soil moisture retrieval algorithm. *IEEE Trans. Geosci. Remote Sens.* **2012**, *50*, 1384–1403. [CrossRef]
6. Entekhabi, D.; Njoku, E.G.; O'Neill, P.E.; Kellogg, K.H.; Crow, W.T.; Edelstein, W.N.; Entin, J.K.; Goodman, S.D.; Jackson, T.J.; Johnson, J. The soil moisture active passive (SMAP) mission. *Proc. IEEE* **2010**, *98*, 704–716. [CrossRef]
7. Hall, C.; Cordey, R. Multistatic Scatterometry. In Proceedings of the International Geoscience and Remote Sensing Symposium, 'Remote Sensing: Moving Toward the 21st Century', Edinburgh, UK, 12–16 September 1988; pp. 561–562.
8. Chew, C.; Shah, R.; Zuffada, C.; Hajj, G.; Masters, D.; Mannucci, A.J. Demonstrating soil moisture remote sensing with observations from the UK TechDemoSat-1 satellite mission. *Geophys. Res. Lett.* **2016**, *43*, 3317–3324. [CrossRef]
9. Camps, A.; Park, H.; Pablos, M.; Foti, G.; Gommenginger, C.P.; Liu, P.-W.; Judge, J. Sensitivity of GNSS-R spaceborne observations to soil moisture and vegetation. *IEEE J. Sel. Top. Appl. Earth Obs. Remote Sens.* **2016**, *9*, 4730–4742. [CrossRef]
10. Katzberg, S.J.; Torres, O.; Grant, M.S.; Masters, D. Utilizing calibrated GPS reflected signals to estimate soil reflectivity and dielectric constant: Results from SMEX02. *Remote Sens. Environ.* **2006**, *100*, 17–28. [CrossRef]
11. Chew, C.; Small, E. Description of the UCAR/CU Soil Moisture Product. *Remote Sens.* **2020**, *12*, 1558. [CrossRef]

12. Chew, C.; Small, E. Soil moisture sensing using spaceborne GNSS reflections: Comparison of CYGNSS reflectivity to SMAP soil moisture. *Geophys. Res. Lett.* **2018**, *45*, 4049–4057. [CrossRef]
13. Ruf, C. Cyclone Global Navigation Satellite System (CYGNSS) Observations and Applications—Part I. In Proceedings of the 99th American Meteorological Society Annual Meeting, Washington, DC, USA, 6–10 January 2019.
14. Al-Khaldi, M.M.; Johnson, J.T.; O'Brien, A.J.; Balenzano, A.; Mattia, F. Time-series retrieval of soil moisture using CYGNSS. *IEEE Trans. Geosci. Remote Sens.* **2019**, *57*, 4322–4331. [CrossRef]
15. Eroglu, O.; Kurum, M.; Boyd, D.; Gurbuz, A.C. High spatio-temporal resolution CYGNSS soil moisture estimates using artificial neural networks. *Remote Sens.* **2019**, *11*, 2272. [CrossRef]
16. Senyurek, V.; Lei, F.; Boyd, D.; Kurum, M.; Gurbuz, A.C.; Moorhead, R. Machine learning-based CYGNSS soil moisture estimates over ISMN sites in CONUS. *Remote Sens.* **2020**, *12*, 1168. [CrossRef]
17. Jia, Y.; Jin, S.; Chen, H.; Yan, Q.; Savi, P.; Jin, Y.; Yuan, Y. Temporal-spatial soil moisture estimation from CYGNSS using machine learning regression with a preclassification approach. *IEEE J. Sel. Top. Appl. Earth Obs. Remote Sens.* **2021**, *14*, 4879–4893. [CrossRef]
18. Zhan, X.; Miller, S.; Chauhan, N.; Di, L.; Ardanuy, P. *Soil Moisture Visible/Infrared Radiometer Suite Algorithm Theoretical Basis Document*; Raytheon Systems Company: Lanham, MD, USA, 2002.
19. Chauhan, N.; Miller, S.; Ardanuy, P. Spaceborne soil moisture estimation at high resolution: A microwave-optical/IR synergistic approach. *Int. J. Remote Sens.* **2003**, *24*, 4599–4622. [CrossRef]
20. Piles, M.; Camps, A.; Vall-Llossera, M.; Corbella, I.; Panciera, R.; Rudiger, C.; Kerr, Y.H.; Walker, J. Downscaling SMOS-derived soil moisture using MODIS visible/infrared data. *IEEE Trans. Geosci. Remote Sens.* **2011**, *49*, 3156–3166. [CrossRef]
21. Choi, M.; Hur, Y. A microwave-optical/infrared disaggregation for improving spatial representation of soil moisture using AMSR-E and MODIS products. *Remote Sens. Environ.* **2012**, *124*, 259–269. [CrossRef]
22. Peischl, S.; Walker, J.P.; Rüdiger, C.; Ye, N.; Kerr, Y.H.; Kim, E.; Bandara, R.; Allahmoradi, M. The AACES field experiments: SMOS calibration and validation across the Murrumbidgee River catchment. *Hydrol. Earth Syst. Sci.* **2012**, *16*, 1697–1708. [CrossRef]
23. Piles, M.; Sánchez, N.; Vall-llossera, M.; Camps, A.; Martínez-Fernández, J.; Martínez, J.; González-Gambau, V. A downscaling approach for SMOS land observations: Evaluation of high-resolution soil moisture maps over the Iberian Peninsula. *IEEE J. Sel. Top. Appl. Earth Obs. Remote Sens.* **2014**, *7*, 3845–3857. [CrossRef]
24. Piles, M.; Petropoulos, G.P.; Sánchez, N.; González-Zamora, Á.; Ireland, G. Towards improved spatio-temporal resolution soil moisture retrievals from the synergy of SMOS and MSG SEVIRI spaceborne observations. *Remote Sens. Environ.* **2016**, *180*, 403–417. [CrossRef]
25. Sánchez-Ruiz, S.; Piles, M.; Sánchez, N.; Martínez-Fernández, J.; Vall-llossera, M.; Camps, A. Combining SMOS with visible and near/shortwave/thermal infrared satellite data for high resolution soil moisture estimates. *J. Hydrol.* **2014**, *516*, 273–283. [CrossRef]
26. Zhao, W.; Li, A. A downscaling method for improving the spatial resolution of AMSR-E derived soil moisture product based on MSG-SEVIRI data. *Remote Sens.* **2013**, *5*, 6790–6811. [CrossRef]
27. Das, N.N.; Entekhabi, D.; Njoku, E.G.; Shi, J.J.; Johnson, J.T.; Colliander, A. Tests of the SMAP combined radar and radiometer algorithm using airborne field campaign observations and simulated data. *IEEE Trans. Geosci. Remote Sens.* **2013**, *52*, 2018–2028. [CrossRef]
28. Srivastava, P.K.; Han, D.; Ramirez, M.R.; Islam, T. Machine learning techniques for downscaling SMOS satellite soil moisture using MODIS land surface temperature for hydrological application. *Water Resour. Manag.* **2013**, *27*, 3127–3144. [CrossRef]
29. Malbéteau, Y.; Merlin, O.; Molero, B.; Rüdiger, C.; Bacon, S. DisPATCh as a tool to evaluate coarse-scale remotely sensed soil moisture using localized in situ measurements: Application to SMOS and AMSR-E data in Southeastern Australia. *Int. J. Appl. Earth Obs. Geoinf.* **2016**, *45*, 221–234. [CrossRef]
30. Malbéteau, Y.; Merlin, O.; Gascoin, S.; Gastellu, J.P.; Mattar, C.; Olivera-Guerra, L.; Khabba, S.; Jarlan, L. Normalizing land surface temperature data for elevation and illumination effects in mountainous areas: A case study using ASTER data over a steep-sided valley in Morocco. *Remote Sens. Environ.* **2017**, *189*, 25–39. [CrossRef]
31. Gleason, S.; Ruf, C.S.; O'Brien, A.J.; McKague, D.S. The CYGNSS level 1 calibration algorithm and error analysis based on on-orbit measurements. *IEEE J. Sel. Top. Appl. Earth Obs. Remote Sens.* **2018**, *12*, 37–49. [CrossRef]
32. Rodriguez-Alvarez, N.; Podest, E.; Jensen, K.; McDonald, K.C. Classifying inundation in a tropical wetlands complex with GNSS-R. *Remote Sens.* **2019**, *11*, 1053. [CrossRef]
33. Clarizia, M.P.; Pierdicca, N.; Costantini, F.; Floury, N. Analysis of CYGNSS data for soil moisture retrieval. *IEEE J. Sel. Top. Appl. Earth Obs. Remote Sens.* **2019**, *12*, 2227–2235. [CrossRef]
34. Carreno-Luengo, H.; Lowe, S.; Zuffada, C.; Esterhuizen, S.; Oveisgharan, S. Spaceborne GNSS-R from the SMAP mission: First assessment of polarimetric scatterometry over land and cryosphere. *Remote Sens.* **2017**, *9*, 362. [CrossRef]
35. Van Dijk, A.I.; Brakenridge, G.R.; Kettner, A.J.; Beck, H.E.; De Groeve, T.; Schellekens, J. River gauging at global scale using optical and passive microwave remote sensing. *Water Resour. Res.* **2016**, *52*, 6404–6418. [CrossRef]
36. Dorigo, W.; Wagner, W.; Hohensinn, R.; Hahn, S.; Paulik, C.; Xaver, A.; Gruber, A.; Drusch, M.; Mecklenburg, S.; van Oevelen, P. The International Soil Moisture Network: A data hosting facility for global in situ soil moisture measurements. *Hydrol. Earth Syst. Sci.* **2011**, *15*, 1675–1698. [CrossRef]

37. Dorigo, W.; Xaver, A.; Vreugdenhil, M.; Gruber, A.; Hegyiova, A.; Sanchis-Dufau, A.; Zamojski, D.; Cordes, C.; Wagner, W.; Drusch, M. Global automated quality control of in situ soil moisture data from the International Soil Moisture Network. *Vadose Zone J.* **2013**, *12*, 1–21. [CrossRef]
38. Gruber, A.; Dorigo, W.A.; Zwieback, S.; Xaver, A.; Wagner, W. Characterizing coarse-scale representativeness of in situ soil moisture measurements from the International Soil Moisture Network. *Vadose Zone J.* **2013**, *12*, 1–16. [CrossRef]
39. Hu, F.; Wei, Z.; Zhang, W.; Dorjee, D.; Meng, L. A spatial downscaling method for SMAP soil moisture through visible and shortwave-infrared remote sensing data. *J. Hydrol.* **2020**, *590*, 125360. [CrossRef]
40. Ranney, K.J.; Niemann, J.D.; Lehman, B.M.; Green, T.R.; Jones, A.S. A method to downscale soil moisture to fine resolutions using topographic, vegetation, and soil data. *Adv. Water Resour.* **2015**, *76*, 81–96. [CrossRef]
41. Busch, F.A.; Niemann, J.D.; Coleman, M. Evaluation of an empirical orthogonal function–based method to downscale soil moisture patterns based on topographical attributes. *Hydrol. Process.* **2012**, *26*, 2696–2709. [CrossRef]
42. Coleman, M.L.; Niemann, J.D. Controls on topographic dependence and temporal instability in catchment-scale soil moisture patterns. *Water Resour. Res.* **2013**, *49*, 1625–1642. [CrossRef]
43. Mascaro, G.; Vivoni, E.R.; Deidda, R. Soil moisture downscaling across climate regions and its emergent properties. *J. Geophys. Res. Atmos.* **2011**, *116*, D22114. [CrossRef]
44. Farr, T.G.; Rosen, P.A.; Caro, E.; Crippen, R.; Duren, R.; Hensley, S.; Kobrick, M.; Paller, M.; Rodriguez, E.; Roth, L. The shuttle radar topography mission. *Rev. Geophys.* **2007**, *45*, RG2004. [CrossRef]
45. Sidhu, N.; Pebesma, E.; Wang, Y.-C. Usability study to assess the IGBP land cover classification for Singapore. *Remote Sens.* **2017**, *9*, 1075. [CrossRef]
46. Zhao, L.; Li, Q.; Zhang, Y.; Wang, H.; Du, X. Normalized NDVI valley area index (NNVAI)-based framework for quantitative and timely monitoring of winter wheat frost damage on the Huang-Huai-Hai Plain, China. *Agric. Ecosyst. Environ.* **2020**, *292*, 106793. [CrossRef]
47. Ho, T.K. The random subspace method for constructing decision forests. *IEEE Trans. Pattern Anal. Mach. Intell.* **1998**, *20*, 832–844.
48. Breiman, L. Random forests. *Mach. Learn.* **2001**, *45*, 5–32. [CrossRef]
49. Chen, T.; Guestrin, C. Xgboost: A scalable tree boosting system. In Proceedings of the 22nd Acm Sigkdd International Conference on Knowledge Discovery and Data Mining, San Francisco, CA, USA, 13–17 August 2016; pp. 785–794.
50. Ke, G.; Meng, Q.; Finley, T.; Wang, T.; Chen, W.; Ma, W.; Ye, Q.; Liu, T.-Y. Lightgbm: A highly efficient gradient boosting decision tree. *Adv. Neural Inf. Process. Syst.* **2017**, *30*, 1–9.
51. Hecht-Nielsen, R. Theory of the backpropagation neural network. In *Neural Networks for Perception*; Elsevier: Amsterdam, The Netherlands, 1992; pp. 65–93.
52. Mirjalili, S. Evolutionary algorithms and neural networks. In *Studies in Computational Intelligence*; Springer: Berlin/Heidelberg, Germany, 2019; Volume 780.
53. Mao, T.; Shangguan, W.; Li, Q.; Li, L.; Zhang, Y.; Huang, F.; Li, J.; Liu, W.; Zhang, R. A Spatial Downscaling Method for Remote Sensing Soil Moisture Based on Random Forest Considering Soil Moisture Memory and Mass Conservation. *Remote Sens.* **2022**, *14*, 3858. [CrossRef]
54. Wakigari, S.A.; Leconte, R. Enhancing Spatial Resolution of SMAP Soil Moisture Products through Spatial Downscaling over a Large Watershed: A Case Study for the Susquehanna River Basin in the Northeastern United States. *Remote Sens.* **2022**, *14*, 776. [CrossRef]
55. Wei, X.; Zhou, Q.; Cai, M.; Wang, Y. Effects of vegetation restoration on regional soil moisture content in the humid karst areas—A case study of Southwest China. *Water* **2021**, *13*, 321. [CrossRef]
56. Tang, R.; Li, Z.-L.; Liu, M.; Jiang, Y.; Peng, Z. A moisture-based triangle approach for estimating surface evaporative fraction with time-series of remotely sensed data. *Remote Sens. Environ.* **2022**, *280*, 113212. [CrossRef]
57. Nadeem, A.A.; Zha, Y.; Shi, L.; Ali, S.; Wang, X.; Zafar, Z.; Afzal, Z.; Tariq, M.A.U.R. Spatial downscaling and gap-filling of SMAP soil moisture to high resolution using MODIS surface variables and machine learning approaches over ShanDian River Basin, China. *Remote Sens.* **2023**, *15*, 812. [CrossRef]
58. Lei, F.; Senyurek, V.; Kurum, M.; Gurbuz, A.C.; Boyd, D.; Moorhead, R.; Crow, W.T.; Eroglu, O. Quasi-global machine learning-based soil moisture estimates at high spatio-temporal scales using CYGNSS and SMAP observations. *Remote Sens. Environ.* **2022**, *276*, 113041. [CrossRef]
59. Kim, H.; Lakshmi, V. Use of cyclone global navigation satellite system (CyGNSS) observations for estimation of soil moisture. *Geophys. Res. Lett.* **2018**, *45*, 8272–8282. [CrossRef]
60. Calabia, A.; Molina, I.; Jin, S. Soil moisture content from GNSS reflectometry using dielectric permittivity from fresnel reflection coefficients. *Remote Sens.* **2020**, *12*, 122. [CrossRef]

remote sensing

MDPI

Article

# Evaluation of the Land GNSS-Reflected DDM Coherence on Soil Moisture Estimation from CYGNSS Data

**Zhounan Dong [1,2] and Shuanggen Jin [1,*]**

1 Shanghai Astronomical Observatory, Chinese Academy of Sciences, Shanghai 200030, China; zndong@shao.ac.cn
2 School of Astronomy and Space Science, University of Chinese Academy of Sciences, Beijing 100049, China
* Correspondence: sgjin@shao.ac.cn or sg.jin@yahoo.com; Tel.: +86-21-34775292

**Abstract:** With the development of spaceborne global navigation satellite system-reflectometry (GNSS-R), it can be used for terrestrial applications as a promising remote sensing tool, such as soil moisture (SM) retrieval. The reflected L-band GNSS signal from the land surface can simultaneously generate coherent and incoherent scattering, depending on surface roughness. However, the contribution of the incoherent component was directly ignored in previous GNSS-R land soil moisture content retrieval due to the hypothesis of its relatively small proportion. In this paper, a detection method is proposed to distinguish the coherence of land GNSS-R delay-Doppler map (DDM) from the cyclone global navigation satellite system (CYGNSS) mission in terms of DDM power-spreading features, which are characterized by different classification estimators. The results show that the trailing edge slope of normalized integrated time-delay waveform presents a better performance to recognize coherent and incoherent dominated observations, indicating that 89.6% of CYGNSS land observations are dominated by the coherent component. Furthermore, the impact of the land GNSS-Reflected DDM coherence on soil moisture retrieval is evaluated from 19-month CYGNSS data. The experiment results show that the influence of incoherent component and incoherent observations is marginal for CYGNSS soil moisture retrieval, and the RMSE of GNSS-R derived soil moisture reaches 0.04 $cm^3/cm^3$.

**Keywords:** GNSS reflectometry (GNSS-R); cyclone global navigation satellite system mission (CYGNSS); coherent scattering; soil moisture

**Citation:** Dong, Z.; Jin, S. Evaluation of the Land GNSS-Reflected DDM Coherence on Soil Moisture Estimation from CYGNSS Data. *Remote Sens.* **2021**, *13*, 570. https://doi.org/10.3390/rs13040570

Academic Editor: Hugo Carreno-Luengo

Received: 7 December 2020
Accepted: 1 February 2021
Published: 5 February 2021



## 1. Introduction

Soil moisture (SM) is an essential parameter for the hydrology and energy cycle. Rapid acquiring and accurate monitoring of terrestrial SM is not only required in the hydrological research but also a significant benefit to water management and agricultural production. Since the L-band microwave has a strong sensitivity to the change of surface SM and can more easily penetrate the atmosphere and vegetation canopy, it has been widely used as the main soil moisture remote sensing frequency band in the satellite-based radiometer and radar missions [1]. Such as the European Space Agency's (ESA) Soil Moisture and Ocean Salinity (SMOS) mission and the National Aeronautics and Space Administration's (NASA) Soil Moisture Active Passive (SMAP) mission, both can provide global SM measurement with the spatial resolution on the order of 40 km and coverage every 2–3 days using carried L-band radiometer. Spaceborne global navigation satellite system-reflectometry (GNSS-R) is an innovative and sustainable low-cost technique with high spatial and temporal resolution [2], which operates as a passive bistatic forward scattering radar. The observe system directly receives the pre-existing signals transmitted by the GNSS satellites reflected off the Earth's surface [3], and the received scattering signals are typically expressed in a delay-Doppler map (DDM) for Earth's surface geophysical parameters retrieval [4], which provide a new paradigm in the land remote sensing to cover the space-time gap of the traditional high-cost dedicated monostatic active or passive satellite missions.

In the past decade, spaceborne GNSS-R has undergone rapid development with successfully deployed satellite missions, such as the UK Technology Demonstration Satellite-1 (TDS-1, launched in July 2014), NASA's cyclone global navigation satellite system (CYGNSS, launched in December 2016), China's Bufeng-1 A/B mission (launched in June 2019) [5]. Although all missions were originally designed for ocean surface wind speed retrieval, they also provided a large number of land observations for terrestrial remote sensing applications, such as soil moisture retrieval, forest biomass estimation, and wetland extent detection [6]. However, there are many differences between GNSS-R land and ocean applications [7]. Before using GNSS-R for geophysical parameters retrieval, the key issue is to determine the scattering mechanisms of observed DDM. Over the sea surface, the surface height standard deviation is at least a significant fraction of the signal wavelength under windy conditions and increases with the wind speed [8,9], so the L-band GNSS signal echoes from the ocean are purely incoherent, which can be well explained by the Z-V model [10]. Compared to the ocean surface, the L-band signal scattering from the land surface is more complicated, and the GNSS signal returns are affected by many factors, such as soil moisture, vegetation, surface roughness, inland water, topographic relief, and soil texture. The DDM generated after noncoherent integration loses phase information, and the land surface small-scale roughness is variant in space and time, which is extremely difficult to be determined. As a result, it is hard to distinguish the coherence of land reflected DDM, which affects its subsequent land applications.

In previous GNSS-R land applications, it has been generally assumed that the coherent component dominates the land scattering field, and the incoherent component is negligible. The coherence defined here refers to reflected signals from the first Fresnel zone arriving at the GNSS-R receiver with similar phase shifts [9]. Many studies have proved that coherent DDM derived reflectivity is sensitive to the change of soil moisture and forest biomass [11–15]. However, due to the sensitivity of coherent and incoherent observation on the land geophysical parameters is different, it is important to distinguish the coherence of observations for quantified parameter retrieval. Theoretical simulations have revealed that the roughness of the land surface was close to 5 cm, where only incoherent scattering will occur [9]. Meanwhile, the effect of topography is independent of surface roughness, and the topographic relief can mitigate the reflectivity [16]. Different DDM observables have been used for GNSS-R sea ice detection based on the difference of coherent reflected signal from the sea ice surface and diffuse scattering from the sea surface [17,18]. However, it is relatively difficult to verify the coherence of current ground CYGNSS data. The coherence of a single complex DDM look can be robustly distinguished based on the differences of coherent and noncoherent integration from the "raw IF" signal [19] because the correlated power of a perfectly coherent signal will increase over the given period from longer integration lengths, while the incoherent will not. Unfortunately, the CYGNSS mission only recorded very few I/F signals limited by its storage capability. Nevertheless, with the help of these I/F signals from the land surface, different estimators have been characterized in the studies for DDM coherence detection [7,20], and the results show that the purely coherent reflection only occurs over the inland water surface in spaceborne GNSS-R observation. The problem is that the differences in estimator performance can lead to different results, and the I/F signal dataset used is too small, lacking sufficient persuasive power. Based on the different assumptions, several SM inversion methods have also been developed, such as spatial averaging, combine linear regression method, machine-learning method, and the global inversion accuracy of SM can reach about 0.05 $cm^3/cm^3$ [21–30].

In this paper, a statistical method is developed to detect the coherence of CYGNSS level-1 DDM from the land. We assume that the delay-Doppler-spreading features of incoherent DDM from the ocean and land scattering are similar, which all present a typical "horseshoe" shape, only the magnitude of the absolute scattering power differs. The defined estimators are used to determine the flag of coherence in terms of known incoherent DDM from the windy ocean surface, and the inversion accuracy of GNSS-R derived soil moisture with high confidence coherent DDM is evaluated and validated.

The paper is organized as follows, Section 2 introduces the scattering theory over the sea surface and smooth soil surface and the definition of the coherent classification estimators based on the difference of typical coherent and incoherent dominated DDM. Section 3 shows the classification performance of different estimators, the distribution characteristics of coherent and incoherent observation over the land surface, and presents soil moisture retrieval results. Section 4 discusses the impact of coherent and incoherent DDM on SM remote sensing applications. Finally, conclusions are summarized in Section 5.

## 2. Methods and Datasets

If the land surface is relatively flat and smooth, the roughness of the scattering region is lower than the scale of the wavelength of the incoming GNSS signal; then the scattering mechanism is different from the diffuse scattering general occurring over the ocean surface. The land-coherent scattering only comes from the first Fresnel zone around the specular point instead of the whole glistening zone. The image theory and Friis transmission equation are used to explain this coherent forward scattering process, and the geophysical characteristics of the reflection surface are indicated by reflectivity [31]. As the surface roughness increases, the contribution of the incoherent component dramatically increases [32]; when the surface roughness approaches the signal wavelength scale, the conditions on the sea surface will recur. Next, we first introduce the scattering model, then present our coherence detection estimators, dataset, and GNSS-R soil moisture retrieval algorithm.

### *2.1. Bistatic Forward Scattering*

The DDM is the function of signal time-delay and Doppler frequency shift from the surface specular point, which implies the mapping relationship of scattering power between space and delay-Doppler domain. When the L-band signal impinges on the rough sea surface, incoherent scattering occurs in most cases and scattering power can be well signified by the Z-V model [6]:

$$P(\hat{\tau},\hat{f}_D) = \frac{T_I^2 P_T \lambda^2}{(4\pi)^3} \iint_A \frac{G_T \sigma^0 G_R}{R_R^2 R_T^2} \Lambda^2(\hat{\tau}-\tau)\mathrm{sinc}^2(\hat{f}_D - f_D)dA \quad (1)$$

$$\chi(\hat{\tau},\hat{f}_D) = \Lambda(\hat{\tau}-\tau)\mathrm{sinc}(\hat{f}_D - f_D) \quad (2)$$

$$\Lambda(\hat{\tau}-\tau) = \begin{cases} (1-\frac{|\hat{\tau}-\tau|}{\tau_c})^2 & |\hat{\tau}-\tau| \le \tau_c \\ 0 & |\hat{\tau}-\tau| > \tau_c \end{cases} \quad (3)$$

$$\mathrm{sinc}(\hat{f}_D - f_D) = \left| \frac{\sin(\pi(\hat{f}_D - f_D)T_I)}{\pi(\hat{f}_D - f_D)T_I} \right|^2 \quad (4)$$

where $P(\hat{\tau},\hat{f}_D)$ is the complex, diffuse scattering power, $P_T$ is the right-hand circular polarized (RHCP) transmitted power of the GNSS satellite, $G_T$ is the transmitter antenna gain, $G_R$ is the GNSS-R receiver antenna gain, $R_R$ is the distance from the receiver to the scatter point over the ocean surface, $R_T$ is the distance from the transmitter to the scatter point, $\lambda$ is the wavelength of the GNSS carrier, $T_I$ is the coherent integration time, $\sigma^0$ is normalized bistatic radar cross-section, $\Lambda(\hat{\tau}-\tau)$ is the correlation function of the GNSS navigation code, $\hat{\tau}$ and $\tau$ are the local replica code in the receiver and received signal time delays, respectively, $\mathrm{sinc}(\hat{f}_D - f_D)$ is the attenuation due to Doppler misalignment, $\hat{f}_D$ and $f_D$ are the local replica code in the receiver and received signal Doppler frequency shift, respectively. The product of the correlation function and the sinc function is called the Woodward ambiguity function (WAF) of the GNSS PRN navigation code. $A$ indicate the glistening zone, $dA$ is the differential area within the glistening zone. The $\sigma^0$ can be further expressed as:

$$\sigma^0 = \frac{\pi|\Re_{LR}(\theta)|^2 \vec{q}^4}{q_z^4} P\left(-\frac{\vec{q}_\perp}{q_z}\right) \tag{5}$$

where $\Re_{LR}$ is the Fresnel reflection coefficient of scattered left-hand circular polarized (LHCP) signal over the sea surface, $\vec{q}$ is the scattering unit vector, $\vec{q}_\perp$ and $q_z$ are horizontal and vertical components, respectively, and $P$ is the probability density function of the sea surface slope. The coherent integration time commonly sets 1ms on the delay Doppler mapping instrument to generate a single DDM look. Due to the diffuse scattering signals over sea surface being relatively weak, to improve the signal-to-noise ratio (SNR) of DDM and reduce the effect of speckle and thermal noise within a coherent integration of a DDM look, there is an extra noncoherent integrated step that takes 1 s, during which the received signal will lose the phase information. Normalized bistatic radar cross-section (NBRCS) has been used as the land surface remote sensing fundamental observable in the backward scatterometer for a very long time, which is also reasonable to be employed in GNSS-R with specific calibration.

Theoretically, real land scattering power consists of coherent and incoherent components. After noncoherent integration, the DDM can be expressed as:

$$\left\langle\left|P(\hat{\tau}, \hat{f}_D)\right|^2\right\rangle = \left\langle\left|P_{coh}(\hat{\tau}, \hat{f}_D)\right|^2\right\rangle + \left\langle\left|P_{incoh}(\hat{\tau}, \hat{f}_D)\right|^2\right\rangle \tag{6}$$

where $\left\langle\left|P_{coh}(\hat{\tau}, \hat{f}_D)\right|^2\right\rangle$ and $\left\langle\left|P_{incoh}(\hat{\tau}, \hat{f}_D)\right|^2\right\rangle$ are coherent and incoherent contributions, respectively. Previous studies focus on land SM retrieval directly assumed that the incoherent item is negligible on Earth's land surface; the received scattering power mainly concentrates from the adjacent region around specular point. According to the image theory and Friis transmission equation, the coherent scattering power coming from the first Fresnel zone can be expressed:

$$P_{coh} = \frac{P_T G_T \lambda^2 G_R}{(4\pi)^2 (R_R + R_T)^2} \Gamma(\theta) \gamma^2 \exp(-(2k\sigma\cos(\theta))^2) \tag{7}$$

where $\Gamma$ is reflectivity, it is the function of the Fresnel reflection coefficient $\Re$ ($\Gamma(\theta) = |\Re_{LR}(\theta)|^2$). $\gamma$ is the transmissivity which indicates the vegetation layer attenuation, it is the function of vegetation opacity depth (VOD) $\tau$($\gamma = \exp(-\tau\sec\theta)$). The exponential term represents signal attenuation caused by surface roughness. $k$ is the wavenumber, and $\sigma$ represents the standard deviation of surface height. The size of the first Fresnel zone is related to the height of the receiver platform and signal incidence angle. For the CYGNSS mission, the diameter of the first Fresnel zone is about 0.5 km. Considering this small area compared to the spatial resolution of CYGNSS DDM pixels, it is reasonable to directly pick the peak power value in the DDM to calculate reflectivity. According to Equation (7), the surface reflectivity can be derived as follows:

$$\Gamma(\theta) = \frac{(4\pi)^2 P_{peak} (R_R + R_T)^2}{\lambda^2 P_T G_T G_R} \tag{8}$$

where $P_{peak}$ is the coherent power with surface roughness and vegetation attenuation correction. Since the velocity of CYGNSS satellites along the track is 7 km/s, the spatial resolution corresponding to the peak power measurement is about 0.5 km $\times$ 7 km. After June 2019, the sampling frequency of the CYGNSS mission has increased to 0.5 Hz, which allows the spatial resolution along the direction of satellite movement to reach 3.5 km.

When the roughness of the land surface is comparable to the scale of GNSS carrier wavelength, incoherent scattering will occur over the land surface. Since there is no reliable high spatial-temporal surface roughness information, it is difficult to determine the coherence of a DDM. The aforementioned assumption indeed ignores two issues for GNSS-R SM inversion. On one hand, if we directly consider that the main contribution of scattering power comes from the coherent component, which implies the influence of the incoherent components in DDM is ignored. On the other hand, when the GNSS-R observation footprint passes through the land surface with large roughness, the purely incoherent signal will be received, the influence of incoherent observations on the SM inversion is ignored as well. There is a big difference between the sensitivity of coherent and incoherent DDM observables to the SM level [9], so it is important to evaluate the influence of the previous assumption. Due to the different scattering mechanisms that happen behind the coherent and incoherent observations, the shape and magnitude of the measured scattering power in the DDM are different. The coherent DDM resembles the WAF itself without delay-doppler spreading [33]. Figure 1a shows a typical land reflected DDM over the winter wheat field. As a comparison, Figure 1b presents the DDM observed over the ocean surface with the 6.6 m/s wind speed near the specular point. The received diffuse scattering signals come from the entire glistening zone with the WAF spreading in the direction of delay and Doppler axis; DDM exhibits the typical "horseshoe" shape.

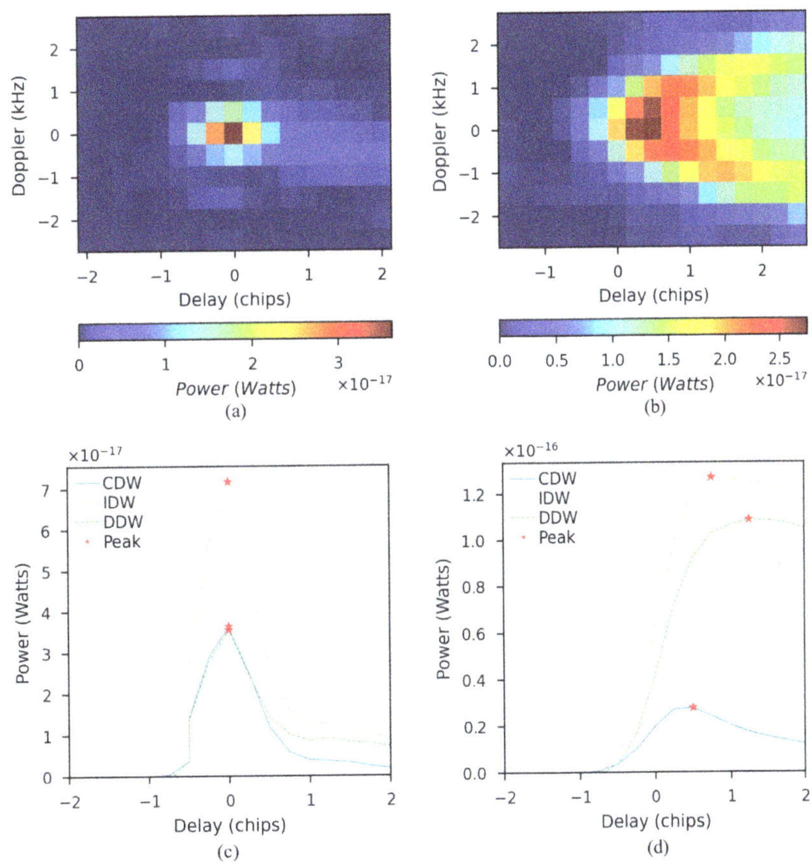


**Figure 1.** Typical scattering-power delay-Doppler map (DDM) and delay waveforms over the land surface (**a**,**c**) and ocean surface (**b**,**d**).

The time-delay waveform (DW) is the 1D representation of DDM; it is also a fundamental observable in the GNSS-R study, which usually includes two types: central Doppler time-delay waveform (CDW) and integrated time-delay waveform (IDW). The CDW is the zero Doppler delay scattering power in the DDM. The IDW is obtained by summing the columns along the Doppler axis of DDM. In our classification method, we also define the deviation of time-delay waveforms (DDW) calculated by IDW subtracting CDW at each time delay bin. It can be noticed from the land reflected DW in Figure 1c that the trailing edge scattering power quickly decreases to the noise floor level after the peak point, and the deviation between CDW and DDW is small. Whereas the scattering power of the trailing edge of DW derived from the sea surface scattered DDM decreases slowly, especially for IDW and DDW, which is shown in Figure 1d, and the peak and trailing edge power of DDW are much larger than CDW. As the land topography changes and the surface roughness increases, which can lead to the intensity of the incoherent field strengthened rapidly, the coherent field weakens according to the conservation of energy. Then, the magnitude and distribution characteristics of DDM gradually approach the sea surface observations. Based on these features, we proposed a method to classify CYGNSS coherent and incoherent observations.

### 2.2. Definition of Classification Estimator

Since the DDM observed from the ocean surface dominated by incoherent scattering and from the relatively flat land surface dominated by coherent scattering are significantly different [17,25], the coherence classification method proposed here is based on the shape and distribution characteristics of power-spreading in the DDM. Here and after, we directly call coherent component dominated DDM as coherent DDM, and incoherent component dominated as incoherent DDM. The whole classification idea is inspired by GNSS-R sea ice detection [17,18], and both are essentially determining the similarity to the coherent model. For the coherent DDM, it resembles the Woodward ambiguity function (WAF) without delay-doppler spreading [33], while incoherent DDM exhibits the typical "horseshoe" shape. The calculation of the defined classification estimators is introduced in the following part in detail. To characterizes the difference in the coherence of DDM, combining the known typical range of estimator values calculated from ocean scattered incoherent DDM, the threshold of coherence can be determined in the CYGNSS land observation. It should be noted that the reference position of defined DDM estimators from the land surface refers to the delay and doppler bin of the DDM peak power. If the DDM observable can be calculated from the DW, the selected window is set to spanning 5 time-delay bins from the peak. For the DDMA calculation, the selected delay/Doppler window is a $5 \times 3$ matrix with the center located on the peak location. The given window size mainly depends on two reasons. On one hand, the position of peak power is not fixed in each CYGNSS DDM, so when the entire DDM is directly used to calculate the estimator, the range of statistical delay and Doppler is different. On the other hand, it is based on the shape of WAF, which is shown in Figure 2. The main coherent reflection power concentrate on this zone. The final objective is to compute the probability density function (PDF) of DDM observables in the land and ocean observations to determine the separation threshold of coherent and incoherent dominated observations.

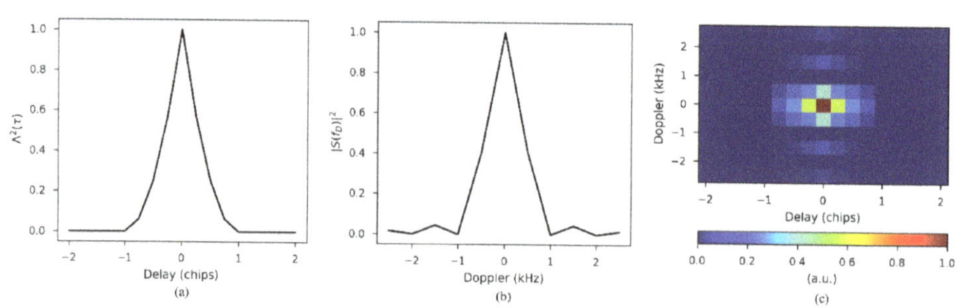


**Figure 2.** The delay spreading function (**a**), Doppler-spreading function (**b**), and Woodward ambiguity function (**c**).

1. TES: It is the trailing edge slope of the normalized DW and determined using the least-squares fitting within the time-delay window to a linear expression:

$$a_{\mathrm{TES}}^{N} = \frac{n\sum_{i=1}^{n}\tau_i P_i^N - \sum_{i=1}^{n}\tau_i \sum_{i=1}^{n} P_i^N}{n\sum_{i=1}^{n}\tau_i^2 - \left(\sum_{i=1}^{n}\tau_i\right)^2} \tag{9}$$

where $n$ = 5 is the number of time-delay bins for the linear fitting. $\tau_i$ is the time-delay of each bin. $P_i^N$ is the normalized scattering power in raw count within the corresponding time-delay bin.

2. TEV: It is the average volume of the normalized DW trailing edge:

$$\overline{P}_{\mathrm{TEV}}^{N} = \frac{1}{n}\sum_{i=1}^{n} P_i^N \tag{10}$$

3. TEV_POW: It is the average absolute scattering power of the DW trailing edge:

$$\overline{P}_{\mathrm{TEV\_POW}} = \frac{1}{n}\sum_{i=1}^{n} P_i \tag{11}$$

where $P_i$ is the scattering power of the corresponding time-delay bin.

4. DDMA: It is the average of the normalized scattering power DDM near its peak:

$$\overline{\sigma}_{\mathrm{DDMA}}^{N} = \frac{1}{nm}\sum_{i=1}^{n}\sum_{j=1}^{m} P_{i,j}^N \tag{12}$$

where $n$ and $m$ indicate the selected size of delay and Doppler window.

5. DDMA_POW: It is the average of the absolute scattering power DDM near the peak:

$$\overline{\sigma}_{\mathrm{DDMA\_POW}} = \frac{1}{mn}\sum_{i=1}^{n}\sum_{j=1}^{m} P_{i,j} \tag{13}$$

6. MF: It is known as the WAF-matched filter (MF) approach, which directly calculates the correlation coefficient of normalized DDM and unitary energy WAF:

$$R_{\mathrm{MF}} = \frac{\left|\left\langle\left\langle\left|P(\hat{\tau},\hat{f}_D)\right|^2\right\rangle,\chi(\hat{\tau},\hat{f}_D)\right\rangle\right|^2}{\left\langle\left\langle\left|P(\hat{\tau},\hat{f}_D)\right|^2\right\rangle,\left\langle\left|P(\hat{\tau},\hat{f}_D)\right|^2\right\rangle\right\rangle\left\langle\chi(\hat{\tau},\hat{f}_D),\chi(\hat{\tau},\hat{f}_D)\right\rangle} \tag{14}$$

### 2.3. *Dataset for Soil Moisture Retrieval*

CYGNSS is part of the NASA Earth system science pathfinder program; it is deployed as the first dedicated spaceborne GNSS-R constellation launched in December 2016. The space segment consists of eight microsatellites orbiting on a non-synchronous near-circular orbit with an inclination of approximately 35° (all spacecraft distributed on the same orbital plane). Each spacecraft is capable of tracking 4 reflections simultaneously, resulting in 32 DDMs per second over the Earth's surface. The standard CYGNSS DDM consists of 17 delay bins with the resolution of a quarter of the GPS C/A chip by 11 Doppler bins with an interval of 500 Hz. Each DDM was processed using 1 ms coherent integration followed by 1000 looks of noncoherent averaging. After July 2019, the noncoherent time was reduced to 0.5 s. The primary objective of CYGNSS is to monitor the wind speed during the evolution of tropical cyclones. The footprint of its measurement covers the critical latitude band between ±38° [34]. At the same time, it also provides substantial land observations within this scope. The work conducting in this paper uses the CYGNSS level-1 DDMs in power analog from published version 2.1 data product ranging from January 2018 to August 2019. General data quality control (QC) in the land application is also utilized, the DDM SNR lower than 2 dB, receiver antenna gain at the specular point direction lower than 0 dB, and specular incidence angle over 65° are screened. To get the purely incoherent DDMs from the ocean surface [9], the corresponding DDMs with ERA5 wind speed greater than 5 m/s are employed. Land and sea surface observations are directly distinguished by the quality flag provided in the CYGNSS level-1 data.

The SMAP dataset used as the land surface reference SM is version 6 level-3 radiometer global daily 36 km equal-area scalable earth grid version 2.0 (EASE-Grid 2.0) soil moisture product within the same period time of the CYGNSS dataset [35], except the data product missing from 20 June to 22 July 2019. The SMAP SM data provides daily descending (a.m.) and ascending passes (p.m.) measurement, including the auxiliary parameters VOD (in the SMAP product indicates vegetation opacity parameter) and surface roughness coefficient (in the SMAP product indicates vegetation_roughness_coefficient), which are used to correct the attenuation of CYGNSS scattering power from the impact of surface roughness and vegetation canopy. In the soil moisture retrieval process, only the recommended data are used without the open water, urban area.

### 2.4. Soil Moisture Retrieval Algorithm

In previous spaceborne GNSS-R soil moisture retrieval studies, the primary methodology is to establish the relationship between GNSS-R derived land surface reflectivity and reference truth SM values, which assumes that coherent component dominates GNSS-R land scattering field. In the theory of surface electromagnetic scattering, the surface reflectivity is the function of the incidence angle of the incoming signal and the Fresnel reflection coefficient; the latter one is mainly affected by the near-surface SM [1]. Figure 3 simulates the relationship between reflectivity and SM at different incidence angles with the solid line, where the semi-empirical Dobson model is used to mapping the relationship between soil moisture and complex permittivity [36]. The surface reflectivity increases monotonously with soil becoming wetter, and the response of reflectivity to the change of SM from 0.0 $cm^3/cm^3$ to 0.7 $cm^3/cm^3$ can reach 10 dB. The effect of the incidence angle on the mapping relationship between SM and reflectivity is negligible when the incidence angle is less than 60°. In our CYGNSS SM inversion experiements, the DDM peak value of coherent scattered power is picked in the CYGNSS level-1 data as the left term of Equation (7). Since the small scale roughness and upwelling vegetation cover can attenuate the scattering signal, the roughness and vegetation correction in Equation (7) directly use the roughness coefficient and VOD parameter provided in SMAP product for individual observation. Although the influence of the signal incidence angle is small, the method proposed in [25] is still used in this work. The effect of the incidence angle correction is represented by the dashed lines in Figure 3 as well.

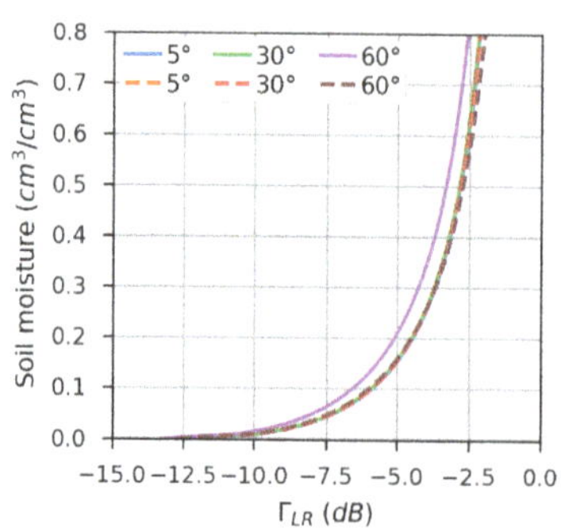


**Figure 3.** The relationship between reflectivity and soil moisture under different incidence angles of GPS signal and the performance of incidence angle correction.

Due to the pseudorandom distribution of CYGNSS measurement, the influence of observation noise, and the spatial difference of surface roughness and vegetation cover at the specular point, currently, it is difficult to directly establish a reliable SM retrieval model at the GNSS-R specular point modeling all these factors, the optimal approach is to improve the SNR of reflectivity using the space–time-averaging method to form the gridded retrieval model [18]. Since the SM reference data used in this work is from the SMAP level-3 version 6 product, the individual CYGNSS reflectivity calculated with Equation (8) within one day will be projected into a global cylindrical 36 km × 36 km EASE-Grid 2.0 grid to align with the reference SM values, the average reflectivity is picked as the grid value. Here, we set a data quality control criterion; if the count of projected reflectivity at the grid is less than three, the corresponding grid observation will be considered invalid on that day. Next, the time matching is used to combine the gridded reflectivity and SMAP SM to establish the training dataset and mask the pixels in the SMAP SM data flagged with inland water and urban areas. Finally, the retrieval model is fitted at each grid. Usually, the variation range of local surface soil moisture is limited in a year; the linear model can achieve high modeling accuracy. Therefore, the training samples are used to fit the linear model between mean reflectivity and reference SM values pixel-by-pixel:

$$SM_{i,j}^{CYGNSS} = a_{i,j}\overline{\Gamma}_{i,j} + b_{i,j} \tag{15}$$

where the $a$ and $b$ are the pending parameters of the model. $i$ and $j$ are the grid location in the 36 km × 36 km EASE-Grid 2.0 grid. $\overline{\Gamma}$ is the grid mean reflectivity after space-time average processing.

## 3. Results and Analysis

### 3.1. Performance Evaluation of DDM Observables

In this work, we assume that coherent and incoherent scattering simultaneously occurs on the land surface in the CYGNSS land observations, and only two scattering cases appear: coherent reflection mainly contributed to DDM or incoherent scattering mainly contributed to DDM. We classify the two cases based on the statistical characteristics of the predefined estimators. Since we have known that ocean surface observation belongs to the latter, the characteristic information of incoherent DDM can be obtained. To evaluate the performance of different classification estimators defined in Section 2.2, the CYGNSS collected land and ocean DDMs in January 2018 are used to calculate the PDF and accumulation distribution function (CDF) of each DDM observable separately. Figure 4 gives the PDF and CDF of TES, TEV, TEV_POW calculated from CDW (top row), IDW (middle row), and DDW (bottom row). It can be found that the performance of the three types of DW-derived classification estimators is different. The PDF of TES between land and ocean observations is separated more and sharper, which means that the classification results of TES are generally better than the other two. TES values from ocean surface scattered signals are generally larger than land observation; its PDF appears on the right side of the figure. The reason is the L-band GNSS signals impinge on the ocean surface always occurring diffuse scattering, the time-delay, and Doppler-spreading cause DW to appear a significant "smearing" feature; in other words, the scattered power of the trailing edge will slowly decrease. In addition, the PDF of land reflected DDM-derived TES is more dispersed than ocean observations. In the first column of Figure 4, the closer TES to the left side of the x-axis, the greater the contribution of the coherent component to the DDM since the DW is much closer to the WAF correlation function. As the roughness of the land surface increases, the contribution of the incoherent component rapidly increases and begins to impact the scattering power of the DW trailing edge, so the TES value gradually approaches the ocean observations, two PDFs finally intersect. The performance of TEV_POW is the worst; the distribution of PDF from the land and ocean observations is overlapped. It can be explained by the fact that the peak value of coherent DM is larger than incoherent DM, while the scattering power of incoherent DW declines slowly after the peak value, the final result is the average of absolute scattering power within 5 time-delay bins between land and ocean DDM derived TEV_POW are close. The performance of TEV is in the middle.

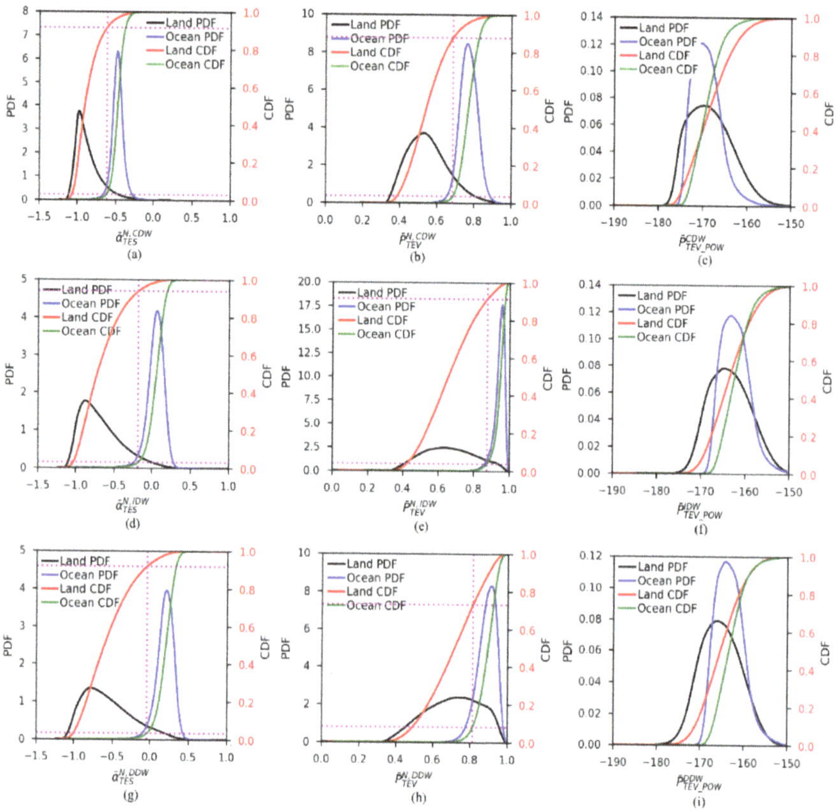


**Figure 4.** Statistic performance of trailing edge slope (TES), the average volume of the normalized time-delay waveform (DW) trailing edge (TEV), average absolute scattering power of the DW trailing edge (TEV_POW) derived from central Doppler time-delay waveform (CDW; **a**–**c**), integrated time-delay waveform (IDW; **d**–**f**), and deviation of time-delay waveform (DDW, **g**–**i**) over land and ocean surface, dataset collected from the cyclone global navigation satellite system (CYGNSS) level-1B in January 2018.

Figure 5 shows the PDF of estimator DDMA, DDMA_POW, and MF derived from the ocean and land DDMs. The performance of DDMA_POW is very close to TEV_POW; the distribution of two PDF almost overlaps, which can be explained by the same reason as TEV_POW. Therefore, we can conclude that it is difficult to determine the coherence of the DDM based on the feature of its absolute power in the given window. In the rest of the paper, we will exclude the absolute power estimators. Here, MF shows the best performance; DDM from land generally has a higher correlation with the WAF in comparison with the ocean, which is in line with the previous assumption.

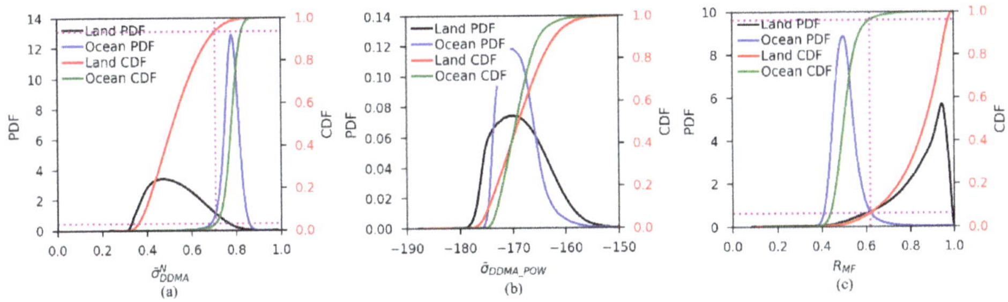


**Figure 5.** Average of the normalized scattering power DDM near its peak (DDMA) (**a**), an average of the absolute scattering power DDM near the peak (DDMA_POW) (**b**), and WAF-matched filter (MF) (**c**) statistic performance over land and ocean area, dataset collected from the CYGNSS level-1B in January 2018.

The performance of different estimators is variant depending on the land surface scattering mechanisms; the classification threshold is determined by the intersection of two PDFs, which is represented by the magenta dotted line in the vertical direction in Figures 4 and 5. The horizontal magenta dotted lines indicate the accumulative probability density of the corresponding estimator computed from the CYGNSS land and ocean surface data, which not only presents the probability of detection (PD) of coherent DDM but also indicates the proportion of the coherent and incoherent data. Table 1 summarizes the classification threshold, PD, the probability of false alarm (PFA), and the probability of error (PE) of each estimator. It can be found that the PD between different observables is small except DDW-derived TEV, and the average PD of all estimators is 89.6%. Among eight estimators, the PD of TES calculated from normalized IDW (NIDW) is the largest, and the PE is the smallest. Comparing all the subgraphs in Figure 4, it can also be found that the PDF of NIDW-derived TES is more separated between land and ocean data. Moreover, it is more concentrated and sharper than normalized CDW, and normalized DDW derived TES. Hence, it is considered the best estimator to detect the coherent and incoherent DDM collected over the land surface in this study. In the rest of the paper, we just use NIDW-derived TES as the classification estimator to recognize the high confidence coherent DDM in the CYGNSS land data for SM retrieval.

**Table 1.** The classification threshold and the probability of different DDM observables.

| Observables | $a_{TES}^{N,CDW}$ | $\overline{P}_{TEV}^{N,CDW}$ | $a_{TES}^{N,IDW}$ | $\overline{P}_{TEV}^{N,IDW}$ | $a_{TES}^{N,DDW}$ | $\overline{P}_{TEV}^{N,DDW}$ | $\overline{\sigma}_{DDMA}^{N}$ | $R_{MF}$ |
|---|---|---|---|---|---|---|---|---|
| Threshold | −0.6191 | 0.6878 | −0.1798 | 0.8726 | −0.0394 | 0.8144 | 0.7053 | 0.6171 |
| PD | 0.9146 | 0.8789 | 0.9379 | 0.9132 | 0.9192 | 0.7307 | 0.9368 | 0.9362 |
| PFA | 0.0284 | 0.0360 | 0.0289 | 0.0376 | 0.0312 | 0.0808 | 0.0310 | 0.0381 |
| PE | 0.0569 | 0.0786 | 0.0455 | 0.0622 | 0.0560 | 0.1751 | 0.0471 | 0.0510 |

### 3.2. Coherent and Incoherent DDM Observations

The coherent and incoherent observation is determined by the threshold of the classification estimator of NIDW-derived TES. Figure 6 shows the average SM values from 9 km EASE-Grid 2.0 SMAP level-3 product, average CYGNSS gridded coherent and incoherent land surface reflectivity with the same projection grid in January 2018. Land coherent DDM can be detected in the entire footprint of the CYGNSS mission, whereas incoherent observations are more likely to occur in high altitude mountainous and hilly terrain.

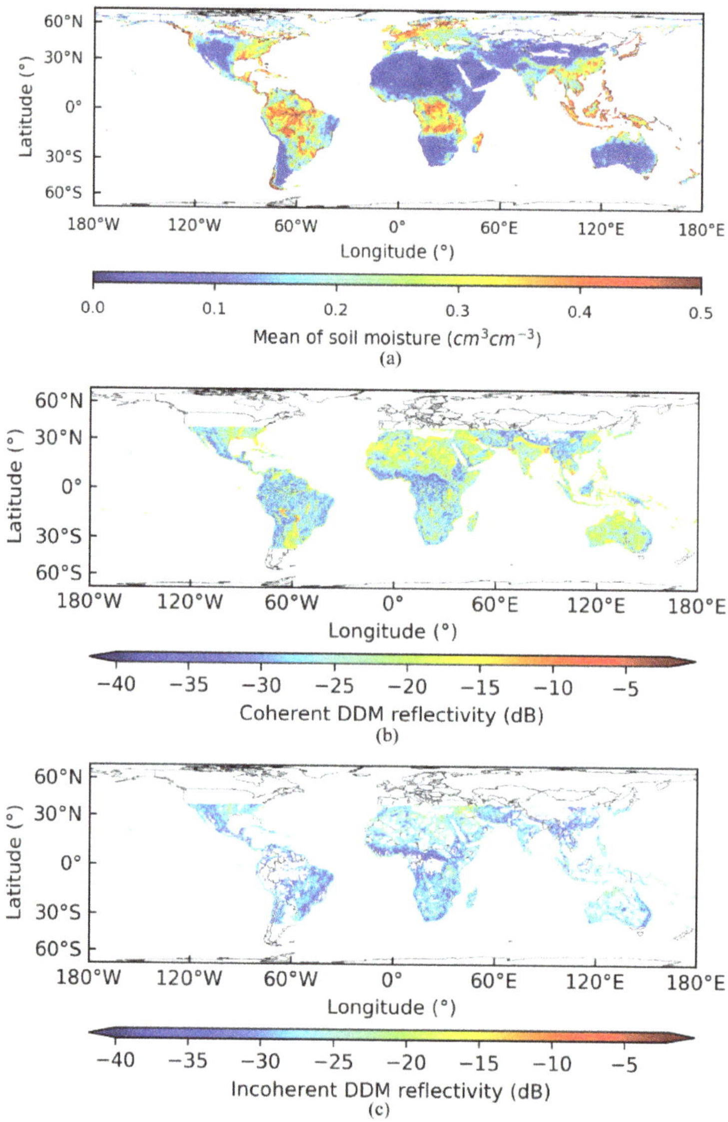


**Figure 6.** The global distribution of monthly average Soil Moisture Active Passive (SAMP) soil moisture (SM) (**a**), coherent reflectivity (**b**), and incoherent observations (**c**) in January 2018.

According to the classification results, the range of coherent reflectivity is from −44 dB to −3 dB; the strongest coherent reflection indeed comes from the inland open water surface, while the area of the tropical rainforest and the arid mountainous area has the lowest reflectivity. It is worthy to note that the GNSS-R reflectivity over tropical dense forest areas is lower than the barren/desert area, which is consistent with previous studies [37]. Nevertheless, compared to the distribution of SMAP SM in Figure 6a, it can be found that the SM values in corresponding areas are high. Meanwhile, incoherent scattering rarely occurs in dense vegetation-covered areas, as Figure 6c shown. However, part of the reason is the applying of QC, which excludes some noisy DDM with lower DDN SNR. However, even we ignore the influence of QC, the count of involved incoherent observations for spatial averaging is still less than four in most of the grids, which is much smaller than the total number of coherent measurements in the same grid. In terms of the International Geosphere-Biosphere Program (IGBP) land cover type parameters provided in the SMAP products, the statistical results also show that the proportion of coherent and incoherent GNSS-R observations over different land cover types is almost the same before and after QC. It confirms that even the dense upwelling vegetation cannot change the scattering mechanism of the land surface, but dense forest canopies will generate a strong attenuation effect on the GNSS-R coherent scattering signals, which may be attributed to vegetation volume scattering. Moreover, many coherent and incoherent overlapped areas can be found in Figure 6b,c; we speculate that the main reason is the spatial distribution of the surface roughness is different within the projected grid, so the coherent and incoherent observations can be collected simultaneously in a grid.

### 3.3. GNSS-R Soil Moisture Retrieval

To analyze the influence of incoherent observations on the GNSS-R land surface SM inversion in the previous SM retrieval method, we compared two retrieval configurations: using 19 months CYGNSS land observations and screened coherent data for retrieval model evaluation with k-fold cross-validation approach, where k = 5. Since the global SM value in most areas of the land is generally small in a year, the PDF of the monthly SMAP SM data in 2018 is presented in Figure 7a, and the maximum probability density of SM is 0.06 $cm^3/cm^3$. To further evaluate the performance of the established SM model over the high-humidity areas, the accuracy of the inversion model is evaluated when the referenced SM value is greater than 0.1 $cm^3/cm^3$.

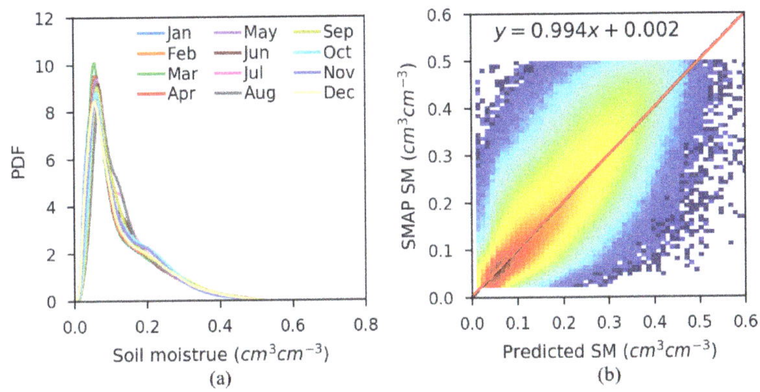


**Figure 7.** Monthly SMAP soil moisture probability density function (**a**) and density scatterplot of GNSS-R derived soil moisture and surface SM reference values in a split of k-fold cross-validation (**b**).

Using the SM inversion method introduced in Section 2.4, Table 2 summarizes the performance of the two models established from two training datasets. When all CYGNSS land observations are used for modeling, the cross-validation model bias, mean absolute error (MAE) and root-mean-square error (RMSE) are −0.0003 $cm^3/cm^3$, 0.0274 $cm^3/cm^3$, and 0.0416 $cm^3/cm^3$, respectively. The inversion results with the distinguished coherent observation training dataset constructed retrieval model show that the bias, MAE, and RMSE are −0.0003 $cm^3/cm^3$, 0.0269 $cm^3/cm^3$, and 0.0408 $cm^3/cm^3$, respectively. The model performance between the two strategies is very close. When the SM reference values are greater than 0.1 $cm^3/cm^3$, the model accuracy of the two methods is 0.0569 $cm^3/cm^3$ and 0.0564 $cm^3/cm^3$, respectively, and the inversion results did not show a big difference. Figure 7b shows the density scatterplot between SMAP reference SM and GNSS-R-derived SM generated from a split of k-fold cross-validation with the coherent observation established model. The red line represents the linear fitting line; the predicted SM shows an overall fairly good agreement with the SMAP SM, all CYGNSS land data retrieved SM show an identical situation. Figure 8 presents the coherent inversion accuracy at each grid pixel with k-fold cross-validation. The analysis shows that CYGNSS incoherent observations will not cause any noticeable SM spatial inversion accuracy differences compared to the coherent results, so it is not given here.

**Table 2.** Soil moisture retrieval model evaluating with k-fold cross-validation (unit: $cm^3/cm^3$).

| Dataset | Total Bias | Total MAE | Total RMSE | SM > 0.1, Bias | SM > 0.1, MAE | SM > 0.1, RMSE |
|---|---|---|---|---|---|---|
| All land observations | −0.0003 | 0.0274 | 0.0416 | −0.0124 | 0.0426 | 0.0569 |
| Coherent observations | −0.0003 | 0.0269 | 0.0408 | −0.0123 | 0.0421 | 0.0564 |

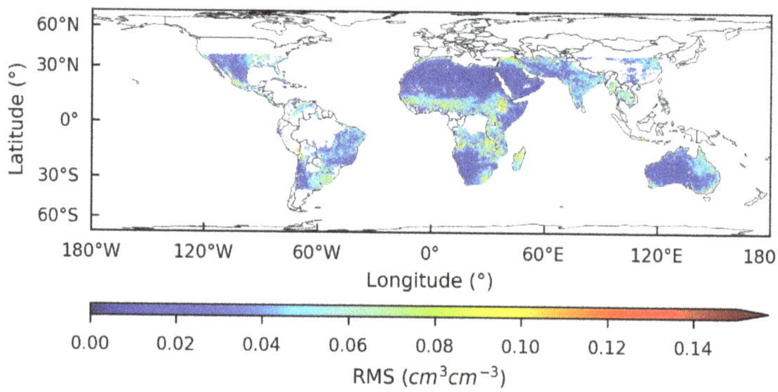


**Figure 8.** Root-mean-square error (RMSE) of CYGNSS retrieved land surface SM at each grid.

## 4. Discussion

GNSS-R coherent and incoherent observations have different sensitivities to the land SM values [9]. According to the classification results with the defined estimator in this study, 6.2% of the measurements in the CYGNSS land observations have a high possibility controlled by the incoherent scattering field. In addition, the PDFs of reflectivity calculated from the land surface coherent and incoherent observations do show distribution differences, as shown in Figure 9. It should be noted that if the DDM scattering power is dominated by incoherent components, NBRCS is commonly picked as the fundamental quantity, which is calculated according to [21]. Since most of the previous studies ignored incoherent scattering, namely the counterpart reflectivity is directly calculated by Equation (8), it is reasonable to use this equation to calculate incoherent reflectivity and analyze their influence on CYGNSS SM retrieval in this paper. The experiments show that extra incoherent observations have no obvious effect on the final CYGNSS SM retrieval with space-time averaging combined with the linear regression method. To further validate this conclusion, the threshold of NIDW-derived TES is set to −0.5 to improve the confidence of discriminated coherent DDM, where the probability of false alarm is only 0.01. It also can be considered that the contribution of the incoherent component is very small in screened coherent observations. At this point, the coherent DDM accounts for 75.8% of CYGNSS land measurements. The bias, MAE, and RMSE of final inversed soil moisture are $-0.0003$ $cm^3/cm^3$, 0.0265 $cm^3/cm^3$, and 0.0403 $cm^3/cm^3$, respectively. The RMSE is reduced by 3.1% compared to the constructed model with assuming all coherent land observations. When the reference SM value is greater than 0.1 $cm^3/cm^3$, the inversion bias is $-0.0145$ $cm^3/cm^3$, MAE values is 0.0416 $cm^3/cm^3$, and RMSE is 0.0558 $cm^3/cm^3$.

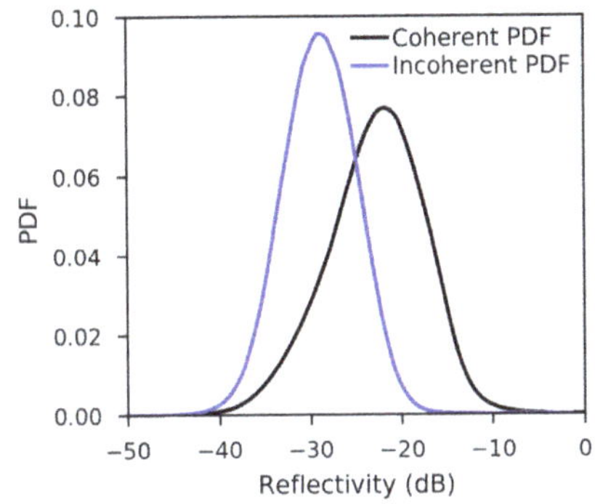


**Figure 9.** The probability density function of coherent and incoherent reflectivity derived from CYGNSS land observations in January 2018.

The inversion accuracy of the aforementioned GNSS-R space-time averaging SM retrieval methods with two different training datasets is similar because the magnitude and number of incoherent reflectivity are smaller when compared to coherent reflectivity, and the spatial average processing will further mitigate its influence. However, there is no doubt that coherence classification methods play a key role in future GNSS-R land detection. The inversion model can be directly established at the individual specular point with improved high-quality and high spatial resolution observations, which provides in the following dedicated spaceborne GNSS-R land remote sensing missions, and also contributes to other land applications, such as inland water system detection, biomass detection, and wetland extent determination. Another noteworthy issue is that the established GNSS-R SM inversion model tends to underestimate the surface soil moisture when the land SM over 0.3 $cm^3/cm^3$, while most of the previous studies also show the same problem. Since most training samples are concentrated in the lower SM range, the regression model is more affected by this part of the data. Therefore, there should be a better weighting strategy to solve this problem in future work.

## 5. Conclusions

This paper presents a classification methodology to distinguish coherent and incoherent DDMs in the CYGNSS land observations. Since the GNSS scattering signals from the windy ocean surface are almost incoherent, while the coherent land DDMs are closer to WAF, six different classification estimators are established based on scattering power-spreading shape and magnitude features over the ocean, and land CYGNSS collected DDMs, which are used to screen the land high confidence coherent component dominated DDMs. The results show that the estimators based on the absolute magnitude features of DDM are difficult to distinguish its coherency, while the estimator indicating shape features performs better. The average proportion of GNSS-R land observations dominated by coherent components is 89.6%. NIDW-derived TES performs best among all defined DDM observables, and its PDFs from the ocean and land DDMs are more separated and sharper, whose detection probability for coherent observations can reach 93.8% with the lowest detection probability of error. The distribution of high-confidence coherent and incoherent surface observation indicates that observations over the dense forest cannot change the surface scattering properties but will greatly weaken the coherent scattering power. Using 19 months of CYGNSS observation data and SMAP SM product for land SM retrieval model validation, the RMSE of model performance with k-fold cross-validation can reach 0.04 $cm^3/cm^3$. Incoherent observations have not seriously impaired the accuracy of CYGNSS soil moisture inversion.

**Author Contributions:** S.J. and Z.D. conceived and designed the experiments and Z.D. performed the experiments and analyzed the data. Both authors contributed to the writing of the paper. Both authors have read and agreed to the published version of the manuscript.

**Funding:** This work was supported by the Strategic Priority Research Program Project of the Chinese Academy of Sciences (Grant No. XDA23040100), and Shanghai Leading Talent Project (Grant No. E056061).

**Acknowledgments:** Authors thank NASA for providing CYGNSS and SMAP data.

**Conflicts of Interest:** The authors declare no conflict of interest.

## References

1. Njoku, E.G.; Entekhabi, D. Passive microwave remote sensing of soil moisture. *J. Hydrol.* **1996**, *184*, 101–129. [CrossRef]
2. Jin, S.; Feng, G.P.; Gleason, S. Remote sensing using GNSS signals: Current status and future directions. *Adv. Space Res.* **2011**, *47*, 1645–1653. [CrossRef]
3. Ruf, C.S.; Chew, C.; Lang, T.; Morris, M.G.; Nave, K.; Ridley, A.; Balasubramaniam, R. A New Paradigm in Earth Environmental Monitoring with the CYGNSS Small Satellite Constellation. *Sci. Rep.* **2018**, *8*, 8782. [CrossRef] [PubMed]
4. Jin, S.; Zhang, Q.; Qian, X. New progress and application prospects of global navigation satellite system reflectometry (GNSS+ R). *Acta Geod. Cartogr. Sin.* **2017**, *46*, 1389–1398.
5. Jing, C.; Niu, X.; Duan, C.; Lu, F.; Di, G.; Yang, X. Sea Surface Wind Speed Retrieval from the First Chinese GNSS-R Mission: Technique and Preliminary Results. *Remote Sens.* **2019**, *11*, 3013. [CrossRef]
6. Zavorotny, V.U.; Gleason, S.; Cardellach, E.; Camps, A. Tutorial on Remote Sensing Using GNSS Bistatic Radar of Opportunity. *IEEE Geosci. Remote Sens. Mag.* **2014**, *2*, 8–45. [CrossRef]
7. Gleason, S.; O'Brien, A.; Russel, A.; Al-Khaldi, M.M.; Johnson, J.T. Geolocation, Calibration and Surface Resolution of CYGNSS GNSS-R Land Observations. *Remote Sens.* **2020**, *12*, 1317. [CrossRef]
8. Comite, D.; Ticconi, F.; Dente, L.; Guerriero, L.; Pierdicca, N. Bistatic Coherent Scattering From Rough Soils With Application to GNSS Reflectometry. *IEEE Trans. Geosci. Remote Sens.* **2019**, 1–14. [CrossRef]
9. Balakhder, A.M.; Al-Khaldi, M.M.; Johnson, J.T. On the Coherency of Ocean and Land Surface Specular Scattering for GNSS-R and Signals of Opportunity Systems. *IEEE Trans. Geosci. Remote Sens.* **2019**, 1–11. [CrossRef]
10. Zavorotny, V.U.; Voronovich, A.G. Scattering of GPS signals from the ocean with wind remote sensing application. *IEEE Trans. Geosci. Remote Sens.* **2000**, *38*, 951–964. [CrossRef]
11. Chew, C.; Shah, R.; Zuffada, C.; Hajj, G.; Masters, D.; Mannucci, A.J. Demonstrating soil moisture remote sensing with observations from the UK TechDemoSat-1 satellite mission. *Geophys. Res. Lett.* **2016**, *43*, 3317–3324. [CrossRef]
12. Camps, A.; Park, H.; Pablos, M.; Foti, G.; Gommenginger, C.P.; Liu, P.W.; Judge, J. Sensitivity of GNSS-R Space-borne observations to soil moisture and vagetation. *IEEE J. Sel. Top. Appl. Earth Obs. Remote Sens.* **2016**, *9*, 4730–4742. [CrossRef]
13. Camps, A.; Llossera, M.V.; Park, H.; Portal, G.; Rossato, L. Sensitivity of TDS-1 GNSS-R Reflectivity to Soil Moisture: Global and Regional Differences and Impact of Different Spatial Scales. *Remote Sens.* **2018**, *10*, 1856. [CrossRef]
14. Zribi, M.; Guyon, D.; Motte, E.; Dayau, S.; Wigneron, J.P.; Baghdadi, N.; Pierdicca, N. Performance of GNSS-R GLORI data for biomass estimation over the Landes forest. *Int. J. Appl. Earth Obs. Geoinf.* **2019**, *74*, 150–158. [CrossRef]
15. Carreno-Luengo, H.; Luzi, G.; Crosetto, M. Biomass Estimation Over Tropical Rainforests Using GNSS-R On-Board The CyGNSS Microsatellites Constellation. In Proceedings of the IGARSS 2019–2019 IEEE International Geoscience and Remote Sensing Symposium, Yokohama, Japan, 28 July–2 August 2019; pp. 8676–8679.
16. Dente, L.; Guerriero, L.; Comite, D.; Pierdicca, N. Space-Borne GNSS-R Signal Over a Complex Topography: Modeling and Validation. *IEEE J. Sel. Top. Appl. Earth Obs. Remote Sens.* **2020**, *13*, 1218–1233. [CrossRef]
17. Alonso-Arroyo, A.; Zavorotny, V.U.; Camps, A. Sea Ice Detection Using U.K. TDS-1 GNSS-R Data. *IEEE Trans. Geosci. Remote Sens.* **2017**, *55*, 4989–5001. [CrossRef]
18. Zhu, Y.; Tao, T.; Yu, K.; Li, Z.; Qu, X.; Ye, Z.; Geng, J.; Zou, J.; Semmling, M.; Wickert, J. Sensing Sea Ice Based on Doppler Spread Analysis of Spaceborne GNSS-R Data. *IEEE J. Sel. Top. Appl. Earth Obs. Remote Sens.* **2020**, *13*, 217–226. [CrossRef]
19. Loria, E.; Brien, A.O.; Gupta, I.J. Detection & Separation of Coherent Reflections in GNSS-R Measurements Using CYGNSS Data. In Proceedings of the IGARSS 2018–2018 IEEE International Geoscience and Remote Sensing Symposium, Valencia, Spain, 22–27 July 2018; pp. 3995–3998.
20. Al-Khaldi, M.M.; Johnson, J.T.; Gleason, S.; Loria, E.; O'Brien, A.J.; Yi, Y. An Algorithm for Detecting Coherence in Cyclone Global Navigation Satellite System Mission Level-1 Delay-Doppler Maps. *IEEE Trans. Geosci. Remote Sens.* **2020**, 1–10. [CrossRef]
21. Clarizia, M.P.; Pierdicca, N.; Costantini, F.; Floury, N. Analysis of CYGNSS Data for Soil Moisture Retrieval. *IEEE J. Sel. Top. Appl. Earth Obs. Remote Sens.* **2019**, *12*, 2227–2235. [CrossRef]
22. Chew, C.C.; Small, E.E. Soil Moisture Sensing Using Spaceborne GNSS Reflections: Comparison of CYGNSS Reflectivity to SMAP Soil Moisture. *Geophys. Res. Lett.* **2018**, *45*, 4049–4057. [CrossRef]
23. Chew, C.; Small, E. Description of the UCAR/CU Soil Moisture Product. *Remote Sens.* **2020**, *12*, 1558. [CrossRef]

24. Kim, H.; Lakshmi, V. Use of Cyclone Global Navigation Satellite System (CyGNSS) Observations for Estimation of Soil Moisture. *Geophys. Res. Lett.* **2018**, *45*, 8272–8282. [CrossRef]
25. M-Khaldi, M.M.; Johnson, J.T.; O'Brien, A.J.; Balenzano, A.; Mattia, F. Time-Series Retrieval of Soil Moisture Using CYGNSS. *IEEE Trans. Geosci. Remote Sens.* **2019**, *57*, 4322–4331. [CrossRef]
26. Jia, Y.; Jin, S.; Savi, P.; Gao, Y.; Tang, J.; Chen, Y.; Li, W. GNSS-R Soil Moisture Retrieval Based on a XGboost Machine Learning Aided Method: Performance and Validation. *Remote Sens.* **2019**, *11*, 1655. [CrossRef]
27. Senyurek, V.; Lei, F.; Boyd, D.; Kurum, M.; Gurbuz, A.C.; Moorhead, R. Machine Learning-Based CYGNSS Soil Moisture Estimates over ISMN sites in CONUS. *Remote Sens.* **2020**, *12*, 1168. [CrossRef]
28. Senyurek, V.; Lei, F.; Boyd, D.; Gurbuz, A.C.; Kurum, M.; Moorhead, R. Evaluations of Machine Learning-Based CYGNSS Soil Moisture Estimates against SMAP Observations. *Remote Sens.* **2020**, *12*, 3503. [CrossRef]
29. Jia, Y.; Jin, S.; Savi, P.; Yan, Q.; Li, W. Modeling and Theoretical Analysis of GNSS-R Soil Moisture Retrieval Based on the Random Forest and Support Vector Machine Learning Approach. *Remote Sens.* **2020**, *12*, 3679. [CrossRef]
30. Yan, Q.; Gong, S.; Jin, S.; Huang, W.; Zhang, C. Near Real-Time Soil Moisture in China Retrieved From CyGNSS Reflectivity. *IEEE Geosci. Remote Sens. Lett.* **2020**, 1–5. [CrossRef]
31. Friis, H.T. A NOTE ON A SIMPLE TRANSMISSION FORMULA. *Proc. Inst. Radio Eng.* **1946**, *34*, 254–256. [CrossRef]
32. Katzberg, S.J.; Garrison, J.L. Utilizing GPS to determine ionospheric delay over the ocean. *NASA Tech. Memo* **1996**, 4750. Available online: https://ntrs.nasa.gov/citations/19970005019 (accessed on 4 February 2021).
33. Clarizia, M.P.; Ruf, C.S.; Jales, P.; Gommenginger, C. Spaceborne GNSS-R Minimum Variance Wind Speed Estimator. *IEEE Trans. Geosci. Remote Sens.* **2014**, *52*, 6829–6843. [CrossRef]
34. Cygnss. *CYGNSS Level 1 Science Data Record*; Version 2.1; NASA Physical Oceanography DAAC: Pasadena, CA, USA, 2017. [CrossRef]
35. O'Neill, E.P.; Chan, S.; Njoku, E.G.; Jackson, T.; Bindlish, R.; Chaubell, J. *SMAP L3 Radiometer Global Daily 36 km EASE-Grid Soil Moisture*; Version 6; NASA National Snow and Ice Data Center Distributed Active Archive Center: Boulder, CO, USA, 2019. [CrossRef]
36. Dobson, M.C.; Ulaby, F.T.; Hallikainen, M.T.; El-rayes, M.A. Microwave Dielectric Behavior of Wet Soil-Part II: Dielectric Mixing Models. *IEEE Trans. Geosci. Remote Sens.* **1985**, *GE-23*, 35–46. [CrossRef]
37. Carreno-Luengo, H.; Luzi, G.; Crosetto, M. Above-Ground Biomass Retrieval over Tropical Forests: A Novel GNSS-R Approach with CyGNSS. *Remote Sens.* **2020**, *12*, 1368. [CrossRef]

*remote sensing*

MDPI

*Article*

# The Correction Method of Water and Fresnel Reflection Coefficient for Soil Moisture Retrieved by CYGNSS

**Qi Wang [1,2], Jiaojiao Sun [3], Xin Chang [4,5,*], Taoyong Jin [5], Jinguang Shang [1,2] and Zhiyong Liu [1,2]**

1 Chengdu Institute of Surveying and Investigation, Chengdu 610081, China; qwangchoice@whu.edu.cn (Q.W.); ybgs@cdkc.cn (J.S.); zhiyong@whu.edu.cn (Z.L.)
2 Urban Informatization Surveying and Mapping Engineering Technology Research Center of Sichuan, Chengdu 610081, China
3 The Third Geographical Information Mapping Institute of Natural Resources Ministry, Chengdu 610100, China; sunjiaojiao@zrzybszztygtx.wecom.work
4 School of Geodesy and Geomatics, Wuhan University, Wuhan 430079, China
5 Hubei Luojia Laboratory, Wuhan 430079, China; tyjin@sgg.whu.edu.cn
* Correspondence: xchang@sgg.whu.edu.cn; Tel.: +86-186-2777-0280

**Abstract:** Spaceborne GNSS-R technology is a new remote sensing method for soil moisture monitoring. Focusing on the significant influence of water on the surface reflectivity of CYGNSS, this paper improved the removal method of water influence according to the spatial resolution of CYGNSS data. Due to the disturbance effect of the incident angle, microwave frequency and soil type on the Fresnel reflection coefficient in surface reflectivity, a normalization method of Fresnel reflection coefficient was proposed after analyzing the data characteristics of variables in the Fresnel reflection coefficient. Finally, combined with the soil moisture retrieval method of linear equation, the accuracy was compared and verified by using measured data, SMAP products and official CYGNSS products. The results indicate that the normalization method of the Fresnel reflection coefficient could effectively reduce the influence of relevant parameters on the Fresnel reflection coefficient, but the normalization effect became worse at large incident angles (greater than 65°). Compared with the official CYGNSS product, the retrieval accuracy of optimized soil moisture was improved by 10%. The method proposed in this paper will play an important reference role in the study of soil moisture retrieval using spaceborne GNSS-R data.

**Keywords:** soil moisture; CYGNSS; normalization method; water removal

**Citation:** Wang, Q.; Sun, J.; Chang, X.; Jin, T.; Shang, J.; Liu, Z. The Correction Method of Water and Fresnel Reflection Coefficient for Soil Moisture Retrieved by CYGNSS. *Remote Sens.* **2023**, *15*, 3000. https://doi.org/10.3390/rs15123000

Academic Editors: Hugo Carreno-Luengo, Dallas Masters, Chun-Liang Lin and Nereida Rodriguez-Alvarez

Received: 8 April 2023
Revised: 20 May 2023
Accepted: 6 June 2023
Published: 8 June 2023



## 1. Introduction

Soil moisture is of great value in understanding plant physiological activities, hydrometeorological processes, global energy exchange and agricultural production [1–3]. The distribution information of accurate soil moisture is not only of great significance for scientific research, but can also serve a number of practical applications. Spaceborne GNSS Reflectometry (GNSS-R) is an emerging remote sensing technology for reflecting soil moisture over a large area due to its advantages of a wide signal source, large data volume, short revisit time, low cost and low power consumption, etc. Its frequency band and high spatial-temporal resolution can effectively compensate for the shortcomings of optical remote sensing, which is easily obscured by clouds, and the low spatial resolution of microwave remote sensing products [4–10].

Since the UK TechDemoSat-1 (TDS-1) and Cyclone Global Navigation Satellite System (CYGNSS) satellites provide spaceborne GNSS-R data for free, the retrieval of soil moisture for spaceborne GNSS-R has gradually become a research hotspot. In order to effectively characterize the relationship between the surface reflectivity of CYGNSS and soil moisture, and thus obtain high accuracy retrieval results for soil moisture, a large number of modeling algorithms have been applied, such as linear models [11–14], machine learning [15–17] and deep learning [18], etc.

The reflection signal carrying water body information will weaken the sensitivity of the surface reflectivity of CYGNSS to soil moisture, and thus reduce the accuracy of soil moisture retrieval. Related studies have shown that a 25 m wide body of water can significantly affect the surface reflectivity of CYGNSS [11]. Therefore, Chew et al. [19], Wan et al. [20] and Zhu et al. [21] performed the removal of the water body effect on CYGNSS observations by means of external data sources such as the Global Surface Water Explorer (GSWE) and the water data of SMAP. The Fresnel reflection coefficient is one of the major component variables of CYGNSS surface reflectivity, and is directly related to the angle of incidence and the soil dielectric constant. The influence of the correlation parameters in the Fresnel reflection coefficient on CYGNSS surface reflectivity can be effectively weakened, which can improve the accuracy of soil moisture retrieval [22]. Al-Khaldi et al. [23] proposed the normalization method of the incident angle to correct the surface reflectivity of CYGNSS, and the results proved that the method could attenuate the effect of the incident angle in the Fresnel reflection coefficient. In addition, this method was also applied to the soil moisture product algorithm of CYGNSS developed by Chew et al. [11,23].

As the sampling frequency of CYGNSS increases, its spatial resolution changes and thus the original method for removing the influence of water bodies will mistakenly pick data carrying valid information. The Fresnel reflection coefficient is composed of the incident angle and the soil dielectric constant, which is directly related to the microwave frequency, soil temperature, soil type composition and soil moisture [24]. Current research has focused on attenuating the effects of incident angle and soil moisture on the Fresnel reflection coefficient, further resulting in a lack of complete analysis of the relevant influences in the Fresnel reflection coefficient and the establishment of a unified model to attenuate the effects of these parameters. Therefore, this paper first improved the removal method of observations affected by water bodies based on the analysis of CYGNSS data. Then, the variable response of the Fresnel reflection coefficient was analyzed in detail and the normalization method was proposed. Finally, the accuracy of soil moisture retrieval using CYGNSS was improved by combining the method of soil moisture retrieval with a linear model. After the introduction, Section 2 describes an overview of the study area and the adopted dataset. Section 3 presents proposed methods and the retrieval method of soil moisture. The results of the proposed method and soil moisture retrieval are displayed and appraised in Section 5. The discussion for results of the study is given in Section 5. Finally, the main conclusions for this study are given in Section 6.

## 2. Study Area and Data Source

### 2.1. Study Area and Ground Measurements

The study area was located in the southern United States, covering 26 states with a geographical range of 75°~122°W and 30°~38°N (Figure 1). The region covers a vast area, with subtropical, temperate continental and tropical climates in the southeast, central and southwest. The southeast is warm and humid, the middle is cold in winter and hot in summer and the southwest has a large annual temperature difference of up to 25 °C. The topography of the entire region is high in the west and low in the east, and there is a clear topographic divide (100°W) in the middle of the region. The western part of this topographic divide is dominated by plateaus and mountains, while the eastern part is plains (Figure 1a). Based on the LC_Type1 (Scheme of global vegetation classification for IGBP) band of the MCD12Q1 product of MODIS, a map of vegetation cover types in the study area was obtained, as shown in Figure 1b.

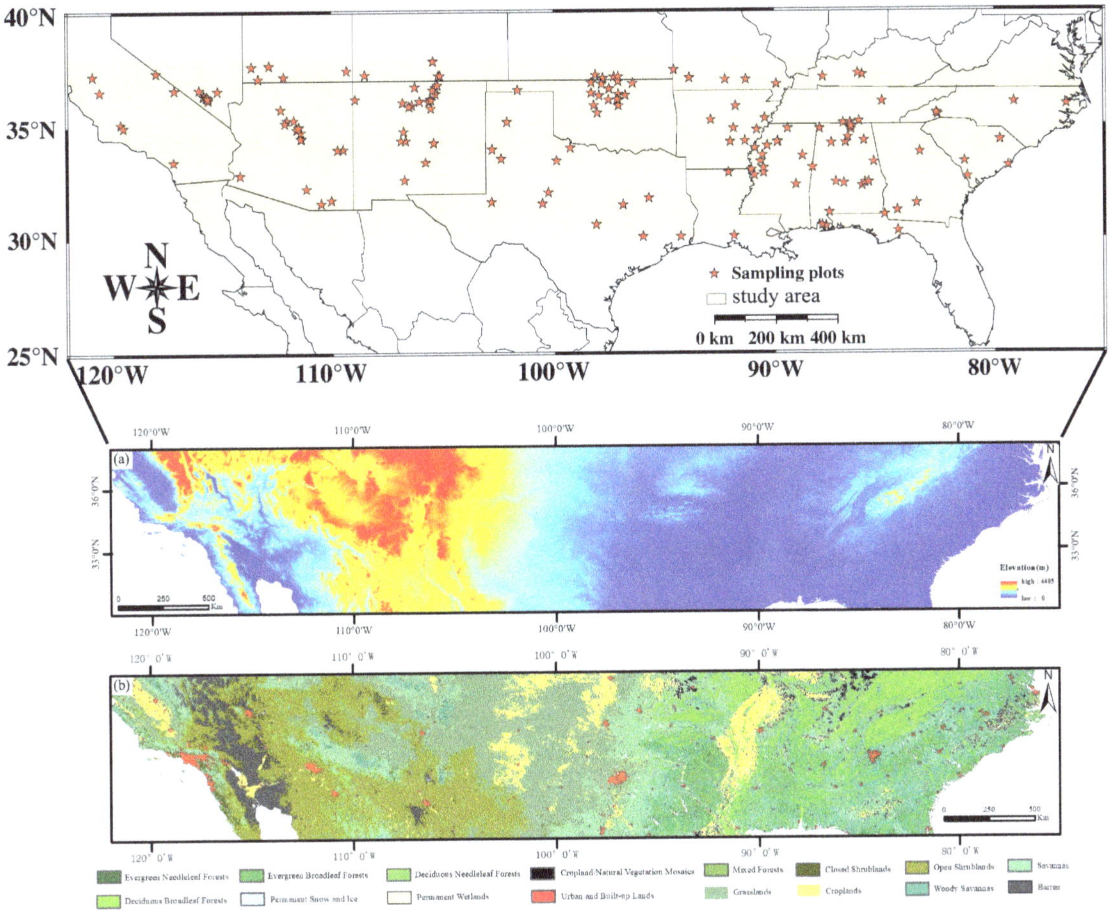


**Figure 1.** Location and measured points of the study area. (**a**) Topographic map, (**b**) soil type map.

The measured soil moisture data used in the study area were freely available to the public in terms of volumetric water content ($m^3/m^3$) and were derived from the International Soil Moisture Observation Network (ISMN; http://ismn.geo.tuwien.ac.at/networks, accessed on 6 November 2021). In this study, four monitoring networks, ARM, SCAN, SNOTEL and USCRN in ISMN, with a total of 160 measured monitoring points, were used from July 2019 to December 2021, with a soil moisture depth of 5 cm. Among these measurement stations of soil moisture, meaningless stations with values less than 0 $cm^3/cm^3$ or greater than 1 $cm^3/cm^3$ were excluded.

### 2.2. CYGNSS Data

CYGNSS is a constellation of eight satellites, each of which has four channels; i.e., thirty-two observations per second can be received. Due to the increase in CYGNSS sampling frequency after July 2019 (i.e., with 500 ms non-coherent accumulation), the resolution of CYGNSS is $3.5 \times 0.5$ km at this time. The CYGNSS data used in this study are the L1-level data of version 3.0 from July 2019 to December 2021, which can be downloaded from the official CYGNSS website (https://cygnss.engin.umich.edu/data-products/ (accessed on 6 November 2021)). To improve the quality of CYGNSS data, standard quality control and empirical quality control were performed [19]. The higher launch power of the GPS Block IIF satellite introduces uncertainty at peak power [25]. Removing these data will reduce the overall data volume by 30%. Therefore, this part of the data was chosen to be retained in this study. The power received by CYGNSS consists of both incoherent and coherent scattering components [26]. Most studies have been based on the assumption that the incoherent component is ignored and only the coherent scattering component is retained [11,12,14,20]. Therefore, based on the land scattering model of GNSS-R, the surface reflectivity $\Gamma(\theta)$ of CYGNSS can be readily derived, following the assumption that the coherent component dominates:

$$\Gamma(\theta) = \frac{P_r(R_t + R_r)^2(4\pi)^2}{P_t G_t G_r \lambda^2} \tag{1}$$

where $P_r$ and $P_t$ are, respectively, the reflected power received by the receiver and the transmitted power. $\lambda$ is the carrier signal wavelength. $R_r$ and $R_t$ are, respectively, the distances from the receiver and transmitter to the specular reflection point. $G_t$ and $G_r$ are the transmitting antenna gain and receiving antenna gain, respectively.

### 2.3. SMAP Data

Soil Moisture Active and Passive (SMAP) is an Earth observation mission carrying an L-band radiometer and radar, which provides soil moisture and freeze–thaw data with a time-resolution of 2–3 days from NASA (https://nsidc.org/data/smap/smap-data.html (accessed on 6 November 2021)). In this study, soil moisture data (L3_SM_P) of SMAP L3 were used, which were obtained by retrieval using the Single Channel Algorithm (SCA) with a spatial resolution of 36 km. The time period was selected from July 2019 to December 2021. SMAP was divided into two types of data: daily 6:00 a.m. (descending orbit) and 18:00 p.m. (ascending orbit). It has been shown that the surface temperature homogenization at 6:00 a.m. is better than that at 18:00 p.m., and the retrieval accuracy is higher than that at 18:00 [27]. In this study, both descending orbit and ascending orbit data were incorporated and averaged to obtain soil moisture data of SMAP.

### 2.4. Water Body Data

The water body data of GSWE came from a 30 m dataset produced based on optical images from Landsat satellites [28], with global coverage. GSWE released several sub-datasets, including Occurrence, Change, Seasonality, Recurrence, etc. Each sub-dataset consisted of a 10° × 10° grid with a total range of 60°S~80°N and 180°W~180°E. The "Seasonality" product was selected for this study, and the individual pixel values were classified on a scale of 1 to 12, indicating how many months of the year were inundated with water. In this case, all values greater than 1 were labeled as water bodies to facilitate the exclusion of observations influenced by water bodies. Since this study was in the southern region of the United States, a total of 12 Seasonality data were selected for the range of 30°N to 50°N and 130°W to 70°W. A Seasonality product from these data was shown in Figure 2. Due to the special and static nature for the time-resolution of Landsat data, the produced GSWE dataset could not be synchronized with the observation time of CYGNSS. However, this dataset could remove the observations affected by water bodies to some extent.

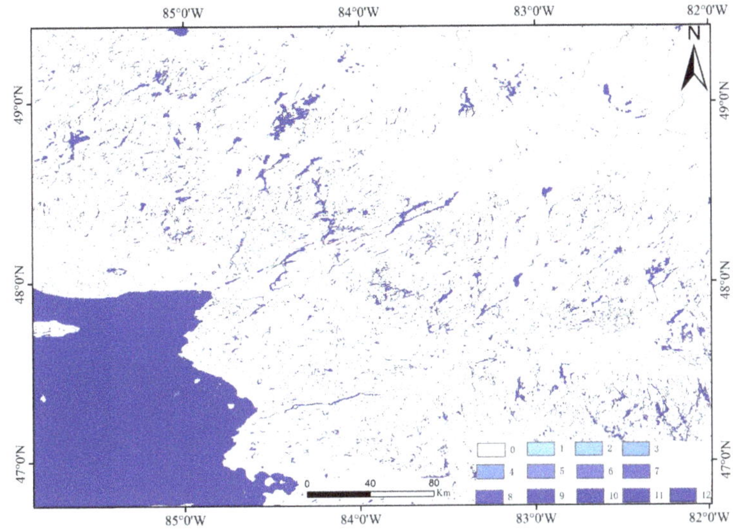


**Figure 2.** The diagram of "Seasonality" product data of GSWE (0 represents no data).

## 3. Methodology

In this study, the procedure for the retrieval of soil moisture using CYGNSS based on proposed methods was illustrated in Figure 3. The first step involved the quality control and processing of data used, such as CYGNSS, SMAP and GSWE. The surface reflectivity derived from CYGNSS was corrected using an improved method of water removal and the normalization method of the Fresnel reflection coefficient. A linear regression equation of soil moisture was established by combining the resampled soil moisture product of SMAP with the corrected surface reflectivity. Finally, the results of soil moisture were obtained by averaging the retrieved soil moisture, and the accuracy was comparatively verified based on the measured data, SMAP products and CYGNSS products.

### *3.1. Removal of Water*

The current solution for observations influenced by water bodies is to exclude observations that carry information about water bodies [19,20]. A square grid of 7 × 7 km to exclude observations influenced by water bodies was designed by Chew et al. [19], based on the "Seasonality" product from GSWE data. According to the research of Chew et al. [19], an improved method for removing observations affected by water bodies in a 3 × 3 km square grid was proposed by analyzing the characteristics of CYGNSS data. The process was as follows:

Step 1: Based on the latitude and longitude of the specular reflection point from CYGNSS, its corresponding location in the "Seasonality" product was searched.

Step 2: A square grid of 3 × 3 km was created, with this corresponding location as the center.

Step 3: When there was a value marked as 1 (i.e., there is a water body) in this 3 × 3 km square grid, then the point was eliminated, i.e., the specular reflection point was eliminated.

Step 4: The above process was repeated to complete the removal of specular reflection points affected by water bodies.

The removal diagram of the observations affected by the water was shown in Figure 4:

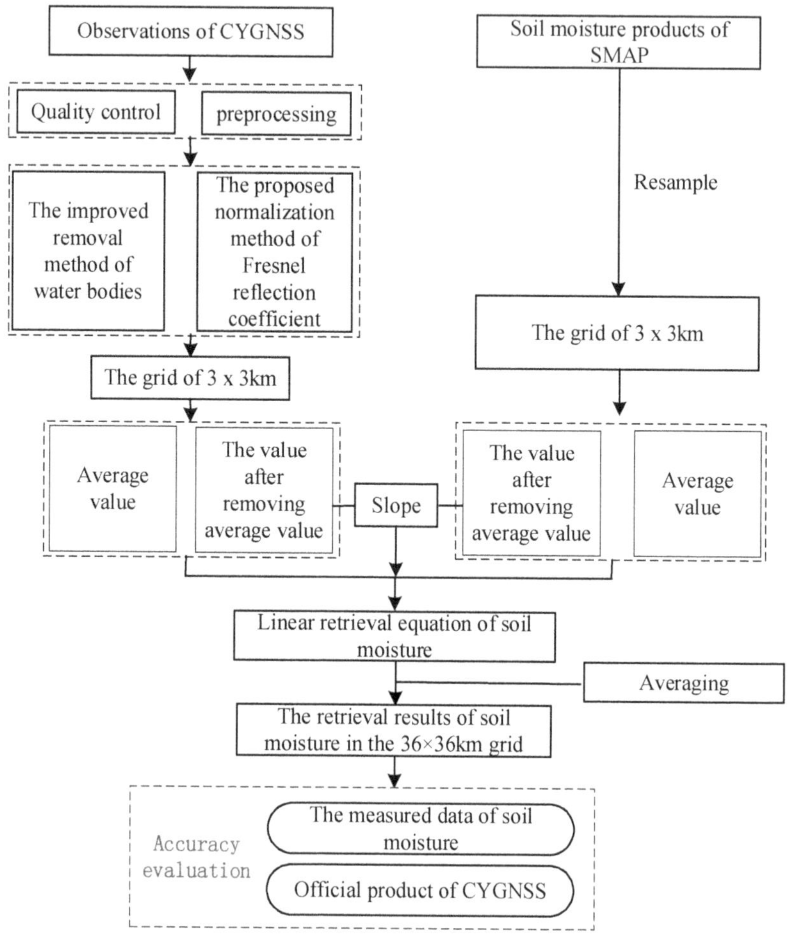


**Figure 3.** Flowchart of this study.

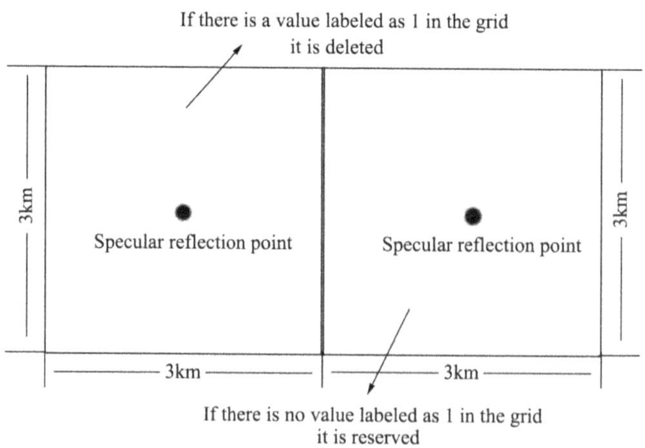


**Figure 4.** Schematic diagram of the removal method for observations affected by water bodies.

### 3.2. The Normalization of the Fresnel Reflection Coefficient

According to the Kirchhoff approximation, the surface reflectivity Γ(θ) of CYGNSS can be further expressed as [29–31]:

$$\Gamma(\theta) = |R_{lr}(\theta)|^2 \gamma \exp(-(2k\sigma\cos\theta)^2) \quad (2)$$

$\gamma$ is the vegetation attenuation term; $\exp(-(2k\sigma\cos\theta)^2)$ is the attenuation term of surface roughness; and $R_{lr}(\theta)$ is the Fresnel reflection coefficient, which is a function of the incident angle θ and the soil dielectric constant ε [32]. ε is calculated by the Dobson model [24], which is adapted to the frequency range of 0.3–1.3 Ghz and 1.4–18 Ghz and consists of microwave frequency, soil temperature, soil type composition and soil moisture:

$$\varepsilon_{soil}^{\alpha} = \left[1 + \frac{p_b}{p_s}(\varepsilon_s^{\alpha} - 1) + m_v^{\beta}\varepsilon_{fw}^{\alpha} - m_v\right]^{\frac{1}{\alpha}} \quad (3)$$

$$\beta = 1.2748 - 0.00519P_{sand} - 0.00152P_{clay} \quad (4)$$

where $P_b$ and $P_s$ are the bulk density of soil and the density of the solid medium in soil, respectively. α is generally 0.65, and $\varepsilon_{fw}$ and $\varepsilon_s$ are the permittivity of free water and solid soil, respectively. $m_v$ is soil moisture. $P_{sand}$ and $P_{clay}$ represent the sand and clay contents of soil (%), respectively.

According to Formulas (3) and (4), the response of these relevant variables to the Fresnel reflection coefficient was shown in Figure 5. Soil type parameters refer to the table of physical parameters published by the Dobson model [24] (Table 1).

**Table 1.** Physical parameters of typical soil types.

| | Sandy Loam | Fertile Land | Silty Loam | Silt Soil |
|---|---|---|---|---|
| $P_{sand}$ (%) | 51.52 | 41.96 | 30.63 | 5.02 |
| $P_{clay}$ (%) | 13.42 | 8.53 | 13.48 | 47.38 |
| $P_{silt}$ (%) | 35.06 | 49.51 | 55.89 | 47.6 |
| $P_s$ | 2.66 | 2.7 | 2.59 | 2.56 |
| $P_b$ (g/cm$^3$) | 1.6006 | 1.5781 | 1.575 | 1.4758 |

From Figure 5a–d, it can be seen that the Fresnel reflection coefficients obtained from different soil moisture values varied greatly under the condition of constant soil temperature and the same soil type. For a constant soil temperature and the same soil moisture, the Fresnel reflection coefficient corresponding to different soil types was also different. The above results indicate that differences in the soil type and soil moisture can lead to changes in the Fresnel reflection coefficient. Figure 5e–h show the response of the Fresnel reflection coefficient for a constant value of soil moisture and different soil temperatures. It can be observed that the changes in Fresnel reflection coefficients obtained from soil temperature differences were small relative to those caused by changes in soil moisture. Of course, differences in the Fresnel reflection coefficient between soil types are always present. Moreover, in Figure 5, the Fresnel reflection coefficient becomes smaller and smaller, with an increasing incident angle regardless of the differences in soil moisture, temperature and soil type, indicating that the incident angle plays a very important role.

In this study, a normalization method of the Fresnel reflection coefficient was proposed to reduce the influence of relevant parameters on the Fresnel reflection coefficient, and thus reduced the surface reflectivity error of CYGNSS caused by the Fresnel reflection coefficient and improved the accuracy of soil moisture retrieval. The process of establishing this method consisted of four steps in total.

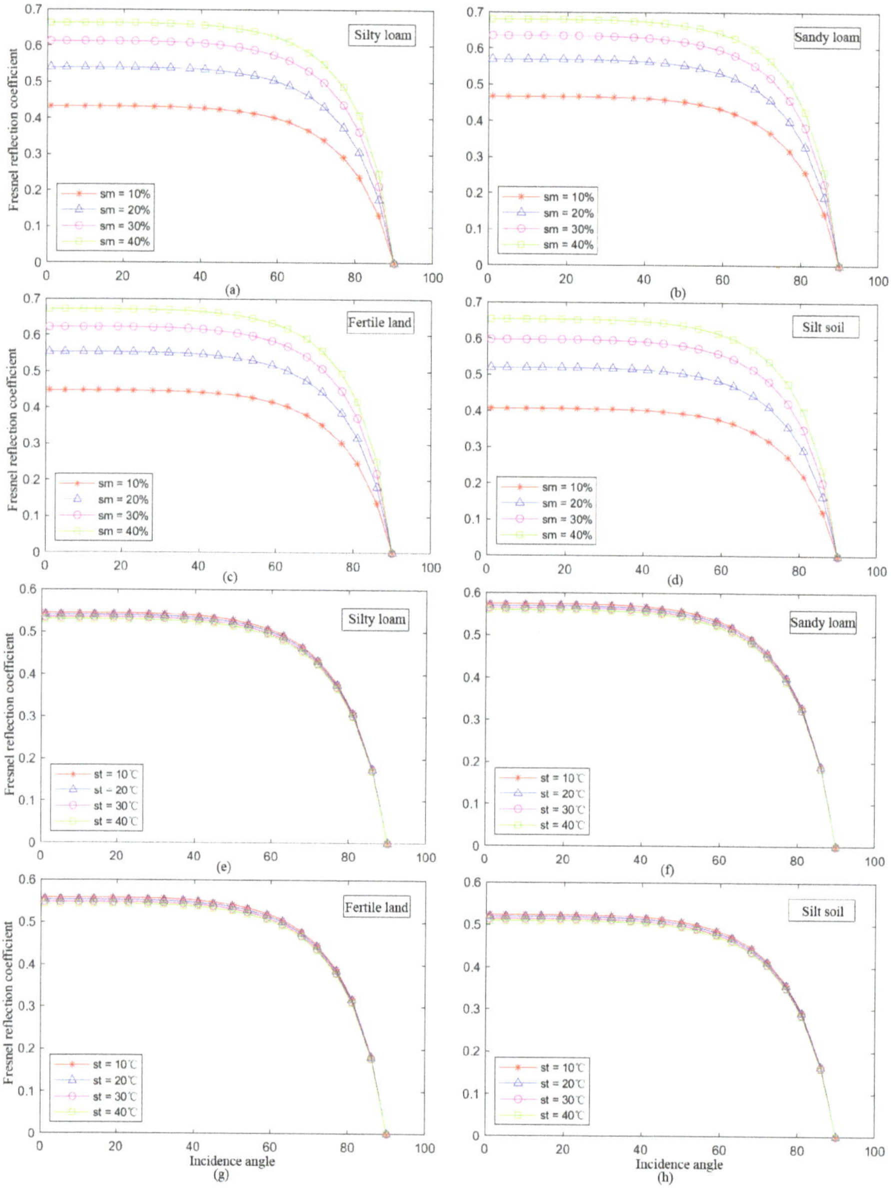


**Figure 5.** Response of the Fresnel reflection coefficient under different conditions: sm is soil moisture, and st is soil temperature ((**a–d**) are response maps with soil temperature as a constant value, (**e–h**) are response maps with soil moisture as a constant value).

Step 1: A database of Fresnel reflection coefficients with incident angle as the independent variable was created by freely combining values in the range of soil moisture, soil temperature, soil type and incident angle. Soil moisture, soil temperature and incident angle were limited to [1, 100], [1, 60] and [1, 90], respectively. The increments for these three parameters were set as 1%, 1 °C and 1°, respectively. Soil types referred to the physical parameters of the Dobson model (Table 1). Microwave frequency was set to 1.57542 GHz.

Through the combination of the above variables, a total of 2,160,000 Fresnel reflection coefficient values in the range of incident angles from 1 to 90° were formed.

Step 2: Based on the combination of variables in Step 1, a database of a total of the values of 2,160,000 Fresnel reflection coefficients with an incident angle of 0° was additionally composed.

Step 3: Based on the database created in Step 1 and Step 2, the correction variable was obtained using the following equation:

$$R_{lr}(\theta)_{cor} = R_{lr}(\theta)/R_{lr}(0) \tag{5}$$

where $R_{lr}(\theta)_{cor}$ is the corrected Fresnel reflection coefficient, $R_{lr}(\theta)$ is the Fresnel reflection coefficient obtained from Step 1 and $R_{lr}(0)$ is the Fresnel reflection coefficient for an incident angle of 0° in Step 2.

Step 4: With incident angle as the independent variable and corrected Fresnel reflection coefficient values of 2,160,000 as the dependent variable, the functional relationship between corrected Fresnel reflection coefficient and incident angle (1–90°) was established. A functional expression for the angle of incidence was as follows:

$$f(\theta) = a \cdot \exp(b \cdot \theta) + c \tag{6}$$

where $a$, $b$ and $c$ are all empirical parameters, which can be obtained by solving the parameters using the least square method.

### 3.3. The Retrieval Algorithm of Soil Moisture

In this study, the soil moisture retrieval algorithm of the linear regression equation proposed by Chew et al. [11] was based on the assumption that the surface reflectivity obtained from CYGNSS data is linearly correlated with the soil moisture of SMAP, and this linear correlation presents spatial variation and does not vary with time. The spatial resolution of CYGNSS is much smaller than the SMAP product, and the surface reflectivity varies depending on land cover and topography. Therefore, to effectively attenuate the effects of vegetation cover and topography, surface reflectivity corrected by the proposed method and soil moisture products of SMAP were sampled on a 3 × 3 km grid using the nearest neighbor sampling method [33–35], and thus soil moisture was retrieved. Among them, the nearest neighbor sampling method did not change the original pixel values of the SMAP products during resampling. The SMAP products of the 36 × 36 km grid were sampled into the 3 × 3 km grid, i.e., all 3 × 3 km grids within a 36 × 36 km grid have the same values. The expression was as follows:

$$SM_{CYGNSS,t} = \beta \cdot (\Gamma_t - \overline{\Gamma_{cal}}) + \overline{SM_{SMAP,cal}} \tag{7}$$

$SM_{CYGNSS,t}$ is the soil moisture retrieved with CYGNSS, $\Gamma_t$ is the corrected surface reflectivity of CYGNSS and $\overline{\Gamma_{cal}}$ and $\overline{SM_{SMAP,cal}}$ are the mean values of corrected surface reflectivity from CYGNSS and soil moisture products of SMAP during the modeling time, respectively. $\beta$ is the slope, representing the slope of the SMAP products and corrected surface reflectivity after removing $\overline{SM_{SMAP,cal}}$ and $\overline{\Gamma_{cal}}$ at the modeling time. If there were less than three sampling points of CYGNSS in a 3 × 3 km grid, the $\beta$ of the grid was not calculated. The value of $\beta$ will be used to estimate the soil moisture during the validation period. Soil moisture retrieval was performed for the 3 × 3 km grid according to the above process, and then the soil moisture result of 3 × 3 km were aggregated and averaged to the 36 × 36 km grid. The mean values for soil moisture of all 3 × 3 km within the 36 × 36 km grid were used as the retrieval values of soil moisture of the final 36 × 36 km grid.

## 4. Results

### *4.1. The Correction Results of the Fresnel Reflection Coefficient*

Figure 6 shows the responses of four soil types at the same soil temperature and different soil moisture after the normalization method of Fresnel reflection coefficient. It can be seen that the effect of differences in the soil moisture and soil type on the Fresnel reflectance coefficient was significantly weakened compared to the previous one, indicating that the surface reflectivity of CYGNSS corrected by the proposed method can significantly attenuate the effect of factors related to the Fresnel term. However, when the incident angle was larger (greater than 65°), the normalization effect became less effective. Excessive incident angles can significantly affect the quality of CYGNSS data [11,23], and this method used should keep the incident angle less than 65°.

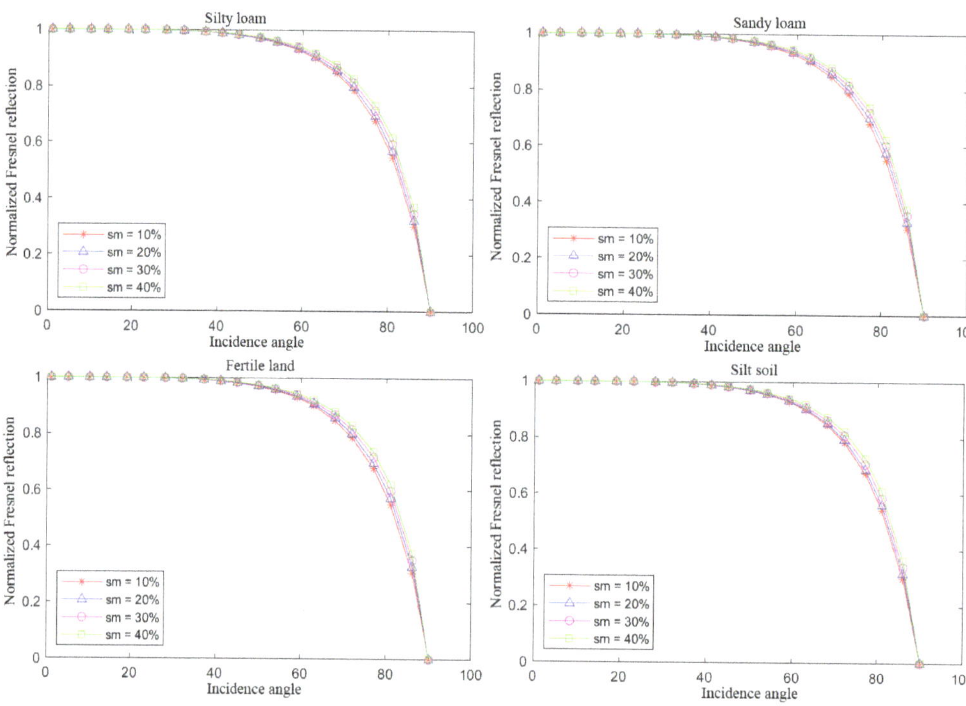


**Figure 6.** The response of normalized Fresnel reflection coefficient for the same soil temperature, different soil moisture and different soil types (sm is soil moisture, and the different lines represent different soil moisture contents).

### *4.2. Estimation of Soil Moisture*

Based on the study area and data, data from July 2019 to December 2020 (modeling time) were used for modeling, and data from 2021 (validation time) were used to retrieve the soil moisture (Figure 7). Accuracy was evaluated in terms of unbiased root-mean-square error (ubRMSE) and Pearson correlation coefficient (R) based on soil moisture data from measured sites. Figure 7 provides the retrieval maps of soil moisture for any time in the four seasons: spring, summer, fall and winter. It can be seen that the western region showed a significantly lower soil moisture phenomenon in four seasons in 2021, while the eastern part was relatively much wetter. Based on the occurrence of the phenomenon of annual drought in the western United States in 2021 (US Drought Monitor, USDM), the retrieval results of soil moisture in this study were more consistent with the actual situation.

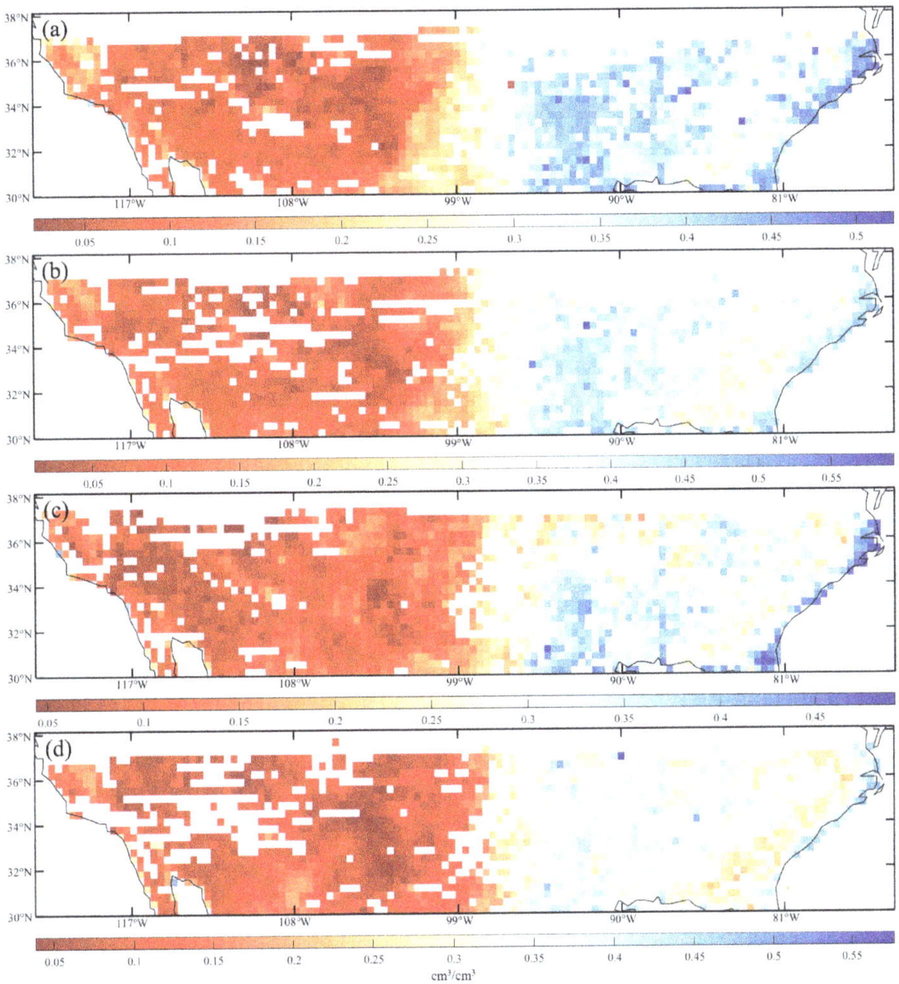


**Figure 7.** Soil moisture maps retrieved using the method of this study for different periods and represent (**a**) 4 January 2021, (**b**) 2 March 2021, (**c**) 4 August 2021 and (**d**) 26 November 2021, respectively.

The sites where the retrieved soil moisture was in good agreement with the measured data are shown in Figure 8. It can be seen that the retrieval results of soil moisture at these four stations were dense, and all of them could capture the changes in measured soil moisture and reflect the changes in low and high values of soil moisture.

### 4.3. Validation of Soil Moisture

The data from the measured sites were based on the results of point measurements, and the soil moisture results were obtained in this study for a 36 × 36 km grid. It is reasonable to use the average value of measured soil moisture in a large area grid as the true value, but there is a lack of such a design for a soil moisture station. Therefore, related studies have used the values of a single measured site to verify the accuracy of soil moisture retrieval [11,14,20]. The accuracy validation of soil moisture results using the retrieval method of this study was executed based on the data measured in the study area. Tables 2 and 3 provide statistics on the accuracy comparison between the SMAP and the soil moisture retrieval results using the method of this study, based on the data of each measured site within the experimental area.

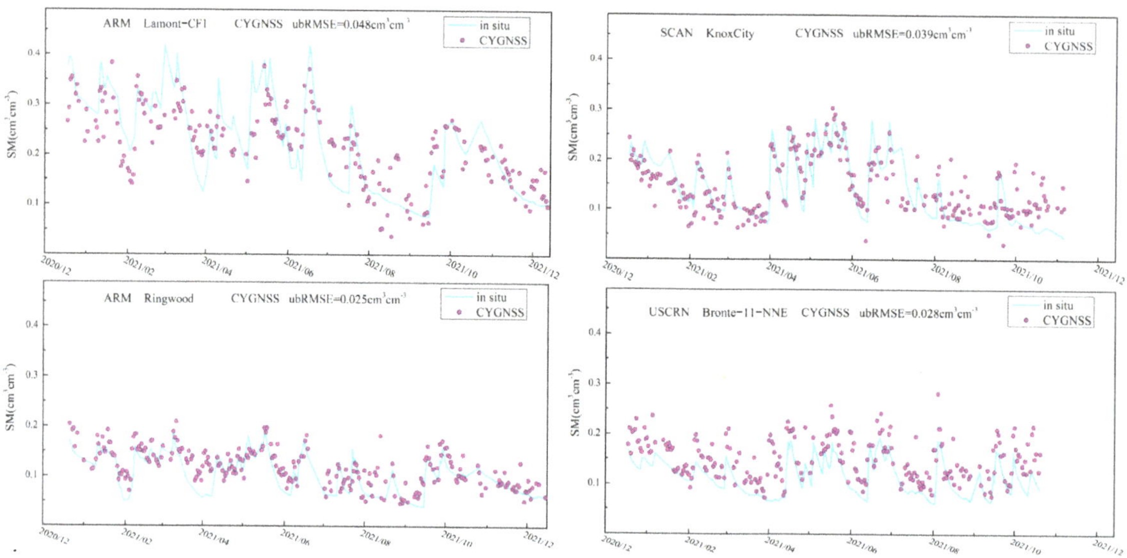


**Figure 8.** Examples of soil moisture retrieved by this study that agree well with the measured data at the measured sites (CYGNSS are the retrieval results of this study, and in situ represents measured soil moisture data; SM is soil moisture).

**Table 2.** Accuracy comparison of SMAP product and this study based on measured site data.

| **Number of Sites** | **ubRMSE ($cm^3/cm^3$)** | | | | | |
|---|---|---|---|---|---|---|
| | **Median** | | **Standard Deviation** | | **Mean** | |
| | CYGNSS | SMAP | CYGNSS | SMAP | CYGNSS | SMAP |
| ALL(160) | 0.057 | 0.051 | 0.026 | 0.024 | 0.061 | 0.056 |
| ARM(15) | 0.633 | 0.050 | 0.011 | 0.008 | 0.058 | 0.049 |
| SCAN(75) | 0.050 | 0.046 | 0.022 | 0.020 | 0.055 | 0.050 |
| SNOTEL(32) | 0.078 | 0.073 | 0.032 | 0.027 | 0.081 | 0.070 |
| USCRN(38) | 0.048 | 0.046 | 0.025 | 0.021 | 0.056 | 0.050 |

**Table 3.** Correlation comparison of soil moisture for SMAP and this study based on measured site data.

| **Number of Sites** | **R** | | | | | |
|---|---|---|---|---|---|---|
| | **Median** | | **Standard Deviation** | | **Mean** | |
| | CYGNSS | SMAP | CYGNSS | SMAP | CYGNSS | SMAP |
| ALL(160) | 0.450 | 0.624 | 0.310 | 0.286 | 0.40 | 0.550 |
| ARM(15) | 0.710 | 0.828 | 0.110 | 0.063 | 0.677 | 0.824 |
| SCAN(75) | 0.457 | 0.624 | 0.278 | 0.288 | 0.380 | 0.558 |
| SNOTEL(32) | 0.241 | 0.366 | 0.291 | 0.264 | 0.180 | 0.300 |
| USCRN(38) | 0.510 | 0.670 | 0.226 | 0.191 | 0.440 | 0.638 |

The average ubRMSE for the retrieval results of soil moisture in this study was 0.061 $cm^3/cm^3$ at 160 measured stations, with an average correlation of 0.4. However, SMAP products presented a better retrieval performance for the soil moisture (ubRMSE = 0.056 $cm^3/cm^3$, R = 0.55). In ubRMSE, the median and mean values of soil moisture retrieved by the method used in this study in the four monitoring networks were worse than those of SMAP, with relatively large standard deviations. The same was true in the correlation. Areas with low or no variation in soil moisture during the validation time resulted in a lower correlation, possibly because the effect of random noise was amplified. The soil moisture values retrieved by CYGNSS may have had more noise than those of SMAP due to the lack of complete coverage of the soil moisture time series in part of the grid.

In order to show the effectiveness of the method in this study, the accuracy indexes for the CYGNSS product (i.e., UCAR/CU) and the results of this study at the common measured sites were systematically counted [19] (Tables 4 and 5). Since the UCAR/CU product was only updated until August 2020, a comparison of the accuracy metric results from this study and the accuracy metric results from the paper by Chew et al. [19] was performed.

**Table 4.** Accuracy comparison for the results of this study and the UCAR/CU product.

| Number of Sites | ubRMSE ($cm^3/cm^3$) | | | | | |
|---|---|---|---|---|---|---|
| | Median | | Standard Deviation | | Mean | |
| | CYGNSS | UCAR/CU | CYGNSS | UCAR/CU | CYGNSS | UCAR/CU |
| ALL(88) | 0.051 | 0.057 | 0.023 | 0.024 | 0.051 | 0.06 |
| SCAN(49) | 0.051 | 0.053 | 0.022 | 0.021 | 0.055 | 0.057 |
| SNOTEL(8) | 0.070 | 0.091 | 0.015 | 0.017 | 0.076 | 0.097 |
| USCRN(31) | 0.044 | 0.054 | 0.023 | 0.024 | 0.051 | 0.057 |

**Table 5.** Correlation comparison for the results of this study and the UCAR/CU product.

| Number of Sites | R | | | | | |
|---|---|---|---|---|---|---|
| | Median | | Standard Deviation | | Mean | |
| | CYGNSS | UCAR/CU | CYGNSS | UCAR/CU | CYGNSS | UCAR/CU |
| ALL(88) | 0.467 | 0.510 | 0.250 | 0.236 | 0.410 | 0.470 |
| SCAN(49) | 0.488 | 0.600 | 0.246 | 0.200 | 0.427 | 0.500 |
| SNOTEL(8) | 0.126 | 0.095 | 0.242 | 0.191 | 0.131 | 0.186 |
| USCRN(31) | 0.500 | 0.470 | 0.224 | 0.233 | 0.440 | 0.457 |

Compared with the UCAR/CU product, the accuracy of the soil moisture results obtained by this study was significantly improved, with the average ubRMSE of SCAN, SNOTEL and USCRN improved by 3.5%, 19% and 10%, respectively. The average ubRMSE at 88 measured sites was improved by 10%. The standard deviation and median had smaller values relative to those of UCAR/CU, indicating that the accuracy of this study remained more stable. In terms of correlation, the results of this study were the same as those of UCAR/CU. Furthermore, the median of the results of two measured networks (SNOTEL and USCRN) was relatively high, and the overall standard deviation was larger.

## 5. Discussion

From Figure 7, there were more blank points in the retrieval map of soil moisture in this study, which may be due to the lack of values in the 3 × 3 km grid, resulting in blank values when 3 × 3 km was aggregated to 36 × 36 km. Therefore, the number of surface reflectivity points of CYGNSS contained in each grid at the 3 km spatial resolution in the modeling time was counted (Figure 9). The number of grids with surface reflectivity points fewer than 20, between 20 and 40, between 40 and 80, and greater than 80 accounted for 40%, 19.7%, 29.3% and 11% of the total number of grids, respectively. As shown in Figure 9, a large number of null values appeared in the central and eastern parts of the study area; the western part presented more grids with fewer surface reflectivity points; grids with fewer than 20 grid points accounted for 40% of the total number of grids in the study area during the modeling time. The absence of values for surface reflectivity points of CYGNSS made the retrieved soil moisture results show the phenomenon of a blank value, probably because some of the surface reflectivity points of CYGNSS were removed by the method of a water body, especially in the central and eastern regions, as well as in the seaward regions, where a large number of water bodies exist. However, after analyzing the results without using the removal method of water bodies, it was found that this method was not the main reason for the appearance of few reflection points and null places, but rather the non-coherent accumulation times of CYGNSS receivers and the irregular distribution of specular reflection points resulted in the sparse spatial distribution of surface reflectivity points of CYGNSS at 3 km spatial resolution and the low distribution of observation points.

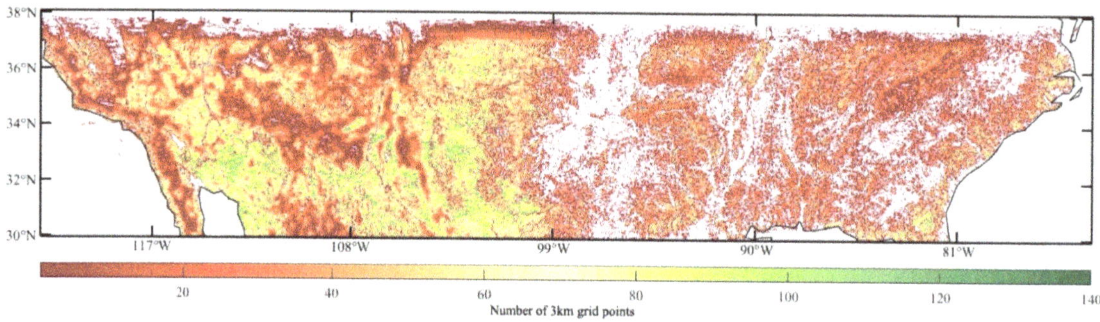


**Figure 9.** The map of CYGNSS points in the 3 km grid during the modeling time.

The slope was calculated based on the SMAP product and the corresponding surface reflectivity of CYGNSS after removing the average value. A temporal resolution of 2~3 days for SMAP and the property of random distribution of CYGNSS would reduce the number of mutually matched values. Meanwhile, if there were fewer than three matching values in the modeling time, the corresponding slope was not calculated. However, the number of grids with fewer than three in this part was 1% of the total number of grids. Therefore, the slope of the linear equation is another reason for the existence of blank values in the retrieval results of soil moisture (Figure 10).

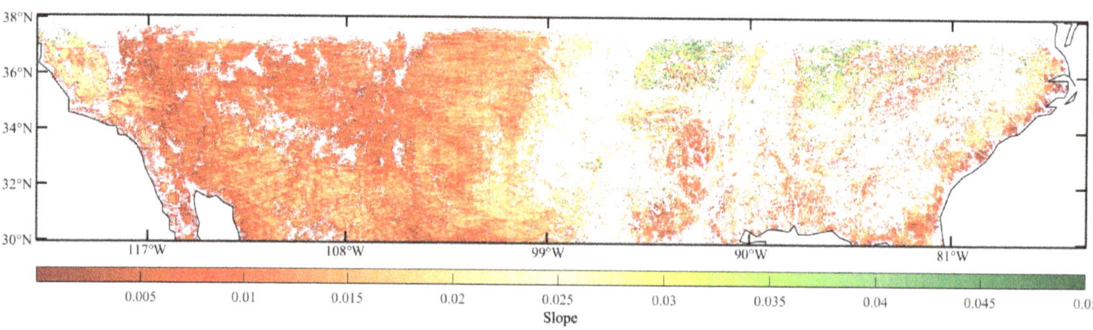


**Figure 10.** Slope map at the 3 km grid during the modeling time.

The soil moisture results obtained in this study varied across the measured soil moisture networks (Tables 2 and 3), and these were projected onto maps for further analysis, as shown in Figures 11 and 12. The ubRMSE for most of the measured sites were below the mean value of 0.061 $cm^3/cm^3$, accounting for about 60% of the total; the rest were maintained between 0.061 $cm^3/cm^3$ and 0.100 $cm^3/cm^3$; the measured sites with an accuracy greater than 0.1 $cm^3/cm^3$ accounted for 7% of the total. The distribution of sites of soil moisture retrieval larger than 0.1 $cm^3/cm^3$ showed that these sites were basically in the forest. The lush vegetation has a significant effect on the reflectance signal, as shown by the lower value of surface reflectivity of CYGNSS, which subsequently reduced the sensitivity to soil moisture. Counting the sites with ubRMSE below 0.061 $cm^3/cm^3$, most of them were in low vegetation areas such as grasslands and wetlands, indicating that the retrieval method in this study still maintains a certain sensitivity to soil moisture in areas with low vegetation cover. In the measured soil moisture network, especially the SNOTEL network, all the results of soil moisture retrieval had poor accuracy performance in this network. The analysis revealed that the sites of the SNOTEL network are mainly located in high altitude and heavily vegetated areas. Due to vegetation and high altitude, the accuracy of the area where the measured network is located was poor. Compared with the correlation between SMAP and the measured data, the correlation between this study and the measured data was mainly distributed between 0.4 and 0.8, indicating that the correlation was weak. The effect of factors such as vegetation and high altitude resulted in little variation in the soil moisture, which may account for the low correlation.

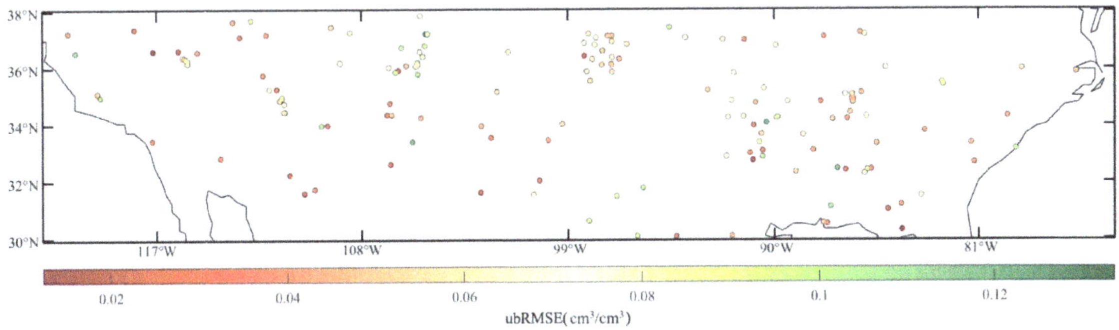


**Figure 11.** The ubRMSE map for the retrieved soil moisture in this study.

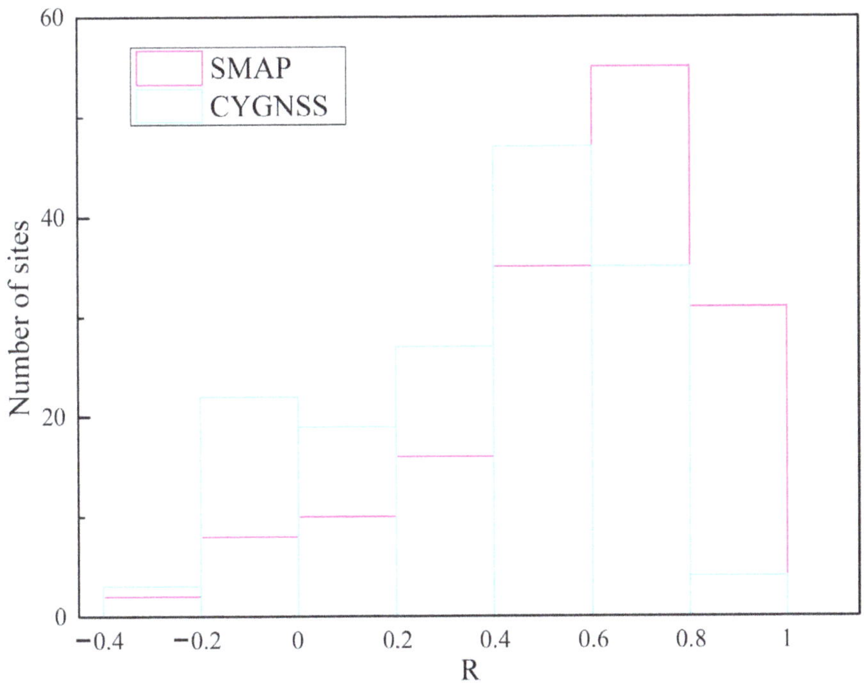


**Figure 12.** The correlation comparison between the SMAP product and retrieved soil moisture.

## 6. Conclusions

With the advantages of wide signal sources, many sampling points and high spatial-resolution for spaceborne GNSS-R, the removal of the influence of water bodies and the correction of the Fresnel reflection coefficient could effectively improve the accuracy of soil moisture retrieved by CYGNSS. According to the problem of water impact, in this paper an improved method to remove the influence of water bodies was proposed. A normalization method for the Fresnel reflection coefficient was proposed to correct the surface reflectivity of CYGNSS data by analyzing the change in the Fresnel reflection coefficient under different influencing factors. Finally, based on the linear algorithm, the results of the retrieved soil moisture were obtained, and the accuracy was compared and verified by the product data and the measurement data.

The improved method of water removal proposed in this paper can effectively remove observations affected by water bodies. The normalization method of the Fresnel reflection coefficient could effectively attenuate the effect of influencing factors on the Fresnel reflection coefficient, but at larger incident angles (greater than 65°) the normalization effect became worse. Compared with the results of the official CYGNSS product, the average ubRMSE of soil moisture retrieved by the method in this paper was improved by 10%, and the correlation was similar overall. Based on all measured data, the average ubRMSE for retrieval results of soil moisture in this paper was 0.061 $cm^3/cm^3$, with an average correlation of 0.4.

The direct removal of observations affected by water bodies is currently the most common approach, which leads to a reduction in the number of sampling points. Retaining observations and studying more efficient removal models are future research topics. The normalization method of the Fresnel reflection coefficient only considers typical soil types. Due to the complexity of soil composition in the natural environment, a unified correction model will be developed in the future by collecting more data.

**Author Contributions:** Q.W., J.S. (Jiaojiao Sun) and J.S. (Jinguang Shang) collected the original data; Q.W. and X.C. jointly designed the study and wrote the manuscript; X.C. and J.S. (Jiaojiao Sun) helped with the revision and discussion; T.J. and Z.L. provided supervision. All authors have read and agreed to the published version of the manuscript.

**Funding:** This work was supported by the National Natural Science Foundation of China (No. 42104037), the China Postdoctoral Science Special Foundation (No. 2022T150487), the Natural Science Foundation of Hubei Province for Distinguished Young Scholars (No. 2022CFA090), the Special Fund of Hubei Luojia Laboratory (No. 220100001), the Natural Science Foundation of Hubei Province of China (No. 2022CFB146), the Key Laboratory of Marine Environmental Survey Technology and Application, the Ministry of Natural Resource (No. MESTA-2020-B007) and the Special Fund of Hubei Luojia Laboratory (No. 220100003).

**Data Availability Statement:** The CYGNSS data used in this study were derived from official CYGNSS website (https://cygnss.engin.umich.edu/data-products/, accessed on 6 November 2021), Soil Moisture Active and Passive (SMAP) data used in this study came from NASA (https://nsidc.org/data/smap/smap-data.html, accessed on 6 November 2021).

**Conflicts of Interest:** The authors declare no conflict of interest.

## References

1. Entekhabi, D.; Rodriguez-Iturbe, I.; Castelli, F. Mutual interaction of soil moisture state and atmospheric processes. *J. Hydrol.* **1996**, *184*, 3–17. [CrossRef]
2. Entekhabi, D.; Asrar, G.R.; Betts, A.K.; Beven, K.J.; Bras, R.L.; Duffy, C.J.; Dunne, T.; Koster, R.D.; Lettenmaier, D.P.; McLaughlin, D.B.; et al. An agenda for land surface hydrology research and a call for the second international hydrological decade. *Bull. Am. Meteorol. Soc.* **1999**, *80*, 2043–2058. [CrossRef]
3. Leese, J.; Jackson, T.; Pitman, A.; Dirmeyer, P. GEWEX/BAHC international workshop on soil moisture monitoring, analysis, and prediction for hydrometeorological and hydroclimatological applications. *Bull. Am. Meteorol. Soc.* **2001**, *82*, 1423–1430. [CrossRef]
4. Shi, J.C.; Du, Y.; Du, J.Y.; Jiang, L.M.; Chai, L.N.; Mao, K.B.; Xu, P.; Ni, W.J.; Xiong, C.; Liu, Q.; et al. Progresses on microwave remote sensing of land surface parameters. *Sci. China Earth Sci.* **2012**, *55*, 1052–1078. [CrossRef]

5. Bu, J.W.; Yu, K.G. Sea surface rainfall detection and intensity retrieval based on GNSS-reflectometry data from the CYGNSS mission. *IEEE Trans. Geosci. Remote Sens.* **2021**, *60*, 5802015. [CrossRef]
6. Zhang, S.C.; Ma, Z.M.; Li, Z.M.; Zhang, P.F.; Liu, Q.; Nan, Y.; Zhang, J.J.; Hu, S.W.; Feng, Y.X.; Zhang, H.B. Using CYGNSS Data to Map Flood Inundation during the 2021 Extreme Precipitation in Henan Province, China. *Remote Sens.* **2021**, *13*, 5181. [CrossRef]
7. Li, X.; Yang, D.; Yang, J.; Zheng, G.; Han, G.; Nan, Y.; Li, W. Analysis of coastal wind speed retrieval from CYGNSS mission using artificial neural network. *Remote Sens. Environ.* **2021**, *260*, 112454. [CrossRef]
8. Guo, F.; Chen, W.J.; Zhu, Y.F.; Zhang, X.H. A GNSS-IR Soil Moisture Inversion Method Integrating Phase, Amplitude and Frequency. *Geomat. Inf. Sci. Wuhan Univ.* **2022**, *1*, 11.
9. Ghiasi, Y.; Duguay, C.R.; Murfitt, J.; van der Sanden, J.J.; Thompson, A.; Drouin, H.; Prévost, C. Application of GNSS interferometric reflectometry for the estimation of lake ice thickness. *Remote Sens.* **2020**, *12*, 2721. [CrossRef]
10. Ghiasi, Y.; Farzaneh, S.; Parvazi, K.; Duguay, C.R. Amplitue Estimation of Dominant Tidal Constituents Using Gnss Interferometric Reflectometry Technique. In Proceedings of the 2021 IEEE International Geoscience and Remote Sensing Symposium IGARSS, Brussels, Belgium, 11–16 July 2021; pp. 8546–8549.
11. Chew, C.C.; Small, E.E. Soil moisture sensing using spaceborne GNSS reflections: Comparison of CYGNSS reflectivity to SMAP soil moisture. *Geophys. Res. Lett.* **2018**, *45*, 4049–4057. [CrossRef]
12. Clarizia, M.P.; Pierdicca, N.; Costantini, F.; Floury, N. Analysis of CYGNSS data for soil moisture retrieval. *IEEE J. Sel. Top. Appl. Earth Obs. Remote Sens.* **2019**, *12*, 2227–2235. [CrossRef]
13. Chen, S.Z.; Yan, Q.Y.; Jin, S.G.; Huang, W.M.; Chen, T.X.; Jia, Y.; Liu, S.C.; Cao, Q. Soil Moisture Retrieval from the CyGNSS Data Based on a Bilinear Regression. *Remote Sens.* **2022**, *14*, 1961. [CrossRef]
14. Yan, Q.Y.; Huang, W.M.; Jin, S.G.; Jia, Y. Pan-tropical soil moisture mapping based on a three-layer model from CYGNSS GNSS-R data. *Remote Sens. Environ.* **2020**, *247*, 111944. [CrossRef]
15. Eroglu, O.; Kurum, M.; Boyd, D.; Gurbuz, A.C. High spatio-temporal resolution CYGNSS soil moisture estimates using artificial neural networks. *Remote Sens.* **2019**, *11*, 2272. [CrossRef]
16. Jia, Y.; Jin, S.G.; Chen, H.L.; Yan, Q.Y.; Savi, P.; Jin, Y.; Yuan, Y. Temporal-spatial soil moisture estimation from CYGNSS using machine learning regression with a preclassification approach. *IEEE J. Sel. Top. Appl. Earth Obs. Remote Sens.* **2021**, *14*, 4879–4893. [CrossRef]
17. Hu, Y.F.; Wang, J.; Li, Z.H.; Peng, J.B. Land Surface Soil Moisture along Sichuan-Tibet Railway Corridor Retrieved by Spaceborne Global Navigation Satellite SystemReflectometry. *Earth Sci.* **2022**, *47*, 2058–2068.
18. Nabi, M.M.; Senyurek, V.; Gurbuz, A.C.; Kurum, M. Deep Learning-Based Soil Moisture Retrieval in CONUS Using CYGNSS Delay–Doppler Maps. *IEEE J. Sel. Top. Appl. Earth Obs. Remote Sens.* **2022**, *15*, 6867–6881. [CrossRef]
19. Chew, C.C.; Small, E.E. Description of the UCAR/CU soil moisture product. *Remote Sens.* **2020**, *12*, 1558. [CrossRef]
20. Wan, W.; Ji, R.; Liu, B.J.; Li, H.; Zhu, S.Y. A two-step method to calibrate CYGNSS-derived land surface reflectivity for accurate soil moisture estimations. *IEEE Geosci. Remote Sens. Lett.* **2020**, *19*, 2500405. [CrossRef]
21. Zhu, Y.F.; Guo, F.; Zhang, X.H. Effect of surface temperature on soil moisture retrieval using CYGNSS. *Int. J. Appl. Earth Obs. Geoinf.* **2022**, *112*, 102929. [CrossRef]
22. Wu, X.R.; Ma, W.X.; Xia, J.M.; Bai, W.H.; Jin, S.G.; Calabia, A. Spaceborne GNSS-R soil moisture retrieval: Status, development opportunities, and challenges. *Remote Sens.* **2020**, *13*, 45. [CrossRef]
23. Al-Khaldi, M.M.; Johnson, J.T.; O'Brien, A.J.; Balenzano, A.; Mattia, F. Time-series retrieval of soil moisture using CYGNSS. *IEEE Trans. Geosci. Remote Sens.* **2019**, *57*, 4322–4331. [CrossRef]
24. Dobson, M.C.; Ulaby, F.T.; Hallikainen, M.T.; El-rayes, M.A. Microwave dielectric behavior of wet soil part II: Dielectric mixing models. *IEEE Trans. Geosci. Remote Sens.* **1985**, *23*, 35–46. [CrossRef]
25. Voosen, P. Satellites see hurricane winds despite military signal tweaks. *Science* **2019**, *364*, 1019. [CrossRef]
26. Camps, A. Spatial Resolution in GNSS-R Under Coherent Scattering. *IEEE Geosci. Remote Sens. Lett.* **2020**, *17*, 32–36. [CrossRef]
27. O'Neill, P.; Chan, S.; Njoku, E.; Jackson, T.; Bindlish, R. *Algorithm Theoretical Basis Document L2 & L3 Soil Moisture (Passive) Data Products*; Jet Propulsion Laboratory: Pasadena, CA, USA, 2014.
28. Pekel, J.F.; Cottam, A.; Gorelick, N.; Belward, A.S. High-resolution mapping of global surface water and its long-term changes. *Nature* **2016**, *540*, 418–422. [CrossRef]
29. Tsang, L.; Newton, R.W. Microwave emissions from soils with rough surfaces. *J. Geophys. Res. Ocean.* **1982**, *87*, 9017–9024. [CrossRef]
30. Carreno-Luengo, H.; Luzi, G.; Crosetto, M. Sensitivity of CyGNSS bistatic reflectivity and SMAP microwave radiometry brightness temperature to geophysical parameters over land surfaces. *IEEE J. Sel. Top. Appl. Earth Obs. Remote Sens.* **2018**, *12*, 107–122. [CrossRef]
31. Carreno-Luengo, H.; Luzi, G.; Crosetto, M. Above-ground biomass retrieval over tropical forests: A novel GNSS-R approach with CyGNSS. *Remote Sens.* **2020**, *12*, 1368. [CrossRef]
32. Zavorotny, V.U.; Voronovich, A.G. Scattering of GPS signals from the ocean with wind remote sensing application. *IEEE Trans. Geosci. Remote Sens.* **2000**, *38*, 951–964. [CrossRef]
33. Cui, C.; Xu, J.; Zeng, J.; Chen, K.; Bai, X.; Lu, H.; Chen, Q.; Zhao, T. Soil moisture mapping from satellites: An intercomparison of SMAP, SMOS, FY3B, AMSR2, and ESA CCI over two dense network regions at different spatial scales. *Remote Sens.* **2018**, *10*, 33. [CrossRef]

34. Kim, H.; Parinussa, R.; Konings, A.G.; Wagner, W.; Cosh, M.H.; Lakshmi, V.; Zohaib, M.; Choi, M. Global-scale assessment and combination of SMAP with ASCAT (active) and AMSR2 (passive) soil moisture products. *Remote Sens. Environ.* **2018**, *204*, 260–275. [CrossRef]
35. Ma, H.; Zeng, J.; Chen, N.; Zhang, X.; Cosh, M.H.; Wang, W. Satellite surface soil moisture from SMAP, SMOS, AMSR2 and ESA CCI: A comprehensive assessment using global ground-based observations. *Remote Sens. Environ.* **2019**, *231*, 111215. [CrossRef]

remote sensing

MDPI

*Article*

# Characterizing Ionospheric Effects on GNSS Reflectometry at Grazing Angles from Space

**Mario Moreno [1,2,*], Maximilian Semmling [1], Georges Stienne [3], Mainul Hoque [1] and Jens Wickert [2,4]**

1 Deutsches Zentrum für Luft-und Raumfahrt, Institut für Solar-Terrestrische Physik (DLR-SO), Kalkhorstweg 53, 17235 Neustrelitz, Germany
2 Institute of Geodesy and Geoinformation Science, Department of GNSS Remote Sensing, Navigation and Positioning, Technische Universität Berlin (TUB), Str. des 17. Juni 135, 10623 Berlin, Germany
3 Laboratoire d'Informatique, Signal et Image de la Côte d'Opale (LISIC), Université Littoral Côte d'Opale (ULCO), 50 rue Ferdinand Buisson, 62228 Calais, France
4 Deutsches GeoForschungsZentrum (GFZ), Wissenschaftpark "Albert Einstein", Telegrafenberg, 14473 Potsdam, Germany
* Correspondence: mario.moreno@dlr.de

**Abstract:** Coherent observations in GNSS reflectometry are prominent in regions with smooth reflecting surfaces and at grazing elevation angles. However, within these lower elevation ranges, GNSS signals traverse a more extensive atmospheric path, and increased ionospheric effects (e.g., delay biases) are expected. These biases can be mitigated by employing dual-frequency receivers or models tailored for single-frequency receivers. In preparation for the single-frequency GNSS-R ESA "PRETTY" mission, this study aims to characterize ionospheric effects under variable parameter conditions: elevation angles in the grazing range (5° to 30°), latitude-dependent regions (north, tropic, south) and diurnal changes (day and nighttime). The investigation employs simulations using orbit data from Spire Global Inc.'s Lemur-2 CubeSat constellation at the solar minimum (F10.7 index at 75) on March, 2021. Changes towards higher solar activity are accounted for with an additional scenario (F10.7 index at 180) on March, 2023. The electron density associated with each reflection event is determined using the Neustrelitz Electron Density Model (NEDM2020) and the NeQuick 2 model. The results from periods of low solar activity reveal fluctuations of up to approximately 300 TECUs in slant total electron content, 19 m in relative ionospheric delay for the GPS L1 frequency, 2 Hz in Doppler shifts, and variations in the peak electron density height ranging from 215 to 330 km. Sea surface height uncertainty associated with ionospheric model-based corrections in group delay altimetric inversion can reach a standard deviation at the meter level.

**Keywords:** GNSS reflectometry; grazing angles; ionospheric delay; ionospheric Doppler shift; NEDM2020 model; NeQuick model; PRETTY mission

**Citation:** Moreno, M.; Semmling, M.; Stienne, G.; Hoque, M.; Wickert, J. Characterizing Ionospheric Effects on GNSS Reflectometry at Grazing Angles from Space. *Remote Sens.* **2023**, *15*, 5049. https://doi.org/10.3390/rs15205049

Academic Editor: Mehrez Zribi

Received: 12 September 2023
Revised: 15 October 2023
Accepted: 17 October 2023
Published: 20 October 2023



## 1. Introduction

The ionosphere, situated between 60 and 2000 km above the Earth's surface, plays a vital role in electromagnetic wave propagation, influenced by solar-radiation-induced ionization [1]. The speed at which the transmitted electromagnetic signals from the GNSS (global navigation satellite system) satellites propagate through the ionosphere depends on the electron density along the line of sight between the satellite and the receiver. Upon traversing the ionosphere, GNSS signals may encounter two distinct forms of perturbations: Firstly, the introduction of an error in the estimated range due to the signal's delay that is proportional to the integrated electron density (slant total electron content—sTEC), and secondly, the occurrence of signal characteristic fluctuations resulting from irregularities in the ionosphere's electron density distribution [2].

The use of GNSS signals, renowned for their global availability and signal propagation characteristics, has been widely investigated and exploited as a powerful tool for ionospheric studies across diverse spatial and temporal scales. Ground-based atmospheric-sounding techniques employing continuously operating reference station (CORS) networks and GNSS receivers, which operate on low Earth orbit (LEO) satellites for the analysis of refracted radio signals via GNSS radio occultation (GNSS-RO), provide key observations for improving global weather forecasts [3]. To further broaden the observations, GNSS reflectometry (GNSS-R) has emerged as a complementary technique that leverages signals reflected off the Earth's surface. This approach not only facilitates the retrieval of reflecting surface properties but also serves as an atmospheric-sounding tool.

In order to understand ionospheric ranging delays within space-borne GNSS-R, simulations are conducted as detailed in [4]. The simulation is based on the Cyclone GNSS (CYGNSS) [5] mission and encompasses different elevation angles, latitudes, and solar activities. The results reveal an inverse relationship between the satellite elevation angle and ionospheric delay, with a larger ionospheric influence at low latitudes. In [6], the impact of scintillation effects on reflectometry has been explored using data from UK TechDemoSat-1 [7]. These effects lead to a degradation of the signal-to-noise ratio that can be utilized for altimetry and scatterometry performance assessments. More recently, studies have been carried out to retrieve the total electron content (TEC) from coherent reflectometry observations. In the work presented in [8], a methodology was introduced for sTEC estimation along the paths of incident and reflected signal rays. This estimation is based on coherent dual-frequency GNSS-R measurements obtained from Spire Global low Earth orbit (LEO) CubeSats. The outcomes have demonstrated a favorable alignment between reflectometry sTEC estimations and the global ionospheric TEC maps (GIM). Furthermore, an algorithm outlined in [9] combines sTEC observation from space-borne reflectometry using CubeSats and data collected from ground-based GNSS stations to generate vertical TEC (vTEC) maps in the Arctic region. Simulations conducted within this study under diverse conditions, involving variations in temporal resolution, solar activity levels, and the number of reflection events, have demonstrated enhanced accuracy in vTEC estimations when coherent GNSS-R observations are incorporated.

In the domain of GNSS-R, it has been empirically established that coherent observations are more frequently observed in the presence of smooth reflecting surfaces, such as sea ice, regions with low sea states, or inland waters, and at low grazing angles [10–12]. Nonetheless, within this range of elevation angles, it is important to note that the trajectories of the LEO GNSS-R rays entail a longer path through the ionosphere. This extended path results in a more pronounced ionospheric impact on the signals themselves. The representation (not to scale) of the LEO GNSS-R configuration along the grazing angle rays' paths and its interaction with the ionosphere are illustrated in Figure 1.

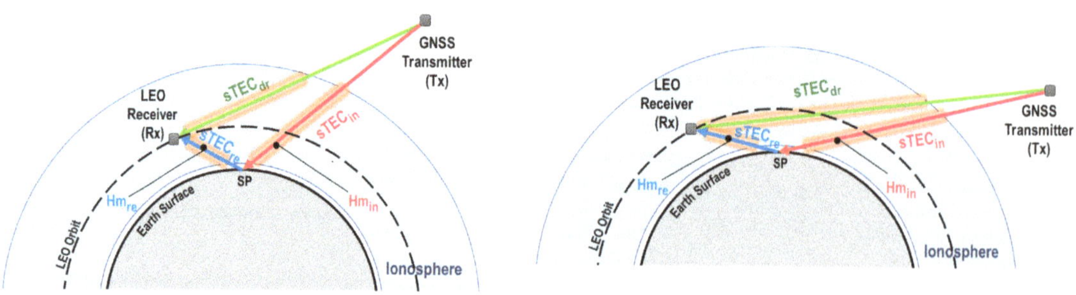


(**a**) (**b**)

**Figure 1.** (**a**) LEO GNSS-R representation at 30° elevation angle at specular point. (**b**) LEO GNSS-R representation at 5° elevation angle. $sTEC_x$ denotes the slant total electron content. Subscripts dr, in, and re correspond to the direct ray (transmitter $T_x$ to receiver $R_x$), incident ray (transmitter to specular point $SP$), and reflected ray (specular point to receiver), respectively. $Hm_x$ represents the peak electron density height for the incident and reflected ray paths.

As described in [13], dual-frequency receivers possess the capability to mitigate these first-order ionospheric effects through the utilization of a linear combination (ionosphere-free) of either code or carrier measurements. Conversely, single-frequency receivers must rely on applying a model to correct for ionospheric refraction, which can introduce delays of several tens of meters. For the Galileo GNSS constellation, the European GNSS Open Service has adopted the Neustrelitz Total Electron Content Model NTCM [14] (NTCM-G) or NeQuick 2 [15] (NeQuick-G) models to provide real-time ionospheric corrections for single-frequency receivers [16].

This study is in preparation for the European Space Agency's GNSS-R CubeSat mission "PRETTY" (passive reflectometry and dosimetry) [17]. The mission's primary goal is to retrieve sea surface height using grazing angle observations. Since PRETTY operates at a single frequency (L5), it requires model-based ionospheric corrections. This study provides a comprehensive characterization of ionospheric effects, at the grazing angle range (5°–30°), considering satellite geometry, latitude-dependent regions, temporal variations, and solar activity. It analyzes variability in the ionospheric group delay, Doppler shift, and peak electron density height. Additionally, the uncertainty in model-based ionospheric corrections for GNSS-R group delay altimetry is assessed.

The analysis is based on utilizing the sTEC obtained from three-dimensional, time-dependent models. To assess model uncertainty, the sTEC values computed using the Neustrelitz Electron Density Model (NEDM2020) [18] are used as a reference and compared with the sTEC retrievals from NeQuick 2. Simulations are conducted to replicate conditions similar to those of the PRETTY mission, utilizing orbit data from the GNSS-R Spire Global Lemur-2 constellation. To provide a comprehensive analysis, the results are categorized into three elevation angle ranges: very-low (5°–10°), low (10°–20°), and mid-low (20°–30°). These categories are further grouped by latitude into three distinct regions: north, tropics, and south. Additionally, this study considers variations in local time and solar activity. Low solar activity (LSA) is represented by F10.7 = 75 in March 2021 and high solar activity (HSA) by F10.7 = 180 in March 2023.

The structure of this paper is outlined as follows: Section 2 presents the GNSS-R data descriptions, reflection events, and ray point settings for the simulations. Section 3 illustrates the methodologies utilized for the determination of parameters such as sTEC, relative ionospheric delay, Doppler shift, and ionospheric piercing points. Subsequently, Section 4 presents the results and analysis of the parameters explained in Section 3. Finally, in Section 5, a discussion of the findings is presented along with the conclusions in Section 6.

## 2. GNSS-R Data and Reflection Events

### 2.1. LEO Data

The LEO data used in this study consist of a total of 1188 reflection events on 1 March 2021, sourced from Spire Global Inc. Currently, the Spire Lemur-2 constellation comprises more than 80 GNSS radio occultation CubeSats, out of which about 30 have been adapted to acquire GNSS reflectometry measurements at grazing angles [19]. The Lemur-2 satellites follow a Sun-synchronous orbit, with altitudes ranging from 400 to 600 km and varying orbit inclinations. This orbital configuration enables them to conduct GNSS-R measurements, encompassing all latitudes of the Earth.

The Spire grazing angle GNSS-R products are collected with a focus on specific regions, including the polar areas, the Gulf of Mexico, and southeast Asia. These regions are selected due to their favorable characteristics, such as sea ice surfaces and calm ocean surfaces, which enable the best performance of coherent reflectometry measurements [10]. Figure 2 displays the track positions of the specular points distributed across both polar regions, as well as in the mid-latitude and tropical regions at different local times. Given the geographical distribution of the events, the dataset has been categorized into three distinct regions: north, covering latitudes between 40°N and 90°N; tropics, spanning latitudes between 40°N and 40°S; and south, covering latitudes between 40°S and 90°S.

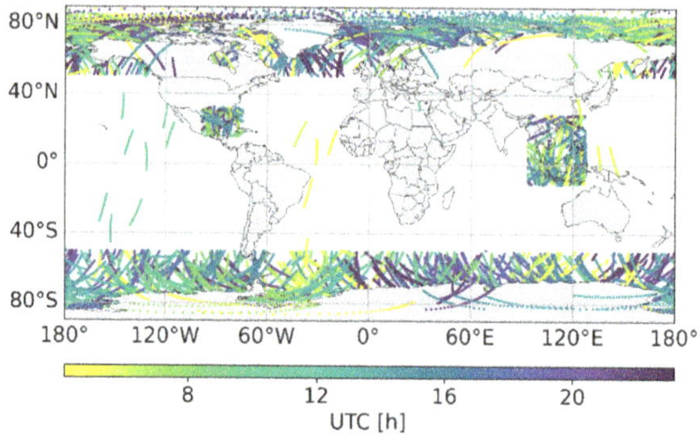


**Figure 2.** Specular point tracks provided by the Spire Lemur-2 CubeSats on 1 March 2021, color-coded according to coordinated universal time (UTC) in hours.

Each Lemur-2 satellite event lasts an average of 4 min, resulting in a total of about 80 h of recorded data. The recording durations vary, with a minimum of 1 min and a maximum of 6 min. Table 1 shows the number of events per region (north, tropics, and south) along with their corresponding durations in minutes.

**Table 1.** Total number of events per region and durations.

| Region | Number of Events | Total Minutes |
|---|---|---|
| North | 474 | 1704 |
| Tropics | 168 | 760 |
| South | 546 | 2335 |
| Total | 1188 | 4798 |

A total of 21 CubeSats from the Spire constellation are evaluated. The metadata include the space vehicle number (SVN) of the Lemur-2 satellite, as well as information about the GNSS satellite and constellation from which the CubeSat receives the reflected signals. For the simulation in this study, the GNSS constellation employed is GPS (global positioning system). Upon analyzing the Spire data, it is found that each Lemur-2 satellite receives the reflected signal from 4 to 19 GPS satellites during different time windows, depending on the positions of the transmitters and the receiver. The Spire SVN and GPS pseudo-random noise code (PRN) are presented in Table 2.

**Table 2.** Space Vehicle Numbers of Lemur-2 CubeSats and GPS Satellite PRNs on 01/03/2021.

| | Spire SVN | GPS PRN |
|---|---|---|
| | 79, 84, 99, 100, 101, 102, 103, 104, 106, 113, 115, 116, 117, 119, 120, 121, 122, 124, 125, 128, 129 | 1, 3, 5, 6, 7, 8, 9, 10, 12, 15, 17, 24, 25, 26, 27, 29, 30, 31, 32 |
| Total | 21 | 19 |

## 2.2. Specular Point Positions and Ray Points

The specular point positions and the ray tracing of the direct, incident, and reflected signals are calculated based on the methodology presented in [11,20]. A geometrical model is employed to characterize specular reflections and determine the specular point position, considering the Earth's surface curvature. For this model, the transmitter ($T_x$) and receiver ($R_x$) positions are needed in an Earth-centered Earth-fixed (ECEF) frame. The $R_x$ position is extracted from the Spire data files. To obtain this position, the Lemur-2 satellites are equipped with a zenith dual-frequency (L1 and L2) antenna, which facilitates precise orbit determination (POD). The $T_x$ position is derived from the broadcasted GPS ephemeris. The Earth's curvature is modeled with an osculating spherical surface with respect to the WGS-84 ellipsoid at a reference specular point. An iterative solution is employed to find the best-fitting sphere that satisfies the condition of equal incident and reflected angles (specular reflection) [20]. The specular point positions are calculated at 10 s intervals on the receiver trajectory. A ray-tracing module is set to compute ray points every 10 km along the three ray paths: $T_x$ to $SP$ (incident), $SP$ to $R_x$ (reflected), and $T_x$ to $R_x$ (direct). The positions of the ray points (latitude, longitude, and ellipsoidal height) are subsequently utilized to obtain the electron density from the ionospheric electron density models. Figure 3 illustrates an example of the electron density retrieval from the NEDM2020 model depicting the change along the specular point tracks every 10 s (blue stars), and the ray points change every 10 km (red dots) along the incident (in), reflected (re), and direct (dr) ray paths.

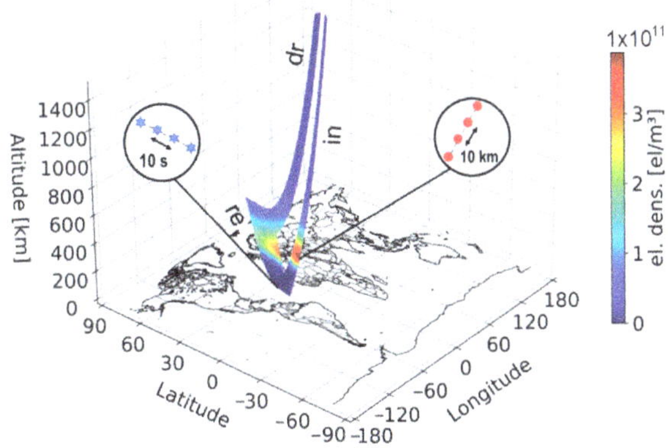


**Figure 3.** Representation of the specular point change along track (blue stars) for the SVN 79 and PRN 30 every 10 s (~45 km) and the ray points every 10 km along the direct, incident, and reflected ray paths (red dots).

Following the ray tracing, a total of 28,790 reflection events are obtained. The total number of reflection events by region is depicted in Figure 4a. Additionally, Figure 4b illustrates the distribution of reflection events concerning the elevation angle by region. Notably, the south pole region exhibits a higher number of events; however, all regions show similar behavior, with a higher concentration of events in the elevation range between 5° and 20°.

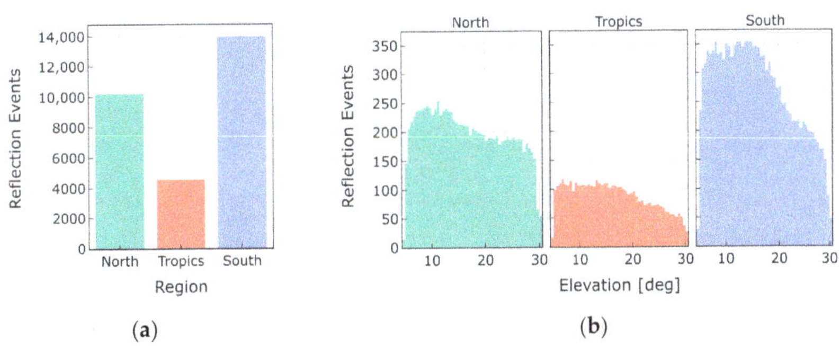


**Figure 4.** (**a**) Total number of reflection events on 1 March 2021 by region. (**b**) The number of reflection events across grazing angles, ranging from 5° to 30°.

## 3. Methodology

### 3.1. Electron Density Models

The electron density in this study is obtained from two three-dimensional and time-dependent electron density models: the Neustrelitz Electron Density Model (NEDM2020) [18] and the NeQuick 2 model [15]. For both models, the input values depending on solar activity are the solar radio flux index F10.7, month, geographic latitude and longitude, height, and universal time (UT). The output obtained is the electron concentration at the specified location and time.

The NeQuick 2 model was developed at the Aeronomy and Radiopropagation Laboratory of The Abdus Salam International Centre for Theoretical Physics (ICTP), Trieste, Italy, and at the Institute for Geophysics, Astrophysics, and Meteorology (IGAM) of the University of Graz, Austria. This model comprises vertical profiles consisting of multiple Epstein layers, and it derives essential electron peak density and height parameters through spatial and temporal interpolation from a comprehensive set of global maps. Consequently, NeQuick 2 incurs significant computational demands in terms of time and processing power [14].

On the other hand, the NEDM2020 model was developed at the German Aerospace Center in the Institute for Solar–Terrestrial Physics (DLR-SO), Neustrelitz, Germany. Including the NTCM model, this model relies on about 100 model coefficients and a set of empirically fixed parameters. Remarkably, the electron density values can be directly computed for any specified location and time without the requirement for the specialized temporal or spatial interpolation of parameters, making it faster than the NeQuick 2 model in terms of computational efficiency [21].

A model comparison of electron density profiles is presented in Figure 5 featuring one example at very low (first row) and mid-low (second row) elevation angle events. The first column displays the electron density per ray mapped along the specular point change using the NEDM2020 model. The subsequent columns (second, third, and fourth) illustrate the electron density profile comparison between the NEDM2020 and NeQuick 2 models for the incident, reflected, and direct rays, respectively.

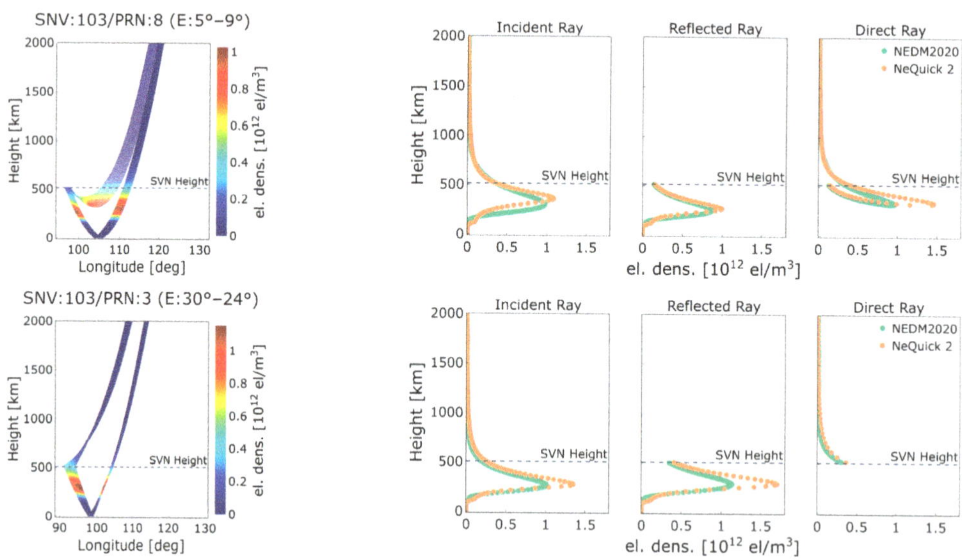


**Figure 5.** Electron density variations along the change and model profile comparison for the incident, reflected, and direct rays at very low and mid-low elevation angles (F10.7 = 75).

### 3.2. Ionospheric Group Delay Computation

The GNSS electromagnetic signal propagation speed in the ionosphere depends on electron density $(N_e)$, which is influenced by daytime ionization and nighttime recombination processes. According to [13], when considering the signal code measurements, the difference between the measured range (using a signal of frequency $f$ in $Hz$) and the Euclidean distance between the satellite and receiver is expressed as follows:

$$\Delta_{gr}^{iono} = +\frac{40.3}{f^2}\int N_e dl \tag{1}$$

$\Delta_{gr}^{iono}$ is the term used for the group ionospheric refraction, and the integral is known as the slant total electron content $(sTEC)$, representing the numerical integration of the electron density along the ray path. $f$ corresponds to the GNSS signal frequency, and in this study, the GPS L1 frequency is 1575.42 MHz. The $sTEC$ is computed for each ray, including the incident $(sTEC_{in})$, reflected $(sTEC_{re})$, and direct $(sTEC_{dr})$ rays, respectively. The $sTEC$ is expressed in total electron content units (TECUs) where one TECU corresponds to $10^{16}$ electrons per square meter $(el/m^2)$. Finally, the group ionospheric delay in meters (for each ray) is obtained from:

$$I_{in,re,dr} = +\frac{40.3 * 10^{16}}{f^2} sTEC_{in,re,dr} \tag{2}$$

As presented in [22], the relative delay between the direct and reflected signals is denoted as $\Delta_p = p_r - p_d$, where $p_r$ is the cumulative path of the incident and reflected rays, while $p_d$ corresponds to the direct path. The relative delay can be influenced by various contributing factors, such as the standard sources of delay within the GNSS signals. Therefore, the extended version of $\Delta_p$ can be written as:

$$\Delta_p = \Delta_{p_{geo}} + \Delta_{p_{trop}} + \Delta_{p_{iono}} + \Delta_{p_{rgh}} + \Delta_{p_{instr}} + n \tag{3}$$

where $\Delta_{p_{geo}}$ represents the relative geometrical delay, and $\Delta_{p_{trop}}$ and $\Delta_{p_{iono}}$ correspond to the relative tropospheric and ionospheric delays, respectively. $\Delta_{p_{rgh}}$ is a bias induced by the surface roughness. The instrumental error is denoted by $\Delta_{p_{instr}}$, and $n$ represents unmodeled errors.

GNSS-R Group Delay Altimetry and Ionospheric Delay Uncertainty

Based on the analysis conducted in [23], the ionospheric delay constitutes a significant component within the error budget associated with GNSS-R ocean surface altimetry retrievals. At elevation angles above 60°, the uncorrected ionospheric delay can reach ~15 m during daytime and ~7 at nighttime. The ionospheric group delay bias propagates to an altimetric bias based on the relation between the height offset $\Delta_h$ and the signal path $\Delta_p$. Consequently, when considering only the ionospheric altimetric error, where $E$ is the elevation angle, it can be expressed as:

$$\Delta h_{iono} = \frac{\Delta_{p_{iono}}}{2 * \sin E} \tag{4}$$

Assuming a relative uncertainty of 30% for the ionospheric delay bias, as established in [20], we introduce normally distributed random errors with a respective standard deviation $(\delta_{iono} = \mathrm{N}(0, 0.3^2))$.

### 3.3. Doppler Shift Computation

The Doppler shift of a GNSS signal is predominantly influenced by the relative velocity between the transmitter satellite and the receiver, along with a common offset that is proportional to the error in the receiver clock's frequency. However, as demonstrated in [24,25], various ionospheric effects, such as changes in the redistribution and density of electrons in the ionosphere, lead to frequency variations in the electromagnetic waves emitted by a stable transmitter. These variations are manifested as the Doppler shift and can be quantified as the time derivative of the phase path of the signal. When considering only the ionospheric delay term in the carrier phase observation model [26], the residual phase path expressed in units of cycles can be given by:

$$\phi = \frac{\Delta_{p_{iono}}}{\lambda} \tag{5}$$

where $\lambda$ is the wavelength of the GPS L1 frequency (0.1905 m). As the Doppler shift $(f_d)$ of a given signal corresponds to the rate of change of its carrier phase over time, it can be computed using the following equation:

$$f_d = \frac{d\phi}{dt} \tag{6}$$

### 3.4. Peak Electron Density Height

Diurnal variations significantly impact the ionosphere, where daytime and nighttime conditions manifest contrasting characteristics. The properties of the ionosphere, such as height, ionized particle concentration, and the presence of distinct layers, change dynamically over time. Regions characterized by high electron densities are designated as the D, E, and F layers. In diurnal cycles, the F layer undergoes separation into two distinct layers termed the F1 and F2 during daytime, while the D layer experiences complete dissipation throughout the nocturnal period [27]. This shifts the height at which the high electron concentration is found. In order to analyze changes in the ionospheric altitude, the height corresponding to the maximum peak of the electron density profile $(Hm)$ is used as the reference point. $Hm$ is obtained for both the incident and reflected rays using the NEDM2020 model.

The LEO GNSS-R space-borne configuration, which enables the simultaneous collection of data from multiple reflections, presents several advantages for ionospheric studies. Firstly, thanks to the fast trajectory change of the LEO satellite, the GNSS-R signal rapidly scans along the ionospheric layers, providing a snapshot view of ionospheric structures [8]. Secondly, the ability to obtain peak electron density points at different locations within a short time interval allows for the mapping of ionospheric structures at varying distances. Assuming the Earth's radius is 6371 km, with a maximum electron density ionospheric shell at a 300 km height, the distance between the incident and reflected $Hm$ points varies depending on the elevation angle as observed in Figure 6.

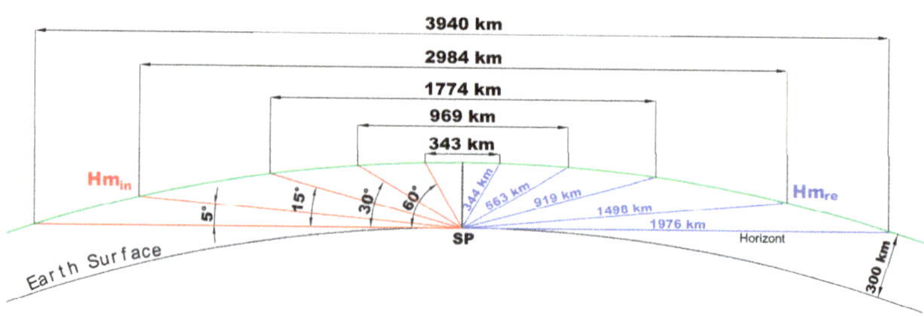


**Figure 6.** Distances between peak electron density height points depending on the elevation angle change at 300 km height.

## 4. Results

### 4.1. Slant Total Electron Content Analysis

The computed sTEC, obtained from the NEDM2020 and NeQuick 2 models, serves as the foundational parameter for the subsequent derivations of the relative ionospheric group delay and Doppler shift. The assessment of the sTEC is presented across different grazing elevation ranges: 5°–10° (very low), 10°–20° (low), and 20°–30°(mid-low), along with the distinct regions of north, tropics, and south. The outcomes of the NEDM2020 and NeQuick 2 sTEC computations during LSA are depicted in Figures 7 and 8, respectively. While discrepancies of up to ~60 TECUs between the two models are noticeable in the tropics region at very low angles for the direct ray, both models consistently demonstrate similar behavior across all analyzed scenarios.

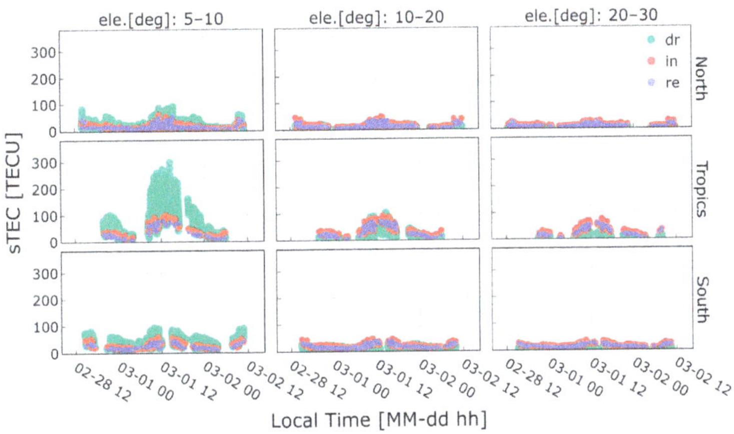


**Figure 7.** sTEC obtained from NEDM2022 model. Color-coded according to the direct (dr), incident (in), and reflected (re) rays.

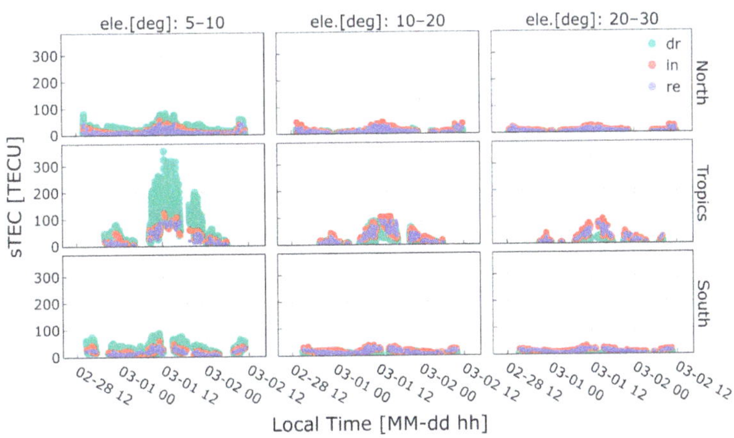


**Figure 8.** sTEC obtained from NeQuick 2 model. Color-coded according to the direct (dr), incident (in), and reflected (re) rays.

The highest sTEC is prominently observed at elevation angles ranging from 5° to 10° within the tropics region, and to a lesser extent in polar regions, but with lower magnitudes, specifically for the direct ray. This behavior occurs because, at such elevation angles, the direct ray traverses a longer path through the ionosphere than the incident and reflected rays. This effect diminishes as the elevation angle increases. At low elevations, the magnitudes of the sTEC are relatively similar for each ray, while at mid-low elevations, the contribution of the incident and reflected rays becomes more prominent in comparison to the direct ray.

Across all scenarios, local time, representing solar radiation, plays a pivotal role in sTEC retrievals. Figures 7 and 8 illustrate how the sTEC values exhibit a progressive increase as the noon-time period approaches, with the highest peaks occurring between 12:00 and 13:00 h. Following sunset, the electron density and consequently the sTEC values gradually decrease accordingly.

Table 3 provides a comparative analysis of both models, presenting the mean and standard deviation values for each ray in the distinct regions. To facilitate interpretation in terms of local time, the events have been categorized into two distinct periods: daytime (DT), spanning from 06:00 to 18:00, and nighttime (NT), encompassing the interval from 18:00 to 06:00. Notably, the range of sTEC magnitude for the direct ray is broader for the NeQuick model computations in the tropics region. However, the NEDM2020 computations consistently yield a higher mean sTEC in most cases except for the direct ray at very low elevations in the tropics during daytime. Particularly higher differences in mean values between the two models are evident in the south region (~6 TECUs), while comparatively smaller differences are observed in the tropics region (~2 TECUs).

**Table 3.** sTEC mean and standard deviation value comparison between NEDM2020 and NeQuick 2 models for F10.7 = 75.

| | | | **sTEC NEDM (TECU)** | | | | | | | | | **sTEC NeQuick 2 (TECU)** | | | | | | | | |
|---|---|---|---|---|---|---|---|---|---|---|---|---|---|---|---|---|---|---|---|---|
| | | | **Ele.: 5°–10°** | | | **Ele.: 10°–20°** | | | **Ele.: 20°–30°** | | | **Ele.: 5°–10°** | | | **Ele.: 10°–20°** | | | **Ele.: 20°–30°** | | |
| | | | **dr** | **in** | **re** | **dr** | **in** | **re** | **dr** | **in** | **re** | **dr** | **in** | **re** | **dr** | **in** | **re** | **dr** | **in** | **re** |
| North | | | | | | | | | | | | | | | | | | | | |
| | DT | mean | 28 | 26 | 23 | 8 | 20 | 18 | 3 | 14 | 13 | 21 | 19 | 17 | 6 | 15 | 13 | 3 | 11 | 9 |
| | | std | 13 | 8 | 11 | 3 | 8 | 7 | 1 | 6 | 5 | 12 | 8 | 9 | 3 | 7 | 6 | 1 | 5 | 4 |
| | NT | mean | 21 | 16 | 13 | 7 | 13 | 11 | 3 | 9 | 8 | 14 | 8 | 7 | 4 | 7 | 5 | 2 | 6 | 4 |
| | | std | 10 | 4 | 3 | 3 | 4 | 2 | 1 | 3 | 1 | 7 | 3 | 2 | 2 | 2 | 2 | 1 | 2 | 1 |
| Tropics | | | | | | | | | | | | | | | | | | | | |
| | DT | mean | 121 | 72 | 63 | 33 | 62 | 52 | 16 | 48 | 39 | 126 | 63 | 60 | 31 | 57 | 49 | 14 | 43 | 37 |
| | | std | 55 | 16 | 11 | 17 | 16 | 12 | 6 | 14 | 10 | 67 | 20 | 19 | 16 | 22 | 17 | 6 | 19 | 14 |
| | NT | mean | 43 | 29 | 20 | 17 | 27 | 18 | 10 | 22 | 15 | 40 | 26 | 16 | 14 | 27 | 17 | 8 | 21 | 16 |
| | | std | 25 | 9 | 7 | 7 | 7 | 6 | 3 | 6 | 4 | 35 | 16 | 12 | 10 | 16 | 11 | 6 | 12 | 10 |
| South | | | | | | | | | | | | | | | | | | | | |
| | DT | mean | 43 | 34 | 32 | 13 | 28 | 25 | 6 | 22 | 18 | 31 | 26 | 25 | 10 | 22 | 20 | 5 | 17 | 14 |
| | | std | 16 | 8 | 6 | 5 | 7 | 5 | 1 | 5 | 4 | 14 | 7 | 5 | 3 | 6 | 4 | 1 | 4 | 3 |
| | NT | mean | 29 | 20 | 16 | 10 | 17 | 14 | 5 | 13 | 11 | 15 | 10 | 7 | 5 | 8 | 6 | 2 | 7 | 5 |
| | | std | 10 | 4 | 3 | 4 | 3 | 2 | 1 | 2 | 2 | 8 | 4 | 3 | 2 | 3 | 2 | 1 | 2 | 2 |

## *4.2. Relative Ionospheric Group Delay Analysis*

The relative ionospheric group delay $\left(\Delta_{p_{iono}}\right)$ denotes the additional delay caused by the ionosphere along the aggregated path of the incident and reflected signals, in comparison to the direct signal. The mitigation of ionospheric delay holds significant importance in reflectometry LEO single-frequency missions, particularly within altimetry applications. The analysis of the relative ionospheric delay follows a similar approach to the sTEC analysis, encompassing the established regions, elevation angle ranges, local time variations, and the change in solar flux index. Figure 9 illustrates the potential ionospheric delays that arise from utilizing the sTEC derived from the NEDM2020 and NeQuick 2 models in conjunction with the GPS L1 frequency and F10.7 = 75.

Consistent with the sTEC analysis outcomes, it is observed that $\Delta_{p_{iono}}$ exhibits greater magnitudes within the tropics region, with the highest values occurring at very low elevation angles for both models. The occurrence of negative values in the relative ionospheric delay is attributed to the dominance of the direct signal contribution in the computation of $\Delta_{p_{iono}}$.

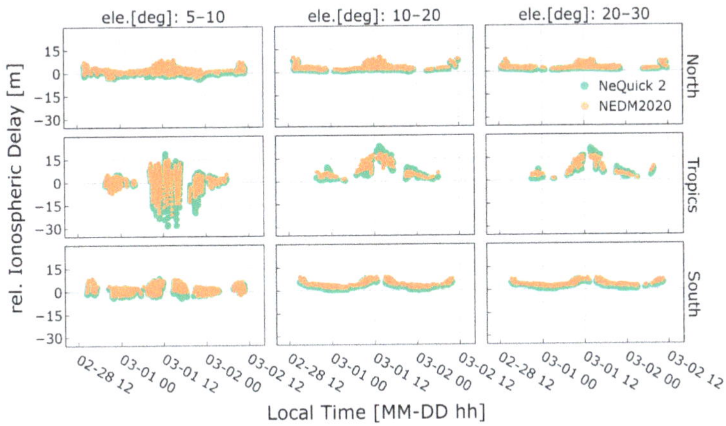


**Figure 9.** Relative ionospheric delay in the distinct regions and grazing angle elevation ranges along with local time variations.

While the outcomes from both models exhibit very similar behavior in terms of relative ionospheric delay, including their dependence on region, elevation angle, and local time, there are noticeable relative differences across the established groups. Taking as a reference the NEDM2020 model, the mean relative difference is computed as follows:

$$\%RD = \frac{mean\left(\left|\Delta_{p_{iono}}^{NeQuick\ 2} - \Delta_{p_{iono}}^{NEDM2020}\right|\right)}{mean\left(\left|\Delta_{p_{iono}}^{NEDM2020}\right|\right)} * 100 \quad (7)$$

Table 4 presents the mean relative differences between low- and high-solar-activity conditions. During LSA, the most significant relative differences occur at very low elevation angles in both the north and south regions during nighttime, showing a notable 64% variation between the two models. This difference decreases as the elevation angle increases. Conversely, during daytime, the differences in the polar regions remain relatively consistent across all scenarios, while variations are more pronounced in the tropics region. During HSA, during nighttime in the north region, the differences can reach up to 98% at very low elevation angles, while in the south region, the differences remain relatively similar when comparing low and high solar activity. In the tropics, an increase in the F10.7 index leads to a higher relative difference between the models during nighttime. However, during daytime, this difference decreases compared to the low-solar-activity condition (F10.7 = 75).

**Table 4.** Mean relative difference in the relative ionospheric delay between NEDM2020 and NeQuick 2 during high and low solar activity.

| | **LSA (F10.7 = 75)** | | | **HSA (F10.7 = 180)** | | |
|---|---|---|---|---|---|---|
| | **Ele.: 5°–10°** | **Ele.: 10°–20°** | **Ele.: 20°–30°** | **Ele.: 5°–10°** | **Ele.: 10°–20°** | **Ele.: 20°–30°** |
| North | | | | | | |
| DT | 30% | 28% | 26% | 59% | 17% | 15% |
| NT | 64% | 49% | 46% | 98% | 38% | 28% |
| Tropics | | | | | | |
| DT | 48% | 17% | 17% | 27% | 16% | 14% |
| NT | 58% | 35% | 38% | 88% | 76% | 48% |
| South | | | | | | |
| DT | 19% | 21% | 23% | 41% | 21% | 24% |
| NT | 64% | 53% | 51% | 66% | 56% | 51% |

The sTEC outcomes obtained from the NEDM2020 model, utilized as the reference model in this study, form the basis for the following analysis. Figure 10 illustrates the ionospheric delay distribution during low solar activity, categorized by elevation angles, regions, and local time distinguishing between daytime and nighttime. At low and mid-low elevation angles, the contribution of each ray to the delay remains relatively similar in magnitude, resulting in positive values for the relative ionospheric delay. Overall, during daytime events, the $\Delta_{p_{iono}}$ is on average 120% greater compared to nighttime events.

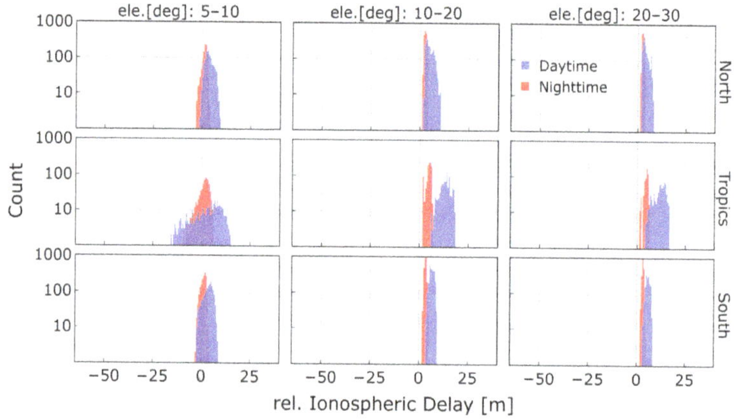


**Figure 10.** Distribution of relative ionospheric delay depending on elevation, daytime (DT), and nighttime (NT) using NEDM2020 sTEC retrievals with F10.7 = 75.

During HSA periods (F10.7 = 180), the relative ionospheric delay range can increase by up to 200% with respect to low-solar-activity periods, as seen in Figure 11. In low- and mid-low-elevation scenarios, the distribution of $\Delta_{p_{iono}}$ behaves similarly to LSA but with higher magnitude values. Notably, in the tropics region at very low elevations, the distribution is more widespread, with relative delays primarily consisting of negative values. This highlights the higher influence of the direct ray on $\Delta_{p_{iono}}$ compared to low-solar-activity periods.

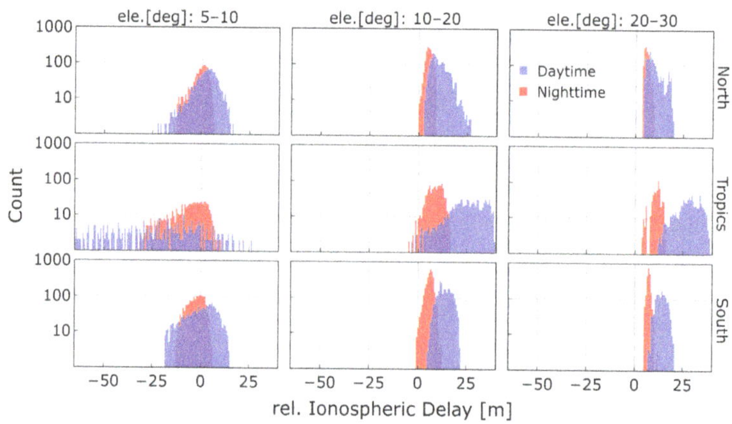


**Figure 11.** Distribution of relative ionospheric delay depending on elevation, daytime (DT), and nighttime (NT) using NEDM2020 sTEC retrievals with F10.7 = 180.

Group Delay Altimetry and Ionospheric Delay Uncertainty Analysis

As a single-frequency GNSS-R mission, PRETTY relies on ionospheric correction models to ensure precise sea surface height measurements, introducing a level of model uncertainty in the correction process. Figure 12 presents the altimetric uncertainty at grazing elevation angles. Figures 10 and 11 depict the distribution of the relative ionospheric delay, showing a noticeable diurnal cycle effect where daytime observations exhibit higher relative ionospheric delays compared to nighttime observations. This diurnal variation is also reflected in the sea surface height uncertainties. Furthermore, it is evident that ionospheric uncertainties have a significantly greater impact on sea height retrievals in the Tropics region, where the general level of ionization is higher. In this geographical area, we observe a higher altimetric uncertainty dispersion, particularly in the mid-low elevation angle regime (during daytime, 0.22 m mean and 4.08 m std), where the combined delay of the incident and reflected rays surpasses that of the direct ray. Consequently, this leads to higher relative delays and, by extension, a more pronounced impact on GNSS-R altimetric retrievals within this specific elevation range and region.

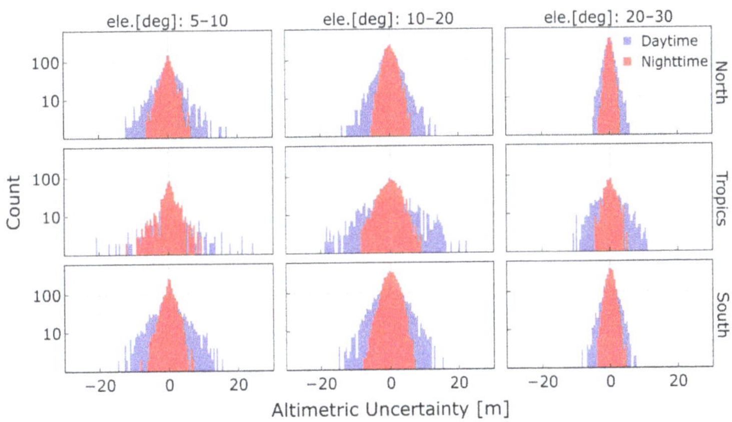


**Figure 12.** Altimetric uncertainty due to uncertainty in ionospheric delay model depending on elevation, daytime (DT), and nighttime (NT) using NEDM2020 during LSA.

### *4.3. Doppler Shift Analysis*

The analysis extends to the Doppler shift observed at the GPS L1 frequency across varying ranges of elevation angles, while considering effects during both day and night periods. Figure 13 illustrates the distribution of the Doppler shift during low solar activity. The electron density variations in grazing angle reflectometry can induce a maximum Doppler shift of $\pm 2$ Hz in the GPS L1 signal during daytime. The attenuation in the Doppler shift demonstrates a strong correlation with diurnal cycles, resulting in a reduction during nighttime periods. This phenomenon can be attributed to the decrease in the rate of electron density changes, which in turn leads to a corresponding decrease in the magnitude of the Doppler shift.

The Doppler shift histograms reveal a symmetrical distribution centered around approximately 0 Hz with a distinct separation in very-low-elevation cases. The distribution is also influenced by the transmitter motion relative to the specular point elevation angle. In Figure 14, it becomes evident that at very low elevation angles, a rising transmitter (ascendant elevation) induces a positive Doppler shift, while a setting transmitter (descendant elevation) results in a negative Doppler shift. However, at higher elevation angles ($20^\circ$ to $30^\circ$), the relationship may vary or even reverse.

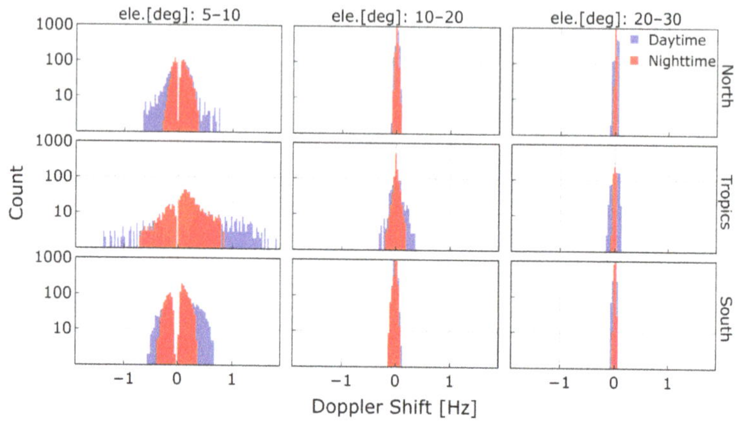


**Figure 13.** Distribution of Doppler shift depending on elevation, daytime (DT), and nighttime (NT) using NEDM2020 sTEC retrievals with F10.7 = 75.

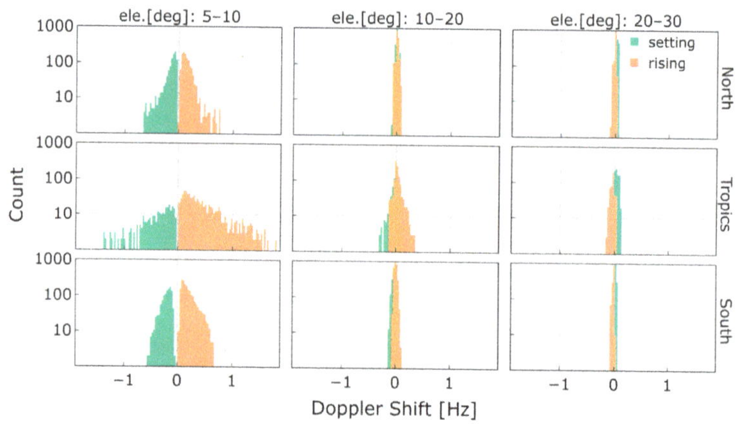


**Figure 14.** Distribution of Doppler shift depending on elevation and rising or setting event using NEDM2020 sTEC retrievals with F10.7 = 75.

The distribution of the Doppler shift for F10.7 = 180 exhibits an increase in dispersion, approximately doubling in all scenarios. In the elevation range of 5°–10° within the tropics region, the range of $f_d$ is more extensive during daytime, reaching maximum values of up to ±4 Hz. The rising and setting event analyses present similar behavior, with negative magnitudes primarily observed during rising events and positive magnitudes during setting events.

### *4.4. Peak Electron Density Height Analysis*

The NEDM2020 model is employed to determine the height at which the maximum electron density peak $Hm$ is observed along the paths of both the incident and reflected rays. This altitude is significant as it represents the point of maximum ionization within the ionosphere that the signals traverse.

From a geometrical standpoint within the grazing GNSS-R configuration, variations in elevation angles directly correspond to changes in the segment of the signal ray that travels along the ionosphere. Furthermore, throughout the diurnal cycle, electron densities within the E and F layers exhibit greater magnitudes during daylight hours compared to nighttime, with the F layer generally obtaining higher electron concentrations. These fluctuations are examined to comprehend the intricate ionospheric interactions that the signals undergo during their propagation. This phenomenon results in variations in the height of the maximum electron density peak, as depicted in Figure 15 for both day and nighttime.

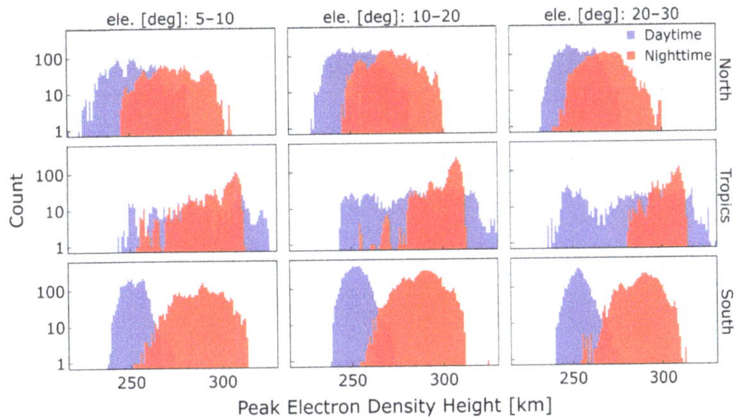


**Figure 15.** Peak electron density height variations depending on elevation ranges, regions, and day and nighttime using NEDM2020 sTEC retrievals with F10.7 = 75.

The overall average of the *Hm* during the LSA period is 270 km. Nevertheless, noticeable variations are evident with respect to daytime and nighttime. In general, during nighttime, the *Hm* is on average 10% higher than during daytime. The tropics region stands out as one of the most dynamically changing areas within the ionosphere. In this zone, the distribution of *Hm* during daytime exhibits a spread ranging from 236 to 326 km, lacking a distinct peak value. However, during nighttime, *Hm* reaches its maximum value at approximately 305 km. This highlights the substantial variations in electron density within this region, particularly during daytime. During HSA, the *Hm* exhibits a consistent increase of 21% across all scenarios.

## 5. Discussion

The analysis provides valuable insights into how ionospheric parameters such as slant total electron content, relative ionospheric delay, Doppler shift, and peak electron density height vary in response to different conditions. These findings are crucial for optimizing the accuracy of space-borne GNSS-R applications, particularly in altimetry, aiding in the development of robust models, and enhancing the interpretation of data acquired through grazing GNSS-R configurations.

Under low-solar-activity conditions (F10.7 = 75), the resulting sTEC values from NEDM2020 and NeQuick 2 reveal that both models exhibit similar behavior across different scenarios. However, it is important to note that while an extensive evaluation of the models is not carried out in this study, differences in the sTEC computations and the relative total delay are observed. Significant differences of ~60 TECUs and up to 64% in relative ionospheric delay are observed in polar regions at very low elevation angles during daytime when comparing NEDM2020 and NeQuick 2. Under high-solar-activity conditions (F10.7 = 180), the relative differences can reach values up to 98%.

Grazing elevation angles, local time, regions, and solar activity emerge as the crucial factors determining ionospheric effects in GNSS-R. The observed elevation angle significantly influences the path traversed by GNSS signals through the ionosphere, while electron density variations rising from ionospheric diurnal cycles and geographical location contribute to fluctuations in the sTEC computation. The sTEC values exhibit a noticeable increase as the elevation angle decreases (very low to mid-low angles) in all regions during both daytime and nighttime.

Daytime events consistently result in higher sTEC values, larger relative ionospheric delay values, and higher Doppler shift magnitudes compared to nighttime events across all regions and elevation angles. The tropics region consistently displays the highest sTEC values across all elevation angle ranges, indicating the presence of higher electron densities. To provide a comprehensive synthesis of the study's findings based on the electron density retrievals from the NEDM2020 model, Tables 5 and 6 provide a summary of the results by presenting the median and standard deviation for each parameter outlined in the Section 4 for F10.7 = 75 and F10.7 = 180, respectively. The parameters provided by the summary tables are the relative ionospheric delay $(\Delta_{p_{iono}})$ in meters, absolute value of Doppler shift $(|f_d|)$ in Hertz, and peak electron density height $(Hm)$ in kilometers.

**Table 5.** Overview of ionospheric parameters from the NEDM2020 sTEC computations during low solar activity (F10.7 = 75).

| | | **Very Low: 5°–10°** | | | **Low: 10°–20°** | | | **Mid-Low: 20°–30°** | | |
|---|---|---|---|---|---|---|---|---|---|---|
| | | $\Delta_{p_{iono}}$ (m) | $\lvert f_d \rvert$ (Hz) | $Hm$ (km) | $\Delta_{p_{iono}}$ (m) | $\lvert f_d \rvert$ (Hz) | $Hm$ (km) | $\Delta_{p_{iono}}$ (m) | $\lvert f_d \rvert$ (Hz) | $Hm$ (km) |
| North | | | | | | | | | | |
| DT | median | 3.10 | 0.133 | 251.7 | 4.35 | 0.022 | 252.3 | 3.33 | 0.026 | 252.2 |
| | std | 1.96 | 0.142 | 11.7 | 1.81 | 0.019 | 10.5 | 1.44 | 0.021 | 8.5 |
| NT | median | 1.38 | 0.097 | 274.0 | 2.61 | 0.009 | 271.9 | 2.27 | 0.011 | 268.3 |
| | std | 1.19 | 0.067 | 12.9 | 0.60 | 0.014 | 11.6 | 0.47 | 0.006 | 9.7 |
| Tropics | | | | | | | | | | |
| DT | median | 3.96 | 0.729 | 287.9 | 13.50 | 0.048 | 287.4 | 12.12 | 0.067 | 285.3 |
| | std | 7.57 | 0.415 | 21.4 | 2.61 | 0.070 | 22.4 | 2.87 | 0.028 | 23.8 |
| NT | median | 1.63 | 0.219 | 302.6 | 4.95 | 0.016 | 305.4 | 4.74 | 0.014 | 305.4 |
| | std | 2.82 | 0.195 | 13.1 | 1.18 | 0.033 | 9.1 | 1.05 | 0.013 | 7.3 |
| South | | | | | | | | | | |
| DT | median | 3.98 | 0.220 | 252.2 | 6.49 | 0.030 | 252.2 | 5.63 | 0.014 | 253.1 |
| | std | 2.26 | 0.138 | 5.2 | 1.15 | 0.021 | 5.2 | 1.02 | 0.022 | 4.9 |
| NT | median | 1.41 | 0.145 | 289.0 | 3.36 | 0.012 | 288.8 | 3.17 | 0.015 | 289.1 |
| | std | 1.27 | 0.073 | 11.2 | 0.48 | 0.021 | 9.8 | 0.40 | 0.007 | 9.2 |

In general, as the F10.7 index increases, notable observations emerge: (1) There is a compensation effect, attributed to the direct signal contribution, leading to a decrease in the median level of relative ionospheric delay as elevation decreases, particularly at very low elevations. (2) The absolute Doppler shift exhibits a substantial increase in median values, scaling up to one order of magnitude, as elevation angles decrease to their lowest. (3) Notably, in tropical regions characterized by higher density peak heights, there is a more pronounced compensation by direct signal contribution in $\Delta_{p_{iono}}$ at the lowest elevations, resulting in negative median delay values.

For a LEO GNSS-R mission employing the GPS L1 frequency, findings show that relative ionospheric delays can reach ~19 m during periods of LSA and ~70 m during HSA, equivalent to about 120 and 430 TECUs. The forthcoming ESA PRETTY mission will pioneer grazing altimetry at the L5 frequency, which, with its longer wavelength (~0.2548 m), is more sensitive to ionospheric group delays. Using 120 and 430 TECUs as benchmarks, relative ionospheric corrections of approximately 35 and 125 m can be expected for group delay altimetry during low and high solar activity.

**Table 6.** Overview of ionospheric parameters from the NEDM2020 sTEC computations during high solar activity (F10.7 = 180).

| | | **Very Low: 5°–10°** | | | **Low: 10°–20°** | | | **Mid-Low: 20°–30°** | | |
|---|---|---|---|---|---|---|---|---|---|---|
| | | $\Delta p_{iono}$ **(m)** | $\lvert f_d \rvert$ **(Hz)** | *Hm* **(km)** | $\Delta p_{iono}$ **(m)** | $\lvert f_d \rvert$ **(Hz)** | *Hm* **(km)** | $\Delta p_{iono}$ **(m)** | $\lvert f_d \rvert$ **(Hz)** | *Hm* **(km)** |
| North DT | median | 2.77 | 0.516 | 306.8 | 9.18 | 0.075 | 307.6 | 8.09 | 0.049 | 307.5 |
| | std | 6.00 | 0.496 | 13.9 | 4.24 | 0.085 | 12.7 | 3.79 | 0.043 | 10.4 |
| NT | median | 0.03 | 0.330 | 333.8 | 5.11 | 0.051 | 331.7 | 5.31 | 0.015 | 327.2 |
| | std | 3.90 | 0.148 | 15.9 | 1.64 | 0.066 | 14.4 | 1.16 | 0.011 | 12.0 |
| Tropics DT | median | −21.45 | 2.329 | 349.5 | 25.56 | 0.246 | 348.4 | 28.45 | 0.139 | 345.2 |
| | std | 24.43 | 0.698 | 26.7 | 9.30 | 0.377 | 27.3 | 5.96 | 0.068 | 28.9 |
| NT | median | −5.21 | 0.817 | 368.1 | 8.82 | 0.144 | 371.9 | 11.55 | 0.022 | 371.9 |
| | std | 9.88 | 0.352 | 15.7 | 3.74 | 0.173 | 10.9 | 2.40 | 0.024 | 8.8 |
| South DT | median | 1.46 | 0.812 | 307.2 | 13.57 | 0.084 | 306.9 | 13.97 | 0.037 | 307.9 |
| | std | 7.58 | 0.333 | 6.5 | 3.39 | 0.109 | 6.4 | 2.62 | 0.043 | 6.0 |
| NT | median | −1.98 | 0.472 | 351.7 | 6.26 | 0.081 | 351.2 | 7.17 | 0.015 | 351.8 |
| | std | 3.99 | 0.146 | 14.1 | 1.84 | 0.092 | 12.3 | 0.92 | 0.015 | 11.4 |

## 6. Conclusions

In this paper, we have analyzed ionospheric effects in GNSS-R at grazing angles. This study encompasses the characterization of slant total electron content, relative ionospheric delay, the influence of ionospheric correction model uncertainties on GNSS-R group delay altimetry retrievals, the Doppler effect, and peak electron density height changes. Various factors have been considered such as satellite geometry, latitude-dependent regions, temporal variations, and solar activity.

When analyzing the results during LSA (low solar activity) and HSA (high solar activity), it becomes evident that as the elevation decreases into the grazing regime below 20°, the median relative ionospheric delay decreases due to the compensation from the direct signal contribution. However, it is important to note that the standard deviation of the delay, especially in terms of the Doppler shift, undergoes a substantial increase. This behavior poses a significant challenge for the model-based correction of ionospheric delay in GNSS reflectometry altimetry at grazing elevation angles.

While model uncertainties do affect group delay sea height estimates it is important to highlight that these effects are not uniform across all GNSS-R observations. Coherent phase observations, for instance, offer a remarkable level of precision, down to the centimeter scale. Along reflection tracks characterized by consistent ionospheric bias, relative altimetry at a centimeter precision level can be achieved. This means that even in the presence of ionospheric delay bias, LEO space-borne GNSS-R systems, as reported in [26], can still provide precise results in the altimetric inversion.

Total electron content, a crucial ionospheric parameter, exhibits complex variations spanning diurnal, monthly, seasonal, and 11-year solar cycles. Extended temporal coverage is essential for deciphering these patterns, especially in dynamic regions allowing analysis of seasonal trends. This study highlights the importance of spatially extended data, particularly in tropical areas with substantial ionospheric variability. Such data is key to comprehending ionospheric parameter evolution across different time scales and regions, influenced by factors like solar activity and geomagnetic storms.

GNSS-R (global navigation satellite system reflectometry) stands as a valuable and complementary remote sensing tool in ionospheric studies, effectively addressing areas not covered by alternative methods. This capacity offers significant contributions to the modeling, prediction, and comprehension of ionospheric effects.

**Author Contributions:** Conceptualization: M.M., M.S. and G.S.; methodology: M.M., M.S. and G.S.; software: M.M. and M.S.; data resources: M.M. and M.S.; writing—original draft preparation: M.M.; writing—review and editing: M.S., G.S., M.H. and J.W.; visualization, M.M.; supervision, M.S., J.W. and M.H. All authors have read and agreed to the published version of the manuscript.

**Funding:** This research was funded by the German Aerospace Center, Institute for Solar-Terrestrial Physics (DLR-SO).

**Data Availability Statement:** Data are available under the authorization of Spire Global Inc.

**Acknowledgments:** The authors would like to thank Spire Global for providing real orbit from Lemur-2 constellation LEO satellites.

**Conflicts of Interest:** The authors declare no conflict of interest. The funders had no role in the design of the study; in the collection, analyses, or interpretation of data; in the writing of the manuscript, or in the decision to publish the results.

## References

1. Hagfors, T. 2.1. The Ionosphere. In *Methods in Experimental Physics*; Meeks, M.L., Ed.; Astrophysics; Academic Press: Cambridge, MA, USA, 1976; Volume 12, pp. 119–135.
2. Dubey, S.; Wahi, R.; Gwal, A.K. Ionospheric Effects on GPS Positioning. *Adv. Space Res.* **2006**, *38*, 2478–2484. [CrossRef]
3. Wickert, J.; Michalak, G.; Schmidt, T.; Beyerle, G.; Cheng, C.Z.; Healy, S.B.; Heise, S.; Huang, C.Y.; Jakowski, N.; Kohler, W.; et al. GPS Radio Occultation: Results from CHAMP. GRACE and FORMOSAT-3/COSMIC. *Terr. Atmos. Ocean. Sci.* **2009**, *20*, 35–50. [CrossRef]
4. Xing, J.; Datta-Barua, S.; Garrison, J.; Ridley, A.; Pervan, B. Relative Ionospheric Ranging Delay in LEO GNSS Oceanic Reflections. *IEEE Geosci. Remote Sens. Lett.* **2015**, *12*, 1416–1420. [CrossRef]
5. Ruf, C.; Gleason, S.; Jelenak, Z.; Katzberg, S.; Ridley, A.; Rose, R.; Scherrer, J.; Zavorotny, V. The NASA EV-2 Cyclone Global Navigation Satellite System (CYGNSS) Mission. In Proceedings of the 2013 IEEE Aerospace Conference, Big Sky, MT, USA, 2–9 March 2013; pp. 1–7.
6. Camps, A.; Park, H.; Foti, G.; Gommenginger, C. Ionospheric Effects in GNSS-Reflectometry From Space. *IEEE J. Sel. Top. Appl. Earth Obs. Remote Sens.* **2016**, *9*, 5851–5861. [CrossRef]
7. Unwin, M.; Jales, P.; Tye, J.; Gommenginger, C.; Foti, G.; Rosello, J. Spaceborne GNSS-Reflectometry on TechDemoSat-1: Early Mission Operations and Exploitation. *IEEE J. Sel. Top. Appl. Earth Obs. Remote Sens.* **2016**, *9*, 4525–4539. [CrossRef]
8. Wang, Y.; Morton, Y.J. Ionospheric Total Electron Content and Disturbance Observations From Space-Borne Coherent GNSS-R Measurements. *IEEE Trans. Geosci. Remote Sens.* **2022**, *60*, 1–13. [CrossRef]
9. Liu, L.; Morton, Y.J.; Wang, Y.; Wu, K.-B. Arctic TEC Mapping Using Integrated LEO-Based GNSS-R and Ground-Based GNSS Observations: A Simulation Study. *IEEE Trans. Geosci. Remote Sens.* **2022**, *60*, 1–10. [CrossRef]
10. Roesler, C.; Wang, Y.; Morton, Y.J.; Nerem, R.S. Coherent GPS Reflections Over Ocean Surface. In Proceedings of the IGARSS 2020–2020 IEEE International Geoscience and Remote Sensing Symposium, Waikoloa, HI, USA, 26 September–2 October 2020; pp. 6218–6221.
11. Moreno, M.; Semmling, M.; Stienne, G.; Dalil, W.; Hoque, M.; Wickert, J.; Reboul, S. Airborne Coherent GNSS Reflectometry and Zenith Total Delay Estimation over Coastal Waters. *Remote Sens.* **2022**, *14*, 4628. [CrossRef]
12. Issa, H.; Stienne, G.; Reboul, S.; Raad, M.; Faour, G. Airborne GNSS Reflectometry for Water Body Detection. *Remote Sens.* **2021**, *14*, 163. [CrossRef]
13. Subirana, J.S.; Zornoza, J.M.J.; Hernández-Pajares, M. *GNSS Data Processing*; Contactivity bv; ESA Communications: Leiden, The Netherlands, 2013; Volume I, Fundamentals and Algorithms; ISBN 978-92-9221-886-7.
14. Hoque, M.M.; Jakowski, N.; Osechas, O.; Berdermann, J. Fast and Improved Ionospheric Correction for Galileo Mass Market Receivers. In Proceedings of the 32nd International Technical Meeting of the Satellite Division of the Institute of Navigation (ION GNSS+ 2019), Miami, FL, USA, 11 October 2019; pp. 3377–3389.
15. Nava, B.; Coïsson, P.; Radicella, S.M. A New Version of the NeQuick Ionosphere Electron Density Model. *J. Atmos. Sol. Terr. Phys.* **2008**, *70*, 1856–1862. [CrossRef]
16. Ionospheric Correction Algorithms | European GNSS Service Centre. Available online: https://www.gsc-europa.eu/support-to-developers/ionospheric-correction-algorithms (accessed on 27 July 2023).
17. Dielacher, A.; Fragner, H.; Koudelka, O. PRETTY—Passive GNSS-Reflectometry for CubeSats. *Elektrotech. Inftech.* **2022**, *139*, 25–32. [CrossRef]
18. Hoque, M.M.; Jakowski, N.; Prol, F.S. A New Climatological Electron Density Model for Supporting Space Weather Services. *J. Space Weather Space Clim.* **2022**, *12*, 1. [CrossRef]
19. Nguyen, V.; Jales, P.; Garbacz, H. Spire Earth Observations for NASA's CSDA Program—Lunch & Learn. 2022. Available online: https://www.earthdata.nasa.gov/s3fs-public/2022-10/2022%2010%2006%20Spire%20Earth%20Observations%20for%20NASA%27s%20CSDA%20Program%20-%20Lunch%20%26%20Learn_0.pdf (accessed on 20 July 2023).

20. Semmling, A.M.; Leister, V.; Saynisch, J.; Zus, F.; Heise, S.; Wickert, J. A Phase-Altimetric Simulator: Studying the Sensitivity of Earth-Reflected GNSS Signals to Ocean Topography. *IEEE Trans. Geosci. Remote Sens.* **2016**, *54*, 6791–6802. [CrossRef]
21. Hoque, M.M.; Jakowski, N.; Cahuasquí, J.A. Fast Ionospheric Correction Algorithm for Galileo Single Frequency Users. In Proceedings of the 2020 European Navigation Conference (ENC), Dresden, Germany, 23–24 November 2020; pp. 1–10.
22. Cardellach, E.; Ao, C.O.; de la Torre Juárez, M.; Hajj, G.A. Carrier Phase Delay Altimetry with GPS-Reflection/Occultation Interferometry from Low Earth Orbiters. *Geophys. Res. Lett.* **2004**, *31*. [CrossRef]
23. Mashburn, J.; Axelrad, P.; Lowe, S.T.; Larson, K.M. Global Ocean Altimetry With GNSS Reflections From TechDemoSat-1. *IEEE Trans. Geosci. Remote Sens.* **2018**, *56*, 4088–4097. [CrossRef]
24. Davies, K.; Watts, J.M.; Zacharisen, D.H. A Study of F 2-Layer Effects as Observed with a Doppler Technique. *J. Geophys. Res.* **1962**, *67*, 601–609. [CrossRef]
25. Jacobs, J.A.; Watanabe, T. Doppler Frequency Changes in Radio Waves Propagating Through a Moving Ionosphere. *Radio Sci.* **1966**, *1*, 257–264. [CrossRef]
26. Cardellach, E.; Li, W.; Rius, A.; Semmling, M.; Wickert, J.; Zus, F.; Ruf, C.S.; Buontempo, C. First Precise Spaceborne Sea Surface Altimetry With GNSS Reflected Signals. *IEEE J. Sel. Top. Appl. Earth Obs. Remote Sens.* **2020**, *13*, 102–112. [CrossRef]
27. Teunissen, P.J.G.; Kleusberg, A. GPS Observation Equations and Positioning Concepts. In *GPS for Geodesy*; Kleusberg, A., Teunissen, P.J.G., Eds.; Lecture Notes in Earth Sciences; Springer: Berlin/Heidelberg, Germany, 1996; pp. 175–217. ISBN 978-3-540-49447-8.

MDPI
St. Alban-Anlage 66
4052 Basel
Switzerland
www.mdpi.com

*Remote Sensing* Editorial Office
E-mail: remotesensing@mdpi.com
www.mdpi.com/journal/remotesensing

www.ingramcontent.com/pod-product-compliance
Lightning Source LLC
Chambersburg PA
CBHW040754250926
51850CB00007BB/1492
*9783036594606*